Sofia Pescarin · Cristiana Barandoni ·
Arthur Clay · George Papadopoulos ·
Irina Crina Anca Sandu
Editors

Chromatic Visions

Exploring Colour in Art, Archaeology and Digital Realities, Part I

Springer

Editors
Sofia Pescarin
CNR ISPC
Florence, Italy

Cristiana Barandoni
CNR ISPC
Florence, Italy

Arthur Clay
HSLU
Lucerne, Switzerland

George Papadopoulos
Foundation for Research and Technology (FORTH)
Heraklion, Greece

Irina Crina Anca Sandu
Munch Museum
Oslo, Norway

ISSN 0302-9743 ISSN 1611-3349 (electronic)
Lecture Notes in Computer Science
ISBN 978-3-032-07791-2 ISBN 978-3-032-07792-9 (eBook)
https://doi.org/10.1007/978-3-032-07792-9

This Springer imprint is published by the registered company Springer Nature Switzerland AG
The registered company address is: Gewerbestrasse 11, 6330 Cham, Switzerland

Lecture Notes in Computer Science 15750

The series Lecture Notes in Computer Science (LNCS), including its subseries Lecture Notes in Artificial Intelligence (LNAI) and Lecture Notes in Bioinformatics (LNBI), has established itself as a medium for the publication of new developments in computer science and information technology research, teaching, and education.

LNCS enjoys close cooperation with the computer science R & D community, the series counts many renowned academics among its volume editors and paper authors, and collaborates with prestigious societies. Its mission is to serve this international community by providing an invaluable service, mainly focused on the publication of conference and workshop proceedings and postproceedings. LNCS commenced publication in 1973.

Preface. PERCEIVE and the Perception of Coloured Collections Through AI and Virtual Experiences

In 2023 the European Commission financed a project dedicated to the conservation and communication of the importance of coloured collections with digital technologies: PERCEIVE (Perceptive Enhanced Realities of Coloured Collections Through Artificial Intelligence and Virtual Experiences https://www.perceive-horizon.eu). The volume "Chromatic Visions" collects the theoretical and scientific work carried out by a consortium of partners experts in the field, during the first year of the project.

PERCEIVE aims at creating a new way to perceive, preserve, curate, exhibit, understand and access coloured Cultural Heritage collections and Digital Artworks to promote their re-appropriation. These collections are, in fact, a priority due to their high fragility, which requires shared methods to preserve and exhibit them, because of the complexity of their study and the significance of being able to properly communicate knowledge about them to future generations. The project started from the twofold need to better preserve and communicate coloured artworks and, at the same time, improve and speed up scientific process results, which can be employed to maximise visitor experiences with a variety of digital-coloured collections including for paintings, antique sculptures, tapestries and garments as well as for born-digital artworks. It aims at advancing the digital capability of scientists and cultural institutions, through a service-based AI architecture and toolkit; and by developing a new design theory for on-site and remote VR/AR/MR experiences, based on "Care", "Accessibility" and "Authenticity" concepts, with and for the creative industries, expanding access to Cultural Heritage and Art outside museums for a wider integration with society.

To be able to reach these goals, the project was developed considering five important "scenarios" that define the context and are used to better focus on specific problems, needs and requirements, and to better identify suitable solutions dedicated to the scientific community, citizens and the creative industries: (1) polychromy, (2) paintings and works on papers, (3) textiles, (4) historical photos and videos, and (5) born-digital artworks.

This first scenario is dedicated to the problem of the *loss of polychromy* in classical sculptures and architectures. Archaeological museums, in fact, often own collections of statues and architectures that were originally coloured. Unfortunately, only tiny traces of these colours are still visible and what is left needs to be preserved carefully. Although it is complex to reconstruct their original appearance, institutions believe that it is also crucial to offer a correct understanding, training visitors' eyes to recognise those original colours, avoiding dangerous ideologies that would support misunderstandings. What matters are the shades of colours, their perception and their semantics. The reliability of digital reconstructions and the authenticity of digital experiences is crucial. Disappearing polychromy is also an amazing occasion to make visitors reflect on the diversity of these shades of colours, developing a new perception on European modern society, through these artworks. The National Archaeological Museum of Naples (MANN) and the Art Institute of Chicago (AIC) have extensively worked on this problem. Unfortunately, examples of polychromy reconstruction are still very few, and this depends also on the long and costly processes required today. These reconstructive studies make use of analytical investigation results, comparisons with similar but better conserved artworks (e.g., painted mummy portraits or terracotta statues) and experts knowledge on ancient techniques and materials. A good example of such a process is the study of the Treu Head by the AIC [1] and the "MANN In Colours" project [2], which since 2018 has aimed to identify, investigate and digitally reconstruct the ancient polychromy of one of the most important collections in the world: the Farnese collection, afterwards extended to the Pompeian collections (i.e., the Venus Lovatelli, Venus Marina, and Venus in Bikini). Although there is still not a common protocol of investigations on ancient polychromy that deals also with the preservation of traces of colour, the above projects are being used by researchers to develop a shared method and to speed up the process thanks to image processing methods (IBM) and Machine Learning approaches (AI). The goal is to obtain, when possible, reliable reconstructions of the original polychromy and, when not enough data are at our disposal, to communicate to the public the scientific process behind the study of residual polychromy. The twofold goals are in fact to train the visitor's eye to recognise the colours, improving their knowledge and understanding, and at the same time to develop a citizen sense of care, through strengthening a sense of wonder for those artworks, without forgetting about the authenticity of the original collection and about the experience itself. Moreover, we believe that there is a strong need to extend civic participation, involving citizens also outside museums (home and street access), developing a new Open Museum concept.

The second scenario faces the problem of colour change in paintings and works on paper. Examples of paintings from the Munch Museum Collection, such as The Scream (ca. 1910) (Woll.M.896) and the Old Man in Warnemünde (Woll.M.491), show how some colours are subjected to this change. For instance, the fading of some cadmium yellows visible in The Scream has been seen as a consequence of the interaction of the paint with moisture and/or light in combination with chlorine compounds, whereas the darkening of some areas of The Old Man in Warnemünde still needs further investigation [3], thus highlighting the many potential causes and complexity. On the other side, the reconstruction of how colours would have been originally perceived is crucial, since their appearance strongly influences our perception of the artworks at the present and also in the future. What are these colours and why do they change over time? How were the colours applied by the artists? What does it mean for a contemporary observer to perceive their colours? The semantics of those paintings as originally conceived by artists or as they may evolve over time would need to be better communicated to the public, letting them experience them in new embodied ways, either at the museum or remotely. These are the questions that this document will try to answer. Moreover, there is a need for curators, conservators and researchers to use "Predictive Systems" that could simulate colour modifications back and forth. These systems could be of high interest also for citizens, who, while playing the role of scientists, could at the same time perceive the digital reconstruction as more authentic, and, ultimately, develop a sense of care for these collections. How an interactive experience could be designed to enhance these concepts will be discussed in more depth in the chapters dedicated to Sense of Care and Authenticity.

The third scenario is to face the problem of colour fading in textiles (dresses and tapestry). We are aware how these materials and their colours are very fragile, suffering rapid degradation, in some cases within a decade. This is particularly true for textiles made of precious metal fibres, such as in the case of embroidery made with silver or gold. Therefore, it is highly necessary to conserve and communicate these collections, which are of inspiration also for the creative industries, when these are appropriately reconstructed. While it is clear how scientific community and museum professionals need to better study the fading process, we need also to explore how we can digitally reconstruct their original appearance and how we can use these reconstructions in digital applications that preserve the authenticity of the original colour.

Also, historical photo collections experience colour degradation and this is what the fourth scenario is focused on. There are two types of assets we believe are more in danger of losing their original appearance: the autochrome and the kodacolour collections, whose colour was made with specific "lost" techniques. How can we set back the clock and digitally acquire / reconstruct them, reproducing the original colour? Most of these photos were made for personal use and not by a professional photographer. They represent therefore personal real-life moments of the last century, costumes, traditions, architectures, that in some cases do not exist anymore. In this case, how can we reconstruct and reproduce their original colour? How can we adopt automatic AI techniques in an ethical way?

The fifth scenario deals with born-digital artwork, created by artists to be perceived by the public "virtually" (either through an AR application or through a screen), in some cases during certain times of the day (night, day, etc.) to make use of real-world blending. While preserving these works and making them accessible for the public we need to consider colour rendering in combination with the physical environment. What is an authentic experience of these artworks and their designed effect?

To reach the various goals, in line with these scenarios, it is important to set a theoretical background for the design of tools, services and experiences that could be used by scientists, curators, citizens, exhibition designers. The following chapters, therefore, set this out according to concepts such as colour, perception, sense of care, authenticity and civic participation in public spaces. They then analyse what has been already done, in line with these concepts. The overall result is meant to develop design guidelines and strategies useful within the above-mentioned scenarios.

As a result of this European Project, PERCEIVE is publishing this volume "Chromatic Visions", which is divided into two parts. Part I is dedicated to the methodology of colour perception, study, simulation, reconstruction and to the theory of caring and authenticity in XR environments. Part II is dedicated to the technological results (tools and services dedicated to colour) and to the design of PERCEIVE prototypes. All contributions have been revised internally and externally by the project external scientific board, anonymously.

Part I has seven chapters, described here briefly.

Perception. This chapter introduces the scientific foundations of colour perception, emphasizing that colour is not an intrinsic property of objects but a perceptual quality resulting from the interaction between light, material surfaces, and the human visual system. It discusses the physiological mechanisms of colour vision, including trichromatic theory, opponent processes, metamerism, and chromatic induction, underscoring how perception is context-dependent and multisensory.

Colours through Time. Focused on the historical evolution of colour theories and artistic practices, this chapter traces how colour has been understood, used, and symbolized from antiquity to the present. It explores key milestones in colour theory—from Democritus to Newton, Goethe, and Munsell—and highlights how artists have used colour for symbolic, emotional, and aesthetic purposes across different periods and cultures. It also reflects on the evolving materiality of pigments and the cultural meanings of colour.

Authenticity. This chapter explores the concept of authenticity in relation to digital cultural heritage and user experience design. It introduces a theoretical framework that includes the three dimensions—Self, Others, and World—that define an authentic experience. Drawing from philosophy, psychology, and HCI, it presents a methodology for designing hybrid and VR experiences that foster subjective authenticity, and proposes an "authenticity index" for future evaluation.

Sense of Care. The chapter develops a design perspective for the "care theory" and addresses emotional, and relational dimensions of digital heritage. It positions Cultural Heritage as a recipient of care and examines how digital experiences—through storytelling, multisensory interaction, and participatory design—can evoke empathy and foster intergenerational responsibility. Cultural Probe Kits and interactive prototypes are presented as tools to investigate and elicit caring behaviours in visitors.

Participation. Focused on civic participation and social cohesion, particularly through public and hybrid cultural experiences, the chapter introduces the concept of the "Open Space Museum", a modular exhibition system for open air public use, designed to increase accessibility and engagement in non-traditional museum settings. It analyses how curiosity and interactivity can trigger active participation, and discusses how cultural heritage engagement can foster democratic values and community resilience.

Colour Fragility. This chapter presents a scenario-based approach to the fragility of colour in Cultural Heritage, analysing five contexts: ancient sculpture, paintings, textiles, historical photography, and born-digital artworks. It addresses the challenges of reconstructing colour with scientific accuracy and authenticity, highlighting the importance of standardised diagnostic workflows, cross-disciplinary collaboration, and the use of analytical, historical, and iconographic data. The discussion emphasizes also the need for open protocols and FAIR-compliant data for future reuse and transparency.

Designing Coloured Experiences. This final chapter synthesises the design methodologies developed in the PERCEIVE project to create engaging user experiences with coloured collections. Based on the well-known PACT framework (People, Activities, Context, Technology) [4], we present here the co-design approach of the project, which made use of instruments such as Cultural Probe Kits [5] and card-based tools, highlighting iterative and user-driven design processes. It connects design strategies with conceptual pillars like care, authenticity, and participation, and proposes actionable pathways for experience design in digital and hybrid heritage contexts.

Acknowledgments. This study was funded by the European Union under the PERCEIVE project (grant agreement Nr. 101061157). The views and opinions expressed are however those of the author(s) only and do not necessarily reflect those of the European Union or the European REA.

Disclosure of Interests. The editors and authors have no competing interests.

References

1. Verri, G., Opper, T., Deviese, T.: The ‘Treu Head’: a case study in Roman sculptural polychromy. British Museum Technical Research Bulletin 4 (2010) 39–54
2. Barandoni, C.: MannInColours. The use of digitals to introduce ancient polychromy to general public. In: Proc. 9th Intl. Conf. on Advanced Computer Information Technologies (ACIT). IEEE (2019)
3. Ferrer, E.J. Strand: Altered blue and red paints: investigations of Old Man in Warnemünde and The Drowned Boy by Edvard Munch (1863–1944). M.Sc. thesis (2019)
4. Benyon, D., Turner, P., Turner, S.: Designing Interactive Systems: People, Activities, Contexts, Technologies. Pearson Education (2005)
5. Gaver, B., Dunne, T., Pacenti, E.: Design: Cultural Probes. Interactions 6(1) 21–29 (1999)

Sofia Pescarin

Organization

Scientific Committee

Francesca Casadio	Art Institute of Chicago, USA
Hartmut Kutzke	Museum of Cultural History, Norway
Catlin Langford	Centre for Contemporary Photography, Australia
Paolo Liverani	University of Florence, Italy
Rebecca Pedrazzi	NotiziArte, Italy
Michael Weinmann	Delft University of Technology, Netherlands
Giovanni Verri	Art Institute of Chicago, USA
Peter Nussbaum	Artist, Austria

PERCEIVE Coordinator

Sofia Pescarin	CNR ISPC, Italy

PERCEIVE Scientific Board

Cristiana Barandoni	CNR ISPC, Italy
Lucia Burgio	V&A, UK
Ivana Cerato	CNR ISPC, Italy
Arthur Clay	HSLU, Switzerland
Brenda Doherty	CNR SCITEC, Italy
Daniele Ferdani	CNR ISPC, Italy
Maria Lucia Giacco	MANN Museum, Italy
Holger Graf	Fraunhofer IGD, Germany
Catlin Langford	Centre for Contemporary Photography, Australia
Donata Magrini	CNR ISPC, Italy
Florent Michel	IMKI, France
Letizia Monico	CNR SCITEC, Italy
Sofia Pescarin	CNR ISPC, Italy
Irina Crina Anca Sandu	MUNCH Museum, Norway
Saptarshi Neil Sinha	Fraunhofer IGD, Germany
Sophia Sotiropoulou	FORTH, Greece
Panos Trahanias	FORTH, Greece
Giorgio Trumpy	NTNU, Norway

Giovanni Verri — Art Institute of Chicago, USA

PERCEIVE Management Board

Sofia Pescarin — CNR ISPC, Italy
Holger Graf — Fraunhofer IGD, Germany
Massimo Osanna — MANN Museum, Italy
Sophia Sotiropoulou — FORTH, Greece
Jon Yngve Hardeberg — NTNU, Norway
Simon Sappa — ANAMNESIA, France

Contents

Perception

Yoko Arteaga[1], Vanessa Bonanno[2], Peter Nussbaum[1], Delfina Sol Martinez Pandiani[3], Sofia Pescarin[2], and Sophia Sotiropoulou[4](✉)

[1] NTNU ColourLab, Gjøvik, Norway
{yoko.arteaga,peter.nussbaum}@ntnu.no
[2] CNR ISPC, Florence, Italy
sofia.pescarin@cnr.it
[3] University of Amsterdam, Amsterdam, Netherlands
d.s.martinezpandiani@uva.nl
[4] FORTH IESL, Heraklion, Greece
sophiaso@eap.gr

Abstract. The following chapter introduces the readers to the complexity of colour perception, emphasizing that colour is not an intrinsic property of objects, but a perceptual quality derived from the interaction between light and HVS, the human visual system. It analyses how light, interacting with material surfaces, influences perception through reflection, transmission, and diffusion. The chapter describes the trichromatic theory, which explains that three types of retinal cones (sensitive to short, medium, and long wavelengths) allow the perception of a wide range of colours. Phenomena like metamerism, where different colours appear similar under specific lighting, are also considered, as well as review Hering's opponent-process theory. Finally, the authors discuss the role of chromatic induction where adjacent colours influence the perception of a given colour, demonstrating how colour perception is a dynamic and contextual process.

Keywords: Perception · Colour Vision Deficiency · Multisensory

1 Colour Perception in Human Vision

The Basics of Colour Perception. Most naturally luminous or illuminated objects have the property of being coloured. This self-evident property of objects is not so obvious at first. The objects around us appear red, green, yellow or multicoloured, not because they are so per se, but because that is how we 'see' them. Colour is not an intrinsic physical property of objects, but rather a psychological property that characterises our visual perception i.e. it is a perceptual quality. The colour that we perceive is clearly due to the physical properties of objects, and it is because of these properties that we see objects in colour. However, colour is essentially different from these properties, and arises because of complex interactions between the light with our visual nervous system in the surrounding space or context. Colour can be derived from a variety of physical, chemical, and biological mechanisms [1]. But there are also cases where we

S. Pescarin et al. (Eds.): *Chromatic Visions*, LNCS 15750, pp. 1–28, 2026.
https://doi.org/10.1007/978-3-032-07792-9_1

have transient colour illusions, that is, we perceive colours that do not result from an external light stimulus but are caused by physiological causes, due to normal reactions of the visual neurons when stimulated by other factors.

Colour and the Interaction of Light with Material Surfaces. When light interacts with a surface, the physical effect (reflected light) is described as a spectral emission distribution resulting from the product of the spectral power distribution curve of the light and the spectral reflectance of the surface. The characteristic absorbance of a material, due to its molecular or crystalline structure, is the most common cause of colour. Colours of natural dyes and inorganic pigments from minerals (e.g. azurite or lapis lazuli), colours of precious stones (e.g. amethyst) and a plethora of rocks in nature or colours of synthetic pigments (e.g. ultramarine), belong to this category. Colour vision, an aspect of visual perception, is the ability of the human visual system (HVS) to perceive differences between visible light with different spectral characteristics. This luminous stimulus is perceived as colour, which our visual system does not analyse spectrally, but "translates" it, in terms of its three independent psychological dimensions: hue, saturation, and brightness.

The perception of colour as a result of the interaction of light with material surfaces, presupposes, and depends on three elements:

1. light and its spectral power distribution.
2. the physical properties of the surfaces which reflect or transmit or diffuse part of the light, depending on the composition of their matter and the microstructure of their surface; and
3. the functions of the observer's visual system, which is capable of distinguishing the spectrally modulated light that is re-emitted from the surface of objects.

However, colour perception is a more complex process that involves the reception of light by photoreceptor cells, the processing of colour information in the visual system, from the retina through the optic nerve to the primary visual cortex (V1) and higher visual areas in the brain, where interpretation of colours is influenced by both physiological and cognitive factors. Before modern neuroscience opened new avenues for understanding the mechanisms of colour perception, psychophysical experiments played a crucial role in the conceptualisation, testing, validation and evolution of colour theories and colorimetry. Psychophysical experiments were and still are the most important tool for the experimental study of visual perception and the mechanisms of colour vision by investigating the psychological response to physical stimuli, i.e. the relationship between perception and external colour stimuli. The psychological dimensions of colour, as perceived by the visual system of an observer with 'normal' colour vision, correspond to the physical dimensions of the light stimulus resulting from the interaction of light with the surface. But how are the psychological and physical dimensions of colour related?

Hue vs Dominant Wavelength. Hue is defined in the CIECAM02 model as "the degree to which a stimulus can be described as similar to or different from stimuli that are described as red, orange, yellow, green, blue, violet" [2]. The hue of a colour is what allows us to differentiate between different colours, without considering their brightness or saturation. On the other hand, dominant wavelength is a method used to describe a colour's hue in terms of its spectral composition. A colour's dominant wavelength is the wavelength of monochromatic light which appears most similar to the colour in question. While hue is a perceptual attribute categorising colours into distinct groups, dominant wavelength is a physical measure describing the wavelength of light most influential in determining the perceived colour. In some cases, the hue and dominant wavelength of a colour may match closely. However, in most cases, colours perceived as the same hue may have different dominant wavelengths due to differences in spectral composition. Newton's discovery that "white" light is complex and consists of a few distinct components that are perceived by the eye as autonomously different colours is undoubtedly a milestone in our understanding of the nature and perception of colour. In the 17th century, Newton demonstrated that when a narrow beam of sunlight passed through a prism, the white light was dispersed into a spectrum. When one of the coloured beams was passed through a second prism, the identical colour persisted, proving these beams are unique and can be used to define colour. These are known as spectral colours, as they are produced by a single wavelength. The visible spectrum ranges from 380 to 780 nm.

Saturation vs Purity of a Chromatic Stimulus. The second perceptual attribute used to describe colour is saturation. Saturation refers to the vividness of a colour relative to its brightness. Normally, this is described as the distance of the colour from a neutral grey of the same brightness level. A fully saturated colour is considered pure and appears vivid and intense, and a fully desaturated colour becomes achromatic. Desaturated colours contain a mixture of the pure colour with white light and appear muted. Purity, also known as chroma or colourfulness, is the dominance of a particular hue in a colour stimulus. It is commonly assessed in comparison to a fully saturated colour of the same hue. For example, a colour with high purity contains the specific wavelength or mixture of wavelengths associated with its hue, and a colour with lower purity will also contain wavelengths associated with other hues.

Value or Brightness vs Integral of the Reflectance Over the Entire Visible Spectrum. The third perceptual attribute that describes colour is value or brightness. It describes how bright or how dark an object appears to an observer, regardless of its hue or saturation. The integral of the reflectance over the entire visible spectrum is the mathematical calculation that gives the total amount of light reflected by an object across the entire visible spectrum. These two concepts are related as the brightness of an object depends on the total amount of light it reflects, but it is not solely determined by it. The perception of brightness also involves factors such as the spectral power distribution of the light, and higher order mechanisms such as the contrast between the object and its background. The HVS perceives the brightness of an object by comparing it with that of a perfectly white object. This quantity is defined as lightness and refers to the colours of objects rather than to light sources, and ranges in the grey scale. The lightness of a reflective surface depends on the percentage of light energy reflected from the surface, but the relationship between lightness and light energy is not linear, but logarithmic, so equal steps in reflectance are perceived as smaller changes in lightness at higher levels of brightness.

Colour Vision and the Trichromatic Theory. The mechanisms involved in human colour perception are both physiological and psychological [3, 4]. The HVS is responsible for this (Fig. 1). When light enters the eye, it is focused on the retina. The photoreceptor layer at the back of the retina is made up of a combination of photosensitive receptors called rods and cones. Cones are mostly concentrated in the central area of the retina, the macula, which contains the fovea. The central fovea contains 100% cones. Cones support the brain process known as photopic vision, which involves colour vision at varying levels of illumination. Humans have three types of cones which have different spectral sensitivity [5], sensitive to short (S), medium (M), and long (L) wavelengths, resulting in trichromatic colour vision [6]. The three types of cone cells (S, M, and L) do not transmit information about the spectral composition of light. However, the wavelength of light determines the probability that a cone will be excited (i.e. absorb photons); and thus, transmit an electrical signal to the ganglia to which it is connected. The spectral sensitivity curves of the cones of the human eye, as defined by the International Commission on Illumination (CIE), based on experimental measurements of a set of observers with normal vision, express precisely the probability that cones are excited by light as a function of wavelength. If we simulate the photosensitive subretinal leaflet of the retina with a detector whose photosensitive surface consists of a mosaic of unit detectors R, G and B in different ratios, then the three colour simulation curves combined correspond to the total spectral sensitivity of the detector system [7].

According to the trichromatic theory, first proposed around 1801 by Thomas Young, an English physician, and refined about 50 years later by the German scientist Hermann von Helmholtz, the human visual system perceives colour as a triad of stimulus values (which can be represented in a three-dimensional colour space comprising the set of perceived colours). The tristimulus values correspond to the degree to which the three types of cone photoreceptors, sensitive to short (S), medium (M), and long (L) wavelengths respectively, are stimulated [8].

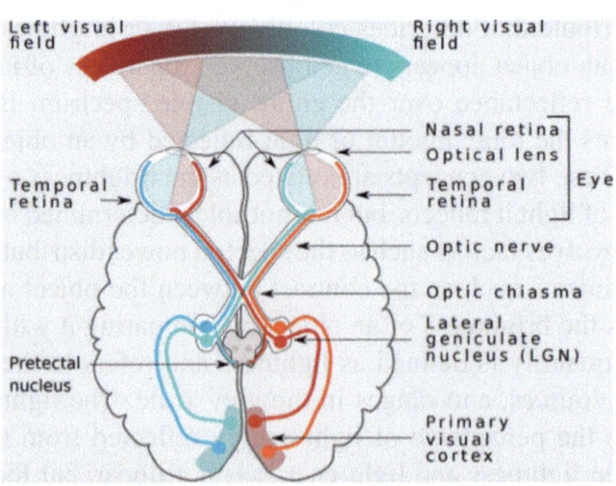

Fig. 1. The human visual system (from Wikimedia Commons, Miquel Perello Nieto)

Although the HVS is defined as trichromatic, its capacity to recognise and differentiate different colours is very extensive. It has been shown that an individual with standard vision can distinguish two hues of around 5 nm spectral difference, and it can achieve a performance peak of around 2 nm at two areas of the chromatic space, specifically in the yellow and blue-green regions [9]. While the perception of colour by the HVS can be very sensitive, it is limited by the encoding of a whole continuous spectrum into three coordinates.

The fact that the mixture of only three primary colours is sufficient to simulate all perceived colours is explained by the ability of the visual system to perceive a multitude of colours, attributing to each distinct colour a unique combination of stimulation of the three types of cones in the retina. The trichromatic theory explains the three most common cases of colour vision deficiency (dyschromatopsia or partial colour blindness) attributed to the absence of one of the three types of cones from the retina. The tri-retinal determination of the colour of a target object is obtained by weighting the spectral power distribution of the light reflected from the object by the three spectral sensitivity curves (colour matching functions) corresponding to each type of cone. This principle was the foundation of modern colorimetry, whose basic principles, laws, and data are derived from psychophysical experiments.

A prerequisite for the validity of psychophysical experimental results is a clear definition and the ability to isolate the physical stimulus (e.g. colour) under investigation, so that only the response to the specific stimulus is measured each time. Accordingly, colorimetric measurements and representations in different colour spaces correspond to the colour as perceived by our visual system isolated from the colour environment. However, the perception of colour in a colour composition or a scene, in a natural space or in a designed environment does not occur in isolation, but as a whole. Colour perception activates all the physiological mechanisms of the visual system: colour induction, colour adaptation and colour constancy.

The trichromatic theory also explains the optical phenomenon of metamerism. Illuminant metamerism occurs when two surfaces of different materials; and therefore, of different spectral reflectance, appear chromatically similar under one light source and chromatically divergent under another light source with a different spectral power distribution. Observer metamerism, occurs when the same material, under the same illumination is perceived as having a different colour by two different observers. The recognition of the colours of a metameric pair is explained by the fact that two colours can have the same trichromatic values for a given illumination and observer, even though they have different spectral reflectivity curves. The phenomenon of metamerism is an obvious proof that our visual system perceives colours without recognising their spectral composition or their physical cause. Pairs of metameric colours are not uncommon in our environment, but we are not always able to recognise them in our everyday experience. It has been observed that there is a great variability in the distribution of cones in the retina, even for non-deficient observers. These variations can happen between individuals, as well as within an individual, between left and right eye, and within the fovea. If the HVS considered these spatial differences, we would perceive colour differences constantly, even when looking directly at an object and then switching to peripheral vision, which is not the case. Therefore, the HVS is capable of extracting properties from objects as well as abstracting itself from variations in the distribution of cones.

The Neural System and Colour Perception. The Theory of Antagonistic Colour. Although the trichromatic theory consistently explains the trichromatic sensitivity of the photosensitive retinal sheet, it fails to explain observed phenomena in colour perception as a conscious subjective (psychological) experience resulting from the overall functioning of the visual system. Investigating the phenomenological dimension of colour perception and in contrast to the supporters of the trichromatic theory, Ewald Hering (1834–1918) developed at the end of the 19th century the opponent process theory (Fig. 2), according to which vision shows colour sensitivity to colours perceived in the following three dimensions: the green-red axis, the blue-yellow axis, and the black-white axis. About a century later, DeValois, Abramov and Jacobs confirmed Herring's theory. Studying the cells of the lateral geniculate nucleus (LGN), they distinguished six cell types. Four types that respond in an antagonistic manner: (R+ G−); or (G+ R−); (B+ Y−); or (B− Y+), and two types of non-antagonistic cells (Wh+ Bl−) or (Bl+ Wh−). The latter two types of non-antagonistic neurons contribute to the distinction of white and black and convey lightness information to the details of an image, regardless of colour. A hundred years after Hering's theory of antagonistic processes was formulated, and thanks to the contribution of other theorists, Leo Hurvich (1910–2009) and Dorothea Jameson (1920–1998) formulated the dual process theory in 1957, which came to bridge the trichromatic theory with the theory of antagonistic colours, proposing a synthesis of the two. According to the two-stage theory, in a first stage the receptors operate trichromatically, and then the output signals of the first stage are input signals of the second stage in which processing takes place based on the theory of competing mechanisms [10, 11]. The two stages of processing, according to the theoretical model of Hurvich and Jameson's theory, take place in the retina, which has been confirmed by subsequent physiological studies [12, 13]. It is widely used, since it is a reasonable simplification of colour vision [14].

Colour Perception as a Function of Lightness and Colour Contrasts. Due to the organization of the synaptic connections and the processing of the signals transmitted through them, the signals emitted by each photoreceptor contain simple light intensity information in the spectral region to which it is sensitive. On the contrary, signals emanating from the ganglia and transmitted through the LGNs contain light intensity difference (contrast) information. This is further processed in subsequent centres of the visual cortex (V1, V2, V4), by so called dual antagonist cells. These neurons respond simultaneously to both spatially competitive signals, between the centre and periphery of their receptive field, and to chromatically competitive signals (bipolar response R+G− or G+R− or R+G− or G+R− or B+Y− or Y+B−). These cells are selective to colour contrasts between neighbouring regions contributing to the perception of the individual elements of the visual scene. In summary, the function of the cone photoreceptors verifies the trichromatic theory, while the function of the ganglion cells and LGN cells, process the signals of the photoreceptors falling within their receptive field, in accordance with the theory of the antagonistic process. Therefore, colour is first determined by the response of the three types of cones and then recoded by the ganglia into a colour contrast signal. The connections of neurons in the primary visual cortex are such (their subfields have an appropriate spatial organization) that the colour coding is of dual antagonism, both chromatic and spatial. Thus, colour is coded not as the absolute colour of a point but as a colour contrast between adjacent areas.

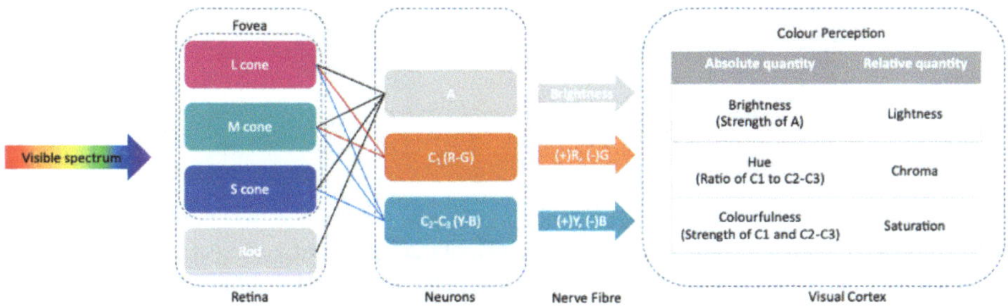

Fig. 2. Opponent process theory

If we relate the physiological functions of visual perception to artistic creation, we could say that a painter does not copy nature, but rather perceives it in terms of relationships/contrasts of lightness and colour, so that they can then capture the 'feeling' they get in their work. Since what we see is determined both by the laws and organisation of the brain, and by the tangible reality of the outside world; art, from another perspective, is a special kind of investigation into the laws governing perception and the aesthetic fulfilment we derive through perception [15].

Chromatic Induction – Interaction of Adjacent Colours. The perception of the colour of a stimulus is predictable based on its spectral composition, under certain conditions of observation and illumination, with the basic assumption that the stimulus is observed in isolation from other colour stimuli. However, in the case of observing a complex chromatic sequence or scene, the perceived hue of a stimulus may change drastically under the influence of other colours in the visual field, especially those immediately adjacent. The change in apparent colour that occurs when an object is observed in the presence of other coloured objects is called chromatic induction. Chromatic induction is a general term that includes individual discrete phenomena such as simultaneous contrast and chromatic adaptation [16]. Chromatic induction is observed not only in the case of strong hue contrasts between adjacent colours, but also in the case of strong contrasts in lightness between adjacent colours. The effect of chromatic induction can be quite significant. For example, in the case of 'simultaneous contrast' the observed colour objects have lightness and hue close to those of their complementary surroundings (Fig. 3). Then, the possible difference in perceived hue tends towards the complementary colour of that which is interacting and causing the chromatic induction. Also, we can say that when perceiving a visual-colour target surrounded by a uniform background, the chromatic components (chromaticity and luminance), which are common to the central colour and the background, do not contribute to the perceived colour of the visual target. That is, the perceived colour results from subtracting the background colour from the observed target colour.

The phenomenon of chromatic induction was known to artists and scientists in the 19th century, with Michel Eugène Chevreul (1786–1889) and Helmholtz being the most

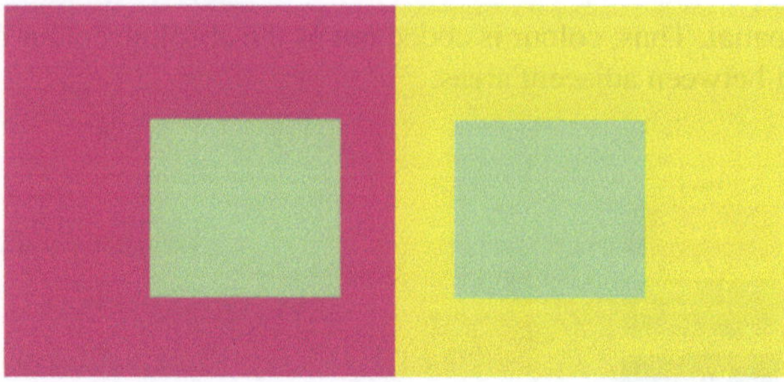

Fig. 3. Simultaneous contrast of colour: when a colour is surrounded by a neighbouring colour, the perception of its colour shifts to its opposite neighbouring colour. Left: An olive green square surrounded by magenta background, shifts in visual perception towards brighter green. Right: An olive green square surrounded by yellow, shifts in visual perception towards darker green (bluer). (Color figure online)

important theoretical representatives. In the 20th century, the most important study of the phenomena of chromatic induction and their application in painting was proposed by Josef Albers (1888–1976) at the Bauhaus School [17]. The interaction or enhanced contrast between adjacent colours is perceived and interpreted by the law of chromatic competition and simultaneous contrast. When colours of a competitive relationship (red-green, blue-yellow) coincide spatially, they cancel each other out (chromatic antagonistic effect), while when they are adjacent to each other they enhance each other (simultaneous contrast effect). Another artist who explores chromatic induction is Carlos Cruz-Diez. In his works *induction chromatique*, he explores complementary colour due to retinal persistence. Cruz-Diez shows that the colour induced on our retina is as real as the colour painted on the surface of an object. Thus, colour is a sensation on the brain of the viewer. Another process which happens along the visual pathway is that of **signal compression**. Considering light aimed at an eye increases in even increments, the resulting visual signal will increase but in diminishing increments. This has been illustrated by Joseph Albers 1963, plate xx-1. For a visually uniform increment, cube-root increments of black should be used [18].

Colour Constancy. Perhaps the most important attribute of colour perception is colour constancy: the fact that the colour of surfaces remains largely unchanged despite any -considerable- variations in the spectral composition of light. Both retinal and cortical factors contribute to colour constancy. Cells in the primary visual cortex (V1) with both spatial and chromatic opponency ('double-opponent' cells) might be important in colour adaptation which significantly supports colour constancy [3].

The 'strategy' used by the brain to ensure colour constancy is not precisely known, but several alternative explanations of the phenomenon have been proposed. Most interpretations converge on the conclusion that we are able to distinguish and identify colour under large variations in the light source thanks to the fact that our visual system responds to colour differences/contrasts rather than to individual colours. In other words, the brain encodes colour as a combination of chromatic and luminance contrast relative to the surrounding environment of the target surface (colour induction mechanism). Thus, given the mechanism of colour perception in relation to its competing environment, the colour we perceive remains constant and independent of the intensity and chromaticity of the lighting conditions. In fact, the visual system seems to be able to "subtract" from the observed colours the intensity and chromaticity of the general illumination, so that it perceives colour as the exclusive result of the reflectance of the surface of objects. It has been argued that chromatic induction can be seen as a consequence of the generalised application of the mechanism of colour constancy. For the perception of the scene, vision is assisted by colour adaptation, so the colour of the background becomes perceived as the prevailing illumination hue and is subtracted from the target colours [18].

Chromatic Adaptation. While the perception of colour can be generalised to standard observers, we have experienced that we do not always see the same colours. An example of this is the viral 2015 picture of #theDress [19–21] (Fig. 4). While some of us saw the dress white and gold, others saw it blue and black. This is because the HVS is constantly inferring queues from the scene and adapting [22]. These adaptations can be spatial, based on the scene and objects in the surrounding, as well as temporal, known as ***chromatic adaptation*** [23].

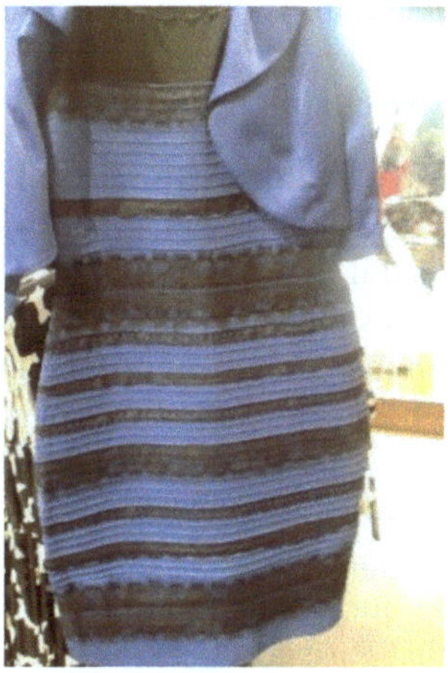

Fig. 4. Original image of a dress people may see as blue/black or white/gold (Wikimedia). (Color figure online)

Chromatic adaptation refers to the ability of the human visual system to adapt to the particular lighting conditions whose spectral composition may significantly vary from absolute white - in order to maintain a 'stable' appearance of the colours of an object. In other words, chromatic adaptation is the ability of the human visual system to compensate the colour of the illumination to consistently perceive the colour appearance of an object. For example, we can still perceive a white sheet of paper as white, under different light sources of different spectral composition and colour temperature. Chromatic adaptation relies on the ability of receptors to 'adapt' - that is to become less responsive - to continuous stimuli of the same wavelength of light. Chromatic adaptation occurs due to the adjustment of the sensitivity of the three types of cone cells working independently in each colour channel.

One of the first mathematical models of chromatic adaptation was proposed by Johannes von Kries (1902), who considered that the regulation of the sensitivity of the three types of cones is independent and linear in each channel. Other mathematical models and studies followed, which converge on some common empirical observations [24]. Firstly, chromatic adaptation is more complete at higher light intensity. Secondly, the light of most light sources, when operated at very high intensity, can be perceived as white. However, chromatic does not work well in the case of light sources of intense chromaticity, such as monochromatic LED sources. Finally, the adaptation ability of blue cones is comparatively lower.

Successive Contrast and Afterimages. Successive contrast occurs when the perception of currently viewed stimuli is modulated by the effect of previously viewed stimuli. The colour afterimages that are observed after the eye has been exposed to coloured light for a period of time can be explained in terms of changes in the relative adaptation of the three types of cones. After a period of exposure to coloured light, the cone type that is relatively weakly stimulated by this light, due to the lack of certain wavelengths, adapts proportionally to darkness. When neutral light is restored, a temporary illusion of a light composed of the 'missing' wavelengths is observed. The colour of an afterimage affects the apparent colour of objects observed after the stimulus in the same field of view, a phenomenon known as successive contrast.

Colour Perception and Cognitive Factors. Moreover, colour perception can be influenced by categorical factors. For example, the study by Berlin and Kay [25] studied how colours would be categorised if there were only 2, 3, 4, and 5 colour terms. Berlin and Kay, also assert that languages can be constrained to up to 11 basic colour terms. More recent studies, have shown that for the English language, these can be augmented to 13 colour terms [26]. It has been shown that it is easier to distinguish two pairs of colours with a 10 nm difference which are at the border between categorical hues, for example blue and purple, than within a hue. However, most recent studies have found this debatable due to methodological reasons [27]. Furthermore, some studies support the Sapir-Whorf hypothesis that our perception of the world depends on language. For instance, research has shown that Russian speakers are faster at discriminating shades of blue for which they have distinct blue terms than non-Russian speakers [28]. This has also been tested in colour naming experiments where two "Russian blues" were used as basic colour terms, showing a significant deviation from English blue [29].

2 Colour Vision Deficiency

The human visual system (HVS) of an individual with normal vision can discern millions of colours under specific viewing conditions. Photoreceptors called cones on the retina of the eye contain pigments that respond to light of varying wavelengths: S-cones for short wavelengths, M-cones for medium wavelengths, and L-cones for long wavelengths, collectively known as trichromacy.

Colour vision deficiency (CVD), commonly known as colour blindness, refers to a condition where individuals have a diminished ability to perceive or distinguish between different colours under normal lighting conditions. It's important to note that CVD is not a form of total blindness; rather, it affects the perception of specific colours [30]. The most prevalent cause of CVD is a defect in the development of the retinal cones responsible for perceiving colour in light and transmitting this information to the optic nerve. These cones, found in the retina of the eye, play a crucial role in the perception of colour. One significant aspect of CVD is its association with sex-linked inheritance. The genes responsible for producing photopigments, which are essential for nor- mal colour vision, are carried on the X chromosome. Therefore, individuals with missing or damaged genes on the X chromosome may experience colour deficiency. Males, who have only one X chromosome, are more likely to express this deficiency compared to females. Statistics indicate that approximately 8% of men and 0.5% of women of Northern European origin are affected by the common form of red-green colour blindness, making it one of the most prevalent types of CVD in the population [31]. Understanding the mechanisms and prevalence of CVD is essential for effective diagnosis, management, and accommodation of individuals with this condition in various aspects of life, including education, employment, and daily activities.

X-linked Recessive Inheritance. Impaired colour vision can arise from two main factors. Firstly, it can be acquired through diseases or medications, such as Plaquenil, used in the treatment of malaria, arthritis, and skin conditions, which can lead to colour blindness. However, the most common cause of impaired colour vision is an inherited defect in the development of the three sets of colour-sensing cones in the eyes. These defects often run in families and follow a specific inheritance pattern. Typically, they occur in every other generation and affect males exclusively. This is because the gene responsible for colour vision deficiency is located on the X chromosome, and males have only one X chromosome (XY), inherited from their mothers. Females, with two X chromosomes (XX), are usually carriers of the defective gene but may not exhibit symptoms of colour deficiency unless the defective gene is present on both X chromosomes. Now, let's consider a family tree: Girls inherit one X chromosome from each parent (mother and father). Boys inherit one X chromosome from the mother and one Y chromosome from the father. If the mother carries the defective gene on one of her X chromosomes, there is a chance she may pass it on to her son, resulting in impaired colour vision. In summary, impaired colour vision in males is primarily due to inheritance from the mother (Fig. 5).

Type of Colour Vision Deficiency (CVD). There are several types of colour vision deficiency (CVD) characterized by varying degrees of impairment in the perception of colours. These types are classified based on the extent of colour perception reduction, ranging from slight to complete absence. Individuals with CVD experience either a slight shift in the sensitivity of cone pigments compared to those with normal vision (known as anomalous trichromacy), or they may have certain cone pigments missing altogether (referred to as dichromacy) [32]. Monochromacy represents total colour blindness, where two or all three cone types responsible for colour vision are absent. Dichromacy denotes partial colour blindness resulting from the absence of one cone type. Anomalous trichromacy refers to a form of moderate colour blindness where all photoreceptors are present, but one type of cone exhibits a shifted sensitivity. In protanomaly, this shift occurs in the L cones, in deuteranomaly it's in the M cones, and in tritanomaly it's in the S cones. Specifically, in protanomalous individuals, the sensitivity function of L cones is shifted closer to that of normal M cones, while in deuteranomaly and tritanomaly, the sensitivity functions are shifted towards longer wavelengths [33] (Fig. 6) Tritanopia/Tritanomaly involves the absence or malfunctioning of the S-cone (blue), Deuteranopia/Deuteranomaly affects the M-cone (green), and Protanopia/ Protanomaly impacts the L-cone (red). Anomaly, in this context, refers to a deviation from the usual rule, type, arrangement, or pattern [31].

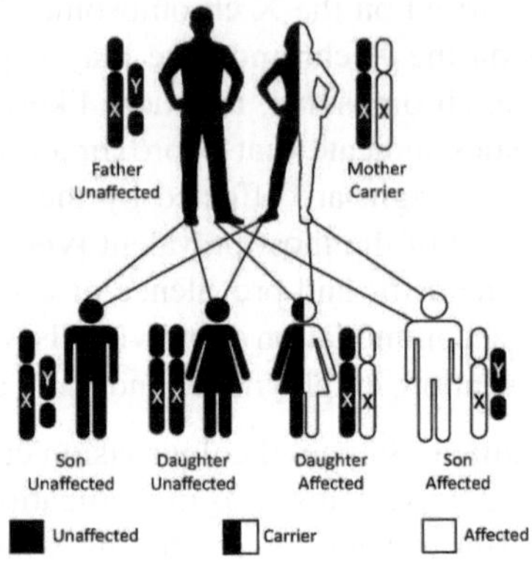

Fig. 5. X-linked recessive inheritance[1].

Confusion Lines. Confusion lines, also known as dichromatic isochromatic lines, are directional lines on a colour chart such as the 1931 xy chromaticity diagram where points lying on them are perceived as the same colour by individuals with a specific type of dichromacy [34]. For instance, protanopes, or those who are "red-blind", have confusion lines associated with them, indicating colours that cannot be distinguished by these individuals, Fig. 7. The confusion lines link a singular point (the copunctal point) with a spectral colour, indicating that colours along these lines would appear identical to a colour vision deficient observer.

Similarly, deuteranopes, or those who are green-blind, exhibit similar con- fusion lines, particularly showing difficulty in distinguishing between red and green colours, hence the term "red-green colour blindness".

While the confusion lines for both types of colour blindness are comparable, there are slight differences between them, with the intersection point often extending beyond the visible colour range, see Fig. 8. This observation underscores the shared challenges in distinguishing certain colours among individuals with red-green colour blindness, while also highlighting nuanced differences in their perception.

Human Colour Vision Test Methods. Colour vision deficiency tests aim to assess the colour discrimination abilities of individuals and determine the type and extent of any colour vision impairment they may have. Dr. Shinobu Ishihara introduced the first test on this subject a century ago, leading to the development of numerous assessments for evaluating colour perception over time. Despite considerable progress in science and

[1] https://commons.wikimedia.org/wiki/file: X − linkedrecessive.svg, accessed April 2024.

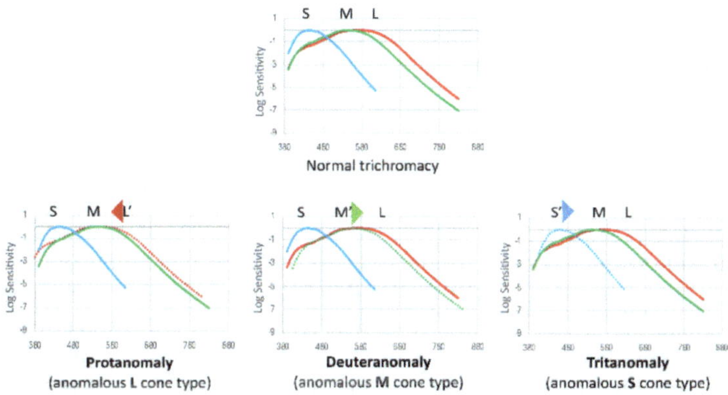

Fig. 6. Anomalous trichromacy refers to moderate colour blindness where all photoreceptors are present, but one cone type exhibits shifted sensitivity, after [33].

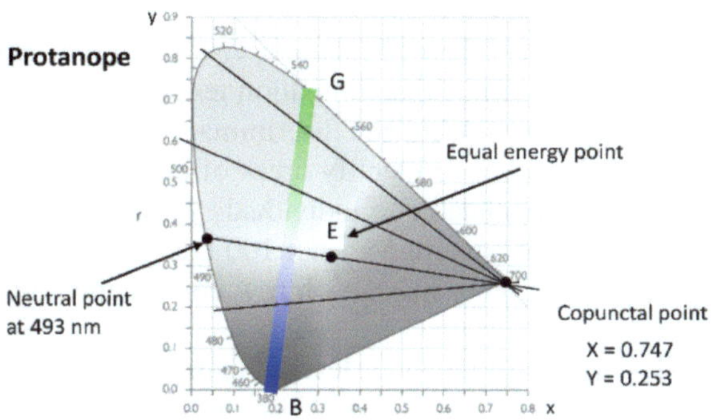

Fig. 7. 1931 xy chromaticity diagram and confusion lines for Protanope.

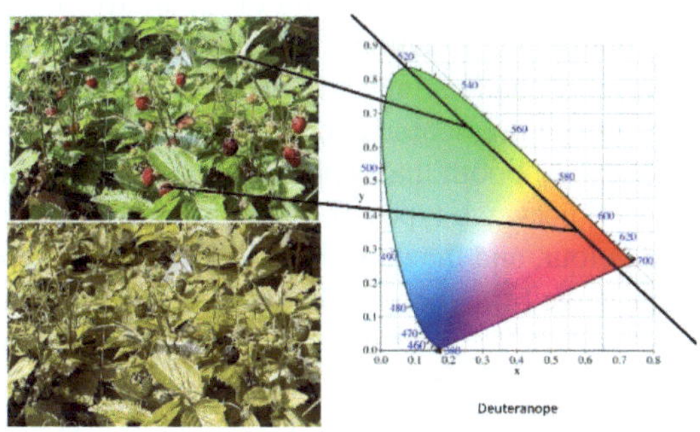

Fig. 8. Schematic illustration demonstrating the concept of confusion for a deuteranope. Both the red berries and the green leaves lie along the same confusion line. (Color figure online)

technology, the Ishihara test continues to be the predominant choice in clinical settings. However, none of the available tests provide a complete evaluation of colour vision.

There are four primary types of colour vision deficiency tests, which include:

- Pseudoisochromatic plates (recognition test)
- Ordering test (arrangement test)
- Anomaloscope
- Lantern tests

While tests utilizing pseudoisochromatic plates like the **Ishihara** [34] test claim to reliably discern between various types of colour defects, inconsistencies in their construction and printing limit their effectiveness to initial screenings only.

In 1954, **Hardy**, **Rand**, and **Rittler** (HRR) [35] introduced a set of pseudoisochromatic plates featuring different symbols such as crosses, triangles, and circles, aimed at identifying different types and severities of CVDs. A revised version of this test, known as the Hardy-Rand-Rittler test (HRR), was later developed. The HRR consists of 24 plates with coloured symbols against a grey background. By adjusting the hue and contrast of the confusion colours in the Fig from strong to weak, it becomes possible not only to differentiate between different types of CVD (protan, deutan, or tritan) but also to assess the severity of the deficiency.

The **Farnsworth-Munsell 100 Hue** [36] colour vision test is a standard assessment tool belonging to the category of hue discrimination tests, also known as arrangement tests. Its primary objective is to evaluate hue discrimination within a constant lightness and chroma curve. This test serves to identify areas of colour confusion and measure colour discrimination ability, applicable to individuals with both normal and impaired colour vision. Typically, the Farnsworth-Munsell 100 Hue Colour Vision Test consists of four rows featuring similar colour hues, each row comprising 25 distinct variations of the hue being assessed.

The **Farnsworth D-15** [37] colour vision test is effective in identifying dichromacies, particularly tritan defects. This test comprises 15 colour plates that must be arranged in the correct colour-coded order, representing a continuum of gradually changing hues. On the other hand, the **Lanthony desaturated D-15** [38] colour vision test also consists of 15 colour discs arranged to form a continuum of gradually changing hues. Like the Farnsworth D-15, the colours in the Lanthony desaturated D-15 test are less saturated. Consequently, this test is more challenging and can detect more subtle colour vision deficiencies. For accurate results, it is essential to conduct both recognition and arrangement colour vision tests in a viewing booth that replicates natural daylight conditions.

An **anomaloscope** [39] uses two types of coloured lights to test colour vision. One light is a single spectral colour, and the other light is a mix of two spectral colours. The person being tested adjusts the brightness of the single colour and the proportion of the two colours in the mix until they appear the same. This helps determine the type and severity of any colour vision deficiency.

Lantern tests [40] are the oldest and simplest confrontation colour test, initially used by the Great North Railway Company in England in 1853 and adopted as a standard colour vision test in France in 1858. An overview of multiple colour vision tests is provided in [41, 42]. Colour deficiency tests typically utilise isolated colours without including edges or altering the spatial arrangement of colours. This methodology ensures that the test focuses solely on assessing pure colour discrimination, minimizing the potential influence of extraneous factors or artifacts. Eschbach et al. [43] suggested that how well colour-deficient individuals perform on colour vision tasks depends on the amount and structure of edge information. Later, Eschbach and Nussbaum suggested adding colour edges to assess how pure colour perception interacts with perception when edges or nearby colours are present [44]. They aimed to maintain similarity to standard tests to avoid the complexity of natural scenes, see Fig. 9. Nowadays, screens are crucial in our lives and make it easy to test colour vision using smartphones. Studies have looked at online colour tests compared to traditional ones [45–47].

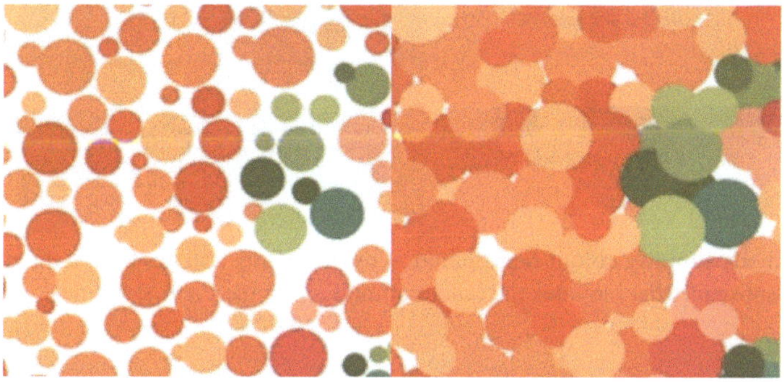

Fig. 9. Standard Ishihara chart (to the left) and a modified version (to the right) that extends the colours so that direct edges occur. This changes the overall colour appearance of the chart.

Simulation. The decrease in sensitivity along the red-green dimension results in a diminished ability to perceive certain colours, depending on the severity of the deficiency. Individuals with colour vision deficiency (CVD) encounter challenges in discerning various visual stimuli, such as navigation cues on maps and signs, interpreting colour nuances in artistic compositions, exploring historical artefacts in archaeology, and immersing themselves in digital environments. This loss of information can be approximated by predicting the ratio of cone stimulation for individuals with CVD, as highlighted in prior research [48]. However, it's important to acknowledge that while such simulations provide insights into the physiological aspects of CVD, they may not fully capture the subjective experience of colour for those with the condition [49, 50]. Green and Nussbaum demonstrated the utilization of ICC[3] profiles1 for implementing diverse transformations aimed at either simulating the visual perception of a colour image for individuals with CVD or enhancing colour discriminability for CVD observers [51].

Thus, while these simulations offer valuable tools for understanding the visual limitations of individuals with CVD, they should be interpreted with caution, recognizing the complex interplay between physiological factors and perceptual experiences.

Art and design professionals often need to check that an image or graphic will be sufficiently clear for a CVD observer, by "simulating" the effect of the colour vision loss. Simulation tools provide designers in graphic and web fields with the ability to preview their designs as they would appear to individuals with colour deficiencies, helping to identify potentially confusing colour combinations. From these simulations emerge guidelines for web accessibility, such as the Techniques for Web Content Accessibility Guidelines (WCAG 2.2)[4]. These guidelines address issues related to colour scheme selection, especially for text against coloured backgrounds across various mediums like documents, software interfaces, and web presentations. Many simulation methods project colours from a three-dimensional colour space onto a subspace containing the previously mentioned colour stimuli along the confusion lines, Fig. 10.

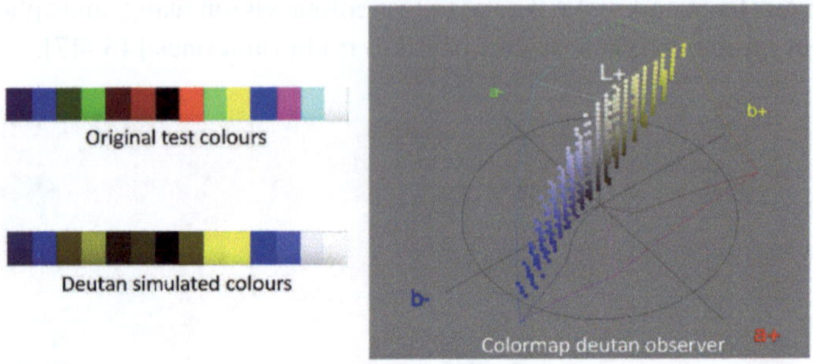

Fig. 10. Methods simplify colour representation by projecting colours from a three-dimensional space onto a subspace that includes the colour stimuli along the confusion lines.

Daltonization. Daltonization refers to the process of improving an image to compensate for missing information perceived by individuals with colour vision deficiency (CVD). Typically, this involves increasing colour contrast between colours that appear distinct to those with normal vision but are (nearly) indistinguishable to those with colour deficiency [48, 52–57]. A work by Green and Nussbaum introduced a daltonisation technique utilizing spectral filtering. They designed and implemented a spectral sharpening filter to enhance spectral reflectance for various colour pairs that individuals with CVD might struggle to distinguish. These reflectance pairs were transformed into colorimetry and displayed on a colour monitor, where observers assessed the visual contrast between the pairs through a psychophysical study. Findings indicate that an appropriate filter can amplify the disparity between red-green pairs [59].

Kotera [58] proposed a technique involving the restoration of spectral information lost to CVD observers through a spectral shift followed by an achromatic-preserving correction. This method begins by identifying spectra using a pseudo-inverse method derived from trichromatic data. In the initial step of the daltonisation process, a simulated version of the original image is generated. This simulated image is then subtracted from the original image in RGB space, resulting in an error image. Subsequently, the error image is added back to the original image, redistributing lost information to the green and blue channels (Fig. 11). Additionally, two main types of daltonisation methods can be distinguished:

Content-Independent and Content-Dependent methods. Content-independent methods do not ensure accurate colour differentiation for dichromats, as they employ a global pixel-based processing approach that disregards image content and spatial distribution of confusing colours. This uniform treatment of the entire image often leads to the creation of new confusing colour pairs or the remapping of distinct colours into indistinguishable combinations. On the other hand, content-dependent methods, including histogram-based, neighbourhood-based, and region-based techniques, consider the initial image gamut, histograms, or pixel location in the image. Although these methods are more complex and computationally intensive, they yield better differentiation of coloured image elements. Simon-Liedtke and Farup assessed well-known daltonisation techniques through a behavioural visual-search approach [32].

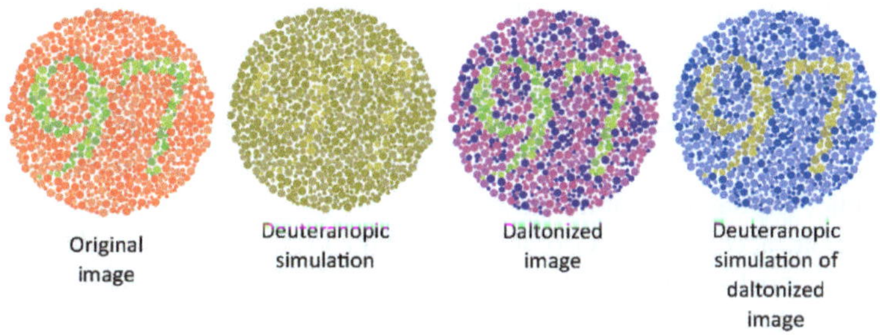

Fig. 11. A simulated version of the original image is created, then subtracted from the original in RGB space to produce an error image. Finally, the error image is added back to the original, restoring the lost information.

Conclusions. In conclusion, this section encompasses a comprehensive exploration of colour vision deficiency (CVD) and its associated topics. The introduction provides an overview of CVD, a condition characterised by difficulties in discerning certain colours due to abnormalities in colour-sensing mechanisms and the necessity of understanding its underlying mechanisms. Subsequently, the genetic basis of CVD is examined, with a focus on X-linked recessive inheritance patterns and the role of mutations in genes located on the X chromosome. The discussion then focuses on various types of CVD, such as protanopia, deuteranopia, and tritanopia, each featuring distinct alterations in colour perception and presenting unique symptomatic challenges. Exploring the concept of confusion lines, which connect colours perceived identically by individuals with CVD, underscores their significance in designing effective colour vision tests and developing accommodation strategies. Further examination involves an overview of methods for assessing colour vision in humans, encompassing pseudoisochromatic plates, anomaloscopes, and hue discrimination tests, along with their diagnostic applications. Lastly, the chapter introduces simulation techniques and daltonisation methods aimed at simulating and enhancing colour vision for individuals with CVD, emphasizing their potential to improve colour perception and overall visual experience in various environments.

3 Colour Perception Beyond Visual Stimuli

3.1 Synaesthesia and Cross-modal Experiences

The term "synaesthesia" is derived from Ancient Greek σύν syn 'together' and αἴσθησις aisthēsis 'sensation' and is the state in which a sensory (e.g. the sound of a note) or conceptual (e.g. a day of the week) stimulus causes, in addition to the corresponding sensation, some other involuntary sensation, a concurrent effect that we characterize as a synaesthetic experience. This is a particular brain state experienced by certain individuals, its "pure form" affects a small proportion, between 0.05% and 4% of the world's population, the so called synaesthetes, who, because of their peculiarity, perceive the world in their peculiar way.

For some reason, in such a state there seems to be an interference between the sensory neurons of different senses, for example, the auditory neurons and the taste neurons. So, while listening to a melody normally activates the auditory cortex, which processes auditory information, it is also possible to simultaneously activate areas of the brain that process information about taste or even texture (touch). For example, listening to a high-pitched melody can be involved in triggering the sense of taste, and the synaesthete tastes an acidic, lemony taste at that moment.

Synaesthesia, according to Grossenbacher and Lovelace [60], is divided into (a) developmental, most commonly manifesting from childhood onwards and remaining relatively stable throughout life, (b) acquired and (c) induced by addictive substances, certain medicines or drugs (e.g., LSD). There are many types of synaesthesia, which are described by the formula x->y, where x is the triggering stimulus or motivator and y is the additional experience, a concurrent effect. Colour synaesthesia is the most common, and it is divided into subcategories such as grapheme->colour to a greater extent (where colourless printed letters, colourless numbers or colourless words are perceived as having an inherent colour), and the sound->colour (either music sound->colour or general sound->colour). Other types include auditory->tactile, sound ->emotion, lexicl->gustatory, etc.

Of particular interest is multisensory synaesthesia, where sounds and specific words evoke tastes, visual patterns and colours. Because of the multiplicity of synaesthesia types, as well as the different synaesthetic experiences of the same category between different individuals, it is not clear in what way one synaesthete understands how another synaesthete perceives things and what his or her aesthetics is. For these and other reasons, the phenomenon has been largely ignored by modern neuroscience and psychology, even though, beyond its intrinsic fascination, its understanding is important for understanding sensory perception and cognitive function [65].

Synaesthesia has been described as the union or synthesis of the senses or also as the most poetic disease. And because synaesthesia is not a painful experience, on the contrary, most synaesthetes know and value their peculiarity, are able to describe it in detail as fascinating, and would certainly not rush to be cured if a therapeutic method could be found. Moreover, Theodor Adorno (1903–1969) pointed out that the appearance of a phenomenon, such as synaesthesia, is often perceived as an abnormality, whereas in fact, it functions as a privilege in the conception and enjoyment of art. According to him, we consider synaesthesia as a perceptual deficiency because we isolate and exaggerate an element that is more or less common to all human beings: the perception of the world through the cooperation of the senses.

Synaesthesia is a particular and curious state of the physiology of perception, but beyond physiological synaesthesia, there is also "metaphorical" synaesthesia. The metaphorical, rhetorical forms of synaesthesia, which are mainly associated with language, concern all of us. Almost all people associate different sensations with metaphorical speech, for example, saying that red is a 'warm' colour, that Helen has a 'sweet' face, or that someone has had a 'hot' argument.

3.2 Synaesthesia in Art

It is believed that synaesthesia is more common in artists or poets than in the rest of the population. Since a key element of creativity is the ability to connect two seemingly unrelated areas or thoughts to reveal a hidden underlying affinity, it is not surprising that some interpretive theories of synaesthesia, such as that of cross-activation or interconnection between cortical areas of the cerebral hemispheres are invoked to explain the creative ability of artists. Synaesthetic art often refers to multisensory experiences such as the visualisation, often with colour, of music, audiovisual art or abstract film. In this sense, synaesthesia in art is seen as the simultaneous perception of two or more stimuli converging in one experience.

The concept of synaesthesia has been used both by artists who were synaesthetes and explored their synaesthetic experiences to create a work of art and by artists who used it in its metaphorical sense, to create synaesthetic associations. Many artists, synaesthetes and non-synaesthetes, have tried to link the different senses, especially the perception of sound and colour, to attempt to create synaesthetic experiences without appealing to a synaesthetic audience, although it is not very clear why non-synesthetes' perception is enhanced by such connections. Twentieth-century artists such as Marc Chagall, Wassily Kandinsky or Henri Matisse, Paul Klee or František Kupka, who sought visual equivalents for concepts borrowed from music, such as rhythm/tempo, harmony, symphony or dissonance, resonance or counterpoint, coupled music and visual art in extraordinary ways. On the other hand, musicians such as Claude Debussy, Gabriel Faure or Rimsky Korsakof expressed the visual world through music. Both paths could have as their axis synaesthesia, in the sense of a connection of the senses, but which ends up as a connection of the arts, or perhaps the concept of "Gesamtkunstwerk" or "Total Art" as proposed by Richard Wagner.

The boundaries between synaesthesia as a neurological phenomenon and its metaphorical use in the context of medicine are clear, but not in art. It is very difficult to distinguish whether a work of art is the result of a synaesthetic experience on the part of the artist, or whether the artist has the perception and the talent to seek forms of expression that deliberately remove the boundaries between the arts, and in this way succeed in expanding the boundaries of the perception of art. Synaesthesia - physiological or metaphorical - is a source of inspiration for visual or musical artists, writers and poets in their art, while synaesthetic works themselves are for the general public a trigger to admire the mysteries of the mind (brain) [62].

And if synaesthesia has been a real driving force or only a source of inspiration in the creation of art that removes the boundaries between sensations and aims at perception as a holistic experience, how can we not consider synaesthesia as a precursor of the current search for artistic expression in the realms of immersive experiences and multisensory engagement? Chromesthesia, also known as sound-colour synaesthesia, is based on the phenomenon in which those who experience it (less than 1 in 2000 people) hear colours and see sounds. Generally, those who experience chromesthesia also have perfect intonation or have perfect pitch, that is, by listening to a note they are able to know exactly which note it is. This is likely because they know which colour each note represents. There are several studies that address and deepen this issue, but, to date, there are few that talk about how the introduction of phenomena such as chromesthesia within interactive experiences related to cultural heritage can positively affect accessibility. In the following section, we will add relevant studies and experiences which Relate Colour to Sound.

Sound to Colour Mapping. In 2017, Dr. Bradley D. Meyer describes about how we can see the world in colour in relation to sound, exploring the relationship between the two and how the sound spectrum can be mapped to colours, both being waves which are measured in frequencies.

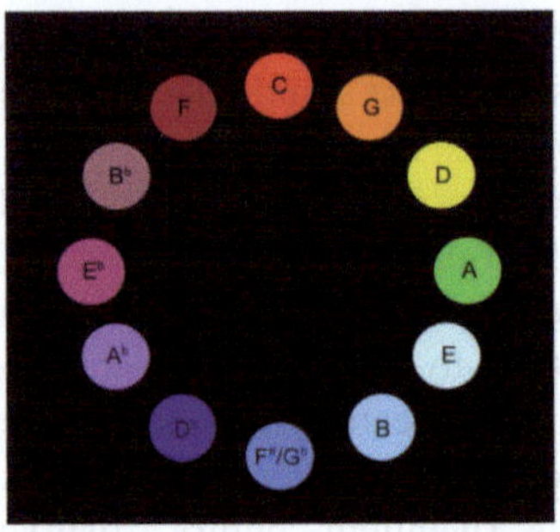

Fig. 12. Scriabin's *circle of fifths*

The idea of how to visualise this relationship had already been described in books on neuropsychology. The author analyses "synaesthesia", a phenomenon experienced by only 1 in 2000 people (e.g., even the well-known composer Scriabin) and there is no single definition. The author describes individuals who can "feel" colours and writes:

- *"they may see a specific piece of music or a chord or scale as being blue or green"*
- *"Interestingly the colours are not necessarily uniform across those with synaesthesia that equates sound to colour: One may see D major as blue, another as green, red, or yellow".*

By mapping the association of colours "felt" by Scriabin with the circle of fifths, it is possible to notice an almost perfect gradient (Fig. 12). The author concludes the article by describing how colours are already widely used in music software (e.g., Iris, SpectraLayers, Metasynth, Audiopaint) to intuitively represent the amplitude or other characteristics of the sound wave.

Soundcolourproject (SOVIS). It is an experiment conceived by Kevin Groat and Derek Torsani with the aim of guaranteeing accessibility to music using visual attributes such as light, colour and texture, thus exploring multisensory accessibility. SOVIS (https://soundcolourproject.com/) can be accessed by having: a device, a microphone, an audio interface, access to the website or by installing the app. It offers different ways of "translating" sound elements into visual elements, such as chromesthesia, chakras, emotion, chromotherapy, etc. The authors propose various applications for use, from musical performances to vocal visualisation and Chromotherapy Healing.

Play a Kandinsky. Google Arts & Culture launched this experiment in the form of a minigame by asking the question "*What if you could hear colour?*" and involving several of the most famous works by Wassily Kandinsky, a well-known painter of the Bauhaus period who probably had chromesthesia, a condition that allowed him to feel his paintings, with each colour and shape connected to a sound, a volume and an intrinsic tone. The project involves a narrative voice and simple instructions for use. All sounds are interpretations of artists who wondered what Kandinsky might have heard while he was painting some of his most famous works. At the end of the online experience, the user is given the opportunity to select which emotions gave rise to the painting he has just seen (Fig. 13).

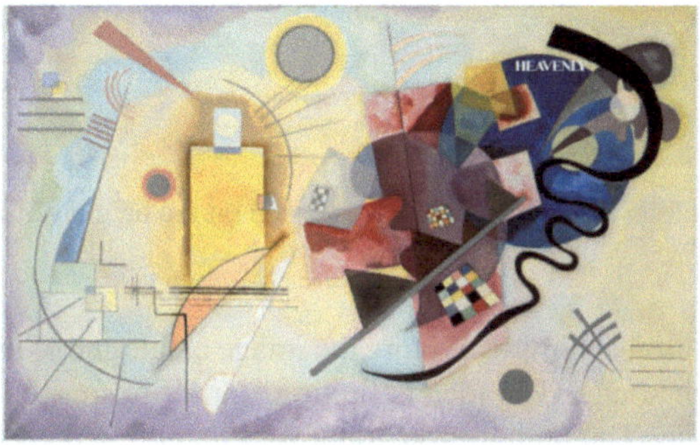

Fig. 13. Play a Kandinsky minigame by Google Art[2]

The Colour of Sound. Flutopedia website has dedicated an article (2010, revised until 2016) dedicated to the nature of sound and colour and to how they are interconnected. The author describes how it is useful in many contexts including treatments for synaesthesia, music education and therapy, and meditation. Basic aspects related to sound (e.g., vowels used during vocalisations, musical intervals) can be connected to colour using pre-existing models, such as a direct relationship (e.g. tone $\Rightarrow$ colour, where each tone has a colour, but not all colours can be mapped to a tone), or an indirect relationship, through the introduction of an intermediary with the purpose of acting as a link between the two. The author mentions the Chakra system, describing how the indirect relationship between colour and tone is affected through the correspondences, precisely, with the Chakras (Fig. 14). On the other hand, the author talks about the correspondences mapped by Scriabin (see Fig. 12), explaining how the composer placed similar colours on notes separated by a perfect fourth and a perfect fifth. At the end of the article, it is included a Clint Goss' pitch-to-colour calculator[3], which provides a visual representation in the

[2] https://artsandculture.google.com/experiment/sgF5ivv105ukhA.

[3] http://flutopedia.com/sound_colour.htm.

form of colour based on the chosen note, frequency standard used, etc. For example, choosing an A4 as Fundamental Note, the resulting colour is #ff4d00 and the calculator provides the following information: the resonant colour of light that is 40 octaves above an A4 at a pitch reference A4 = 436 has a frequency of 479.39 THz. This colour has a wavelength of 625.37 nm. The equivalent RGB colours (shown above) that approximate this colour of light are #ff4d00, based on a modified version of Dan Bruton's colour approximation algorithm.

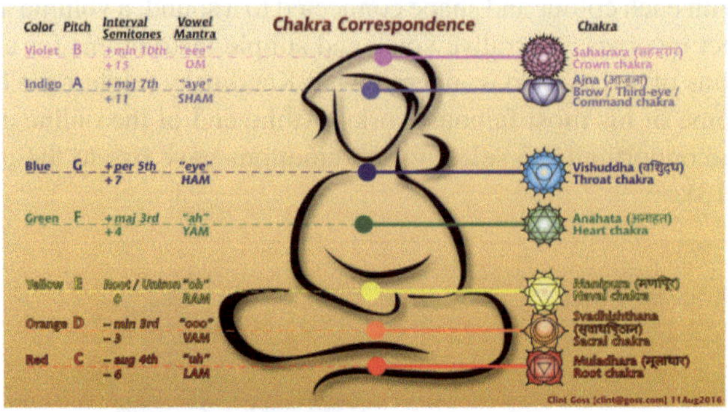

Fig. 14. Colour-tone correspondences based on the height of the Chakras

80 Hz Project. This project aims to explore the image catalogue of the New South Wales State Library through sound. The idea is linking sounds to images to suggest how the latter could sound like. The authors use "data sonification", already used in astrology and oceanography as a model for understanding large amounts of data: this allows those who do not have the ability to see the images to appreciate them in an innovative way. The project has produced an installation that has a curved wooden structure covered with anodised aluminium shingles. Inside the structure, a central mechanism displays a selection of paintings on a reel (similar to an Instagram feed). Visitors can turn a handle to select an image and listen to its dedicated soundscape. The audio is multi-channel in order to create an immersive experience by resonating within the structure, created specifically to generate reverberations. The translation of images into sound elements took into account visual data such as: complexity, colour, tone and face detection and metadata such as date, location and subject. The compositions are computer-generated (Fig. 15).

Jukepix (2018). In 2018, the authors of this project faced the challenge about how to transform paintings into music and proposed a cross-modality model to transform images into multitrack music based on Deep Convolutional Generative Adversarial Network (DCGAN), and unsupervised tasks. The proposed model is trained on a classical music dataset and an impressionist painting dataset and can be applied to transfer impressionist paintings into two-track classical music. Using music evaluation methods, the harmonicity of the generated music can be confirmed. The studio uses the "MuseGAN" project, which allows you to generate music of a length of 4 bars starting from a random noise.

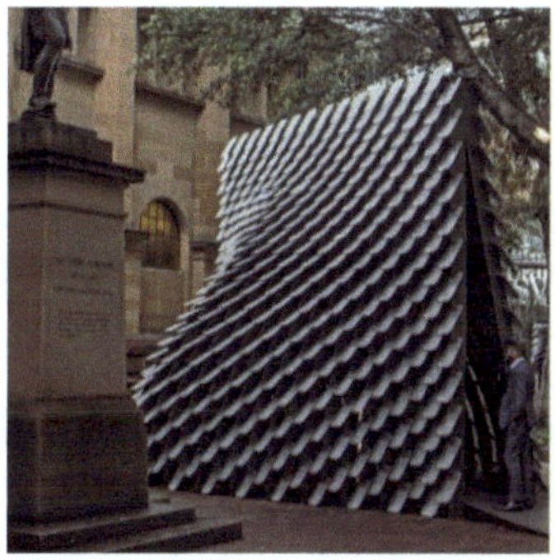

Fig. 15. The wooden structure of the 80Hz project[4]

The model proposed in the Jukepix study allowed AI to establish a method of transformation between two art forms (Fig. 16). In the future, the authors hope to improve their model by combining solo instrument tracks with orchestral tracks to generate concert music [63].

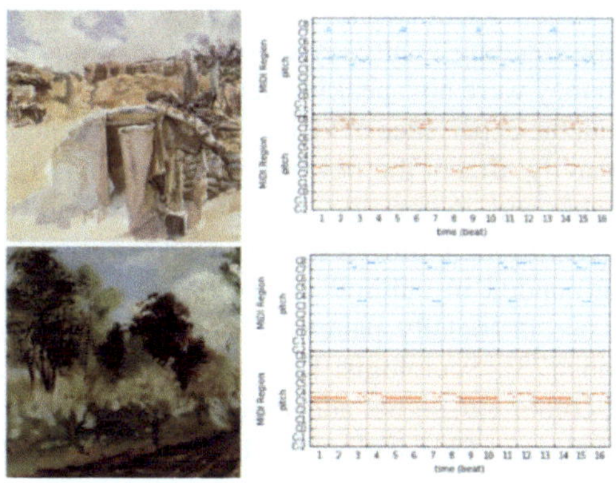

Fig. 16. Paintings and corresponding musical segments[5]

GANs N' Roses (2022). The project, accessible on GitHub[6], aims to make art even more usable by allowing users to experience it through other senses, in their case hearing, transforming a painting into a melody. Firstly, the picture under analysis is introduced into a convolutional neural network (CNN) which can predict possible emotions (20 different ones) aroused in visitors. Secondly, databases containing melodies associated with the same categories of emotions are defined in order to create a system for subsequently translating the images. Finally, the machine learning models created by the team are able to generate new sound content (Cobo 2022, Salu GitHub).

[4] https://www.youtube.com/watch?v=vXyrvWKzMVI.

[5] https://www.youtube.com/watch?v=AURD8NN-6iQ.

[6] https://github.com/GANs-N-Roses.

3.3 The Perception of the Artistic Message Through Colours

As anticipated, archaeological evidence shows that even modern humans' ancestors already used a variety of pigments [64–66], some for symbolic purposes. Why did humans, from the beginning, colour an already colourful world? Paleoanthropologists believe that, while there could be many reasons, at the core was a communication function [67] and the relation we have with colour is connected to the way we survive in this world [68]. Overall, the human vision system's goal is to seize from the continually changing information reaching the brain the most fundamental data, and there are certain kinds of knowledge, such as the colour of a surface, that can only be acquired through it [69]. According to Zeki, the functioning of the brain, and its mechanisms of perception, are governed by an active and continuous process of seeking the essential and an effort to distinguish the eternal from the desperately ephemeral. The essential, however, is nothing other than the pre-existing (Platonic) Idea within us, which, always according to Zeki, is identified through the archive of forms and states that the brain has recorded. And does not the same happen with art, as well? Is it not from this archive that the creations of artists emerge? For this reason, to understand art perception as well as to perceive colour in art, it is essential to consider perception as an inner and holistic operation of the brain. Additionally, the experience of colour is a subjective, relative, and unstable process. Josef Albers called colour "the most relative medium in art", and Henry Matisse said that "seeing is already a creative operation, one that demands an effort" [70]. Although they were not speaking neurologically, these statements express a neurological truth about the experience of vision. In fact, the context in which a perceived object is presented heavily affects the perception of colour [15]. Additionally, colour perception is an unstable and contestable phenomenon shaped by social and material factors [71]. Colour perception, in this view, is a social experience [72]. The experience of colour is deeply grounded in cultural meaning. For anthropologist Victor Turner (1920–1983), colour use is socially patterned, and reflective of basic life-and-death processes and emotions [73]. Sociologist Rose-Greenland sees colour perception also as socially patterned: colour can signal adherence to gender norms, political alliances, age, religion and marital status [74]. Colour has long-lasting cultural meanings. Colours even affect mood, which explains the careful attention that institutions pay to principles of colour harmony in public spaces [75]. The cultural currency of colour allows for a conceptualization of colour as material. Rose-Greenland finds three reasons for doing so: she sees that to trade on immaterial sensations, we need to materialise them; second, she believes that materialising colour perception draws attention to the importance of temporality, specifically the instability of colour; finally, materialization allows for the use of analytical parameters that comprise cultural objects [71].

Acknowledgments. This study was funded by the PERCEIVE project that received funding from the European Union's Horizon research and innovation programme under grant agreement No 101061157.

Authors Contributions. Yoko Arteaga and Sophia Sotiropoulou contributed Sect. 1; Peter Nussbaum contributed to Sect. 2; Sophia Sotiropoulou and Vanessa Bonanno contributed Sect. 3; Delfina Sol Martinez Pandiani and Sofia Pescarin contributed to Sect. 4.

Disclosure of Interests. The authors have no competing interests to declare that are relevant to the content of this article.

References

1. Nassau, K.: The Physics and Chemistry of Colour: The Fifteen Causes of Colour. Wiley, New York (1983)
2. Moroney, N., Fairchild, M., Hunt, R., Li, C.: The CIECAM02 colour appearance model. In: Color Imaging Conference, vol. 23, pp. 1–13 (2002)
3. Gegenfurtner, K.R.: Cortical mechanisms of colour vision. Nat. Rev. Neurosci. **4**(7), 563–572 (2003). https://doi.org/10.1038/nrn1138
4. Shevell, S.K., Kingdom, F.A.A.: Color in complex scenes. Annu. Rev. Psychol. **59**(1), 143–166 (2008). https://doi.org/10.1146/annurev.psych.59.103006.093619
5. Kevan, P.G., Backhaus, W.G.: Colour vision: ecology and evolution in making the best of the photic environment. In: Backhaus, W., Kliegl, R., Werner, J. (eds.) Colour Vision. Perspectives from Different Disciplines, pp. 1–23. Springer, New York (1998)
6. Smith, V.C., Pokorny, J.: Spectral sensitivity of the foveal cone photopigments between 400 and 500 nm. Vision. Res. **15**(2), 161–171 (1975)
7. Wyszecki, G., Stiles, W.S.: Colour Science: Concepts and Methods, Quantitative Data and Formulae, 2nd edn. Wiley, New York (2000)
8. Stockman, A., MacLeod, D.I.A., Johnson, N.E.: Spectral sensitivities of the human cones. J. Opt. Soc. Am. A **10**(12), 2491–2500 (1993). https://doi.org/10.1364/JOSAA.10.002491
9. Pokorny, J., Smith, V.C.: Wavelength discrimination in the presence of added chromatic fields. J. Opt. Soc. Am. **60**(4), 562–567 (1970). https://doi.org/10.1364/JOSA.60.000562
10. Kries, V.: Beitrag zur physiologie der gesichtsempfindung. Arch. Anat. Physiol. **2**, 505–524 (1878)
11. Judd, D.B.: Fundamental studies of color vision from 1860 to 1960. Proc. Natl. Acad. Sci. U.S.A. **55**(6), 1313–1330 (1966). https://doi.org/10.1073/pnas.55.6.1313
12. Svaetichin, G.: Receptor mechanisms for flicker and fusion. Acta Physiol. Scand. Suppl. **39**(134), 47–54 (1956)
13. Hurvich, L.M., Jameson, D.: An opponent-process theory of colour vision. Psychol. Rev. **64**(6), 384–404 (1957)
14. Daw, N.: How Vision Works: The Physiological Mechanisms Behind What We See. Oxford University Press, Oxford (2012). https://doi.org/10.1093/acprof:oso/9780199751617.001.0001
15. Zeki, S.: Inner Vision: An Exploration of Art and the Brain. Oxford University Press, Oxford (2002)
16. Walraven, J.: Discounting the background—the missing link in the explanation of chromatic induction. Vision. Res. **16**(3), 289–295 (1976). https://doi.org/10.1016/0042-6989(76)90112-7
17. Albers, J.: Interaction of Color. 50th anniversary edn. Yale University Press, New Haven (2013)
18. Stevens, S.S.: To honor Fechner and repeal his law: a power function, not a log function, describes the operating characteristic of a sensory system. Science **133**(3446), 80–86 (1961). https://doi.org/10.1126/science.133.3446.80
19. Walraven, J., Benzschawel, T.L., Rogowitz, B.E.: Colour-constancy interpretation of chromatic induction. In: Stiles-Wyszecki Memorial Proceedings. AIC Interim Meeting, pp. 269–273. Muster-Schmidt, Göttingen (1987)
20. Aston, S., Hurlbert, A.: What #theDress reveals about the role of illumination priors in color perception and color constancy. J. Vis. **17**(9), 4 (2017). https://doi.org/10.1167/17.9.4

21. Brainard, D.H., Hurlbert, A.C.: Colour vision: understanding #TheDress. Curr. Biol. **25**(13), R551–R554 (2015). https://doi.org/10.1016/j.cub.2015.05.020
22. Jameson, D., Hurvich, L.M.: Color adaptation: sensitivity, contrast, after-images. In: Jameson, D., Hurvich, L.M. (eds.) Visual Psychophysics. Handbook of Sensory Physiology, vol. 7/4, pp. 568–581. Springer, Berlin, Heidelberg (1972). https://doi.org/10.1007/978-3-642-88658-4_22
23. Monnier, P., Shevell, S.K.: Large shifts in color appearance from patterned chromatic backgrounds. Nat. Neurosci. **6**(8), 801–802 (2003). https://doi.org/10.1038/nn1099
24. Lee, H.-C.: Introduction to Color Imaging Science. Cambridge University Press, Cambridge (2005)
25. Berlin, B., Kay, P.: Basic Colour Terms: Their Universality and Evolution. University of California Press, Berkeley and Los Angeles (1969)
26. Mylonas, D., MacDonald, L.: Augmenting basic colour terms in English. Color. Res. Appl. **41**(1), 32–42 (2016). https://doi.org/10.1002/col.21944
27. Witzel, C.: Misconceptions about colour categories. Rev. Phil. Psych. **10**(3), 499–540 (2019). https://doi.org/10.1007/s13164-018-0404-5
28. Winawer, J., Witthoft, N., Frank, M.C., Wu, L., Wade, A.R., Boroditsky, L.: Russian blues reveal effects of language on color discrimination. Proc. Natl. Acad. Sci. U.S.A. **104**(19), 7780–7785 (2007). https://doi.org/10.1073/pnas.0701644104
29. Paramei, G.V., Griber, Y.A., Mylonas, D.: An online color naming experiment in Russian using Munsell color samples. Color. Res. Appl. **43**(3), 358–374 (2018). https://doi.org/10.1002/col.22190
30. Cole, B.L.: The handicap of abnormal colour vision. Clin. Exp. Optom. **87**(4–5), 258–275 (2004). https://doi.org/10.1111/j.1444-0938.2004.tb05056.x
31. Simunovic, M.P.: Colour vision deficiency. Eye **24**(5), 747–755 (2010). https://doi.org/10.1038/eye.2009.251
32. Simon-Liedtke, J.T., Farup, I.: Evaluating color vision deficiency daltonization methods using a behavioural visual-search method. J. Vis. Commun. Image Represent. **35**, 236–247 (2016). https://doi.org/10.1016/j.jvcir.2015.12.014
33. Celebi, E., Lecca, M., Smolka, B. (eds.): Color Image and Video Enhancement. Springer, Cham (2015). https://doi.org/10.1007/978-3-319-09363-5
34. Fry, G.A.: Dichromatic confusion lines and color vision models. Optom. Vis. Sci. **63**(12), 933–940 (1986). https://doi.org/10.1097/00006324-198612000-00001
35. Birch, J.: Efficiency of the Ishihara test for identifying red-green colour deficiency. Ophthalmic Physiol. Opt. **17**(5), 403–408 (1997)
36. Huna-Baron, R., Glovinsky, Y., Habot-Wilner, Z.: Comparison between Hardy–Rand–Rittler 4th edition and Ishihara color plate tests for detection of dyschromatopsia in optic neuropathy. Graefes Arch. Clin. Exp. Ophthalmol. **251**(2), 585–589 (2013). https://doi.org/10.1007/s00417-012-2073-x
37. Zahiruddin, K., et al.: Effect of illumination on colour vision testing with Farnsworth-Munsell 100 Hue Test: customized colour vision booth versus room illumination. Korean J. Ophthalmol. **24**(3), 159–165 (2010). https://doi.org/10.3341/kjo.2010.24.3.159
38. Hovis, J.K., Ramaswamy, S., Anderson, M.: Repeatability indices for the Farnsworth D-15 test. Vis. Neurosci. **21**(3), 449–453 (2004). https://doi.org/10.1017/S0952523804213402
39. Good, G.W., Schepler, A., Nichols, J.J.: The reliability of the Lanthony Desaturated D-15 Test. Optom. Vis. Sci. **82**(12), 1054–1059 (2005). https://doi.org/10.1097/01.opx.0000192351.63069.4a
40. Melamud, A., Hagstrom, S., Traboulsi, E.: Color vision testing. Ophthalmic Genet. **25**(3), 159–187 (2004). https://doi.org/10.1080/13816810490498341

41. Dain, S.J., Casolin, A., Long, J., Hilmi, M.R.: Color vision and the railways: Part 1. The Railway LED Lantern Test. Optom. Vis. Sci. **92**(2), 138–146 (2015). https://doi.org/10.1097/OPX.0000000000000460
42. French, A., Rose, K., Cornell, E., Thompson, K.: The evolution of colour vision testing. Aust. Orthoptic J. **40**(2), 7–15 (2008)
43. Fanlo Zarazaga, A., Gutiérrez Vásquez, J., Pueyo Royo, V.: Revisión de los principales test clínicos para evaluar la visión del color. Arch. Soc. Esp. Oftalmol. **94**(1), 25–32 (2019). https://doi.org/10.1016/j.oftal.2018.08.006
44. Eschbach, R., Morgana, S., Quaranta, A., Bonanomi, C., Rizzi, A., et al.: Feeling edgy about colour blindness. In: Electronic Imaging XIX: Displaying, Processing, Hardcopy, and Applications, p. 99 (2014)
45. Eschbach, R., Nussbaum, P.: Examining spatial attributes for color-deficient observers. Electron. Imaging **33**(16), 308-1–308-10 (2021). https://doi.org/10.2352/ISSN.2470-1173.2021.16.COLOR-308
46. Khizer, M.A., Ijaz, U., Khan, T.A., et al.: Smartphone color vision testing as an alternative to the conventional Ishihara booklet. Cureus (2022). https://doi.org/10.7759/cureus.30747
47. Fliotsos, M.J., Zhao, J., Pradeep, T., Ighani, M., Eghrari, A.O.: Testing a popular smartphone application for colour vision assessment in healthy volunteer subjects. Neuro-Ophthalmology **45**(2), 99–104 (2021). https://doi.org/10.1080/01658107.2020.1817947
48. Rozhkova, G., Belokopytov, A., Gracheva, M., Ershov, E., Nikolaev, P.: A simple method for comparing peripheral and central color vision by means of two smartphones. Behav. Res. **55**(1), 38–57 (2022). https://doi.org/10.3758/s13428-021-01783-3
49. Brettel, H., Viénot, F., Mollon, J.D.: Computerized simulation of color appearance for dichromats. J. Opt. Soc. Am. A **14**(10), 2647 (1997). https://doi.org/10.1364/JOSAA.14.002647
50. Viénot, F., Brettel, H., Ott, L., M'Barek, A.B., Mollon, J.D.: What do colour-blind people see? Nature **376**(6536), 127–128 (1995). https://doi.org/10.1038/376127a0
51. Green, P.: Why simulations of colour for CVD observers might not be what they seem. In: Eschbach, R., Marcu, G.G., Rizzi, A. (eds.) Electronic Imaging 2015, San Francisco, USA, p. 939511 (2015)
52. Green, P., Nussbaum, P.: Colour vision deficiency transforms using ICC profiles. Electron. Imaging **28**(20), 1–5 (2016). https://doi.org/10.2352/ISSN.2470-1173.2016.20.COLOR-328
53. Anagnostopoulos, C.-N., Tsekouras, G., Anagnostopoulos, I., Kalloniatis, C.: Intelligent modification for the daltonization process of digitized paintings. In: International Conference on Computer Vision Systems: Proceedings (2007)
54. Kim, H.-J., Jeong, J.-Y., Yoon, Y.-J., Kim, Y.-H., Ko, S.-J.: Color modification for color-blind viewers using the dynamic color transformation. In: 2012 IEEE International Conference on Consumer Electronics (ICCE), Las Vegas, USA, pp. 602–603 (2012). https://doi.org/10.1109/ICCE.2012.6162036
55. Huang, J.-B., Tseng, Y.-C., Wu, S.-I., Wang, S.-J.: Information preserving color transformation for protanopia and deuteranopia. IEEE Signal Process. Lett. **14**(10), 711–714 (2007). https://doi.org/10.1109/LSP.2007.898333
56. Machado, G.M., Oliveira, M.M.: Real-time temporal-coherent color contrast enhancement for dichromats. Comput. Graph. Forum **29**(3), 933–942 (2010). https://doi.org/10.1111/j.1467-8659.2009.01701.x
57. Viénot, F., Brettel, H., Mollon, J.D.: Digital video colourmaps for checking the legibility of displays by dichromats. Color. Res. Appl. **24**(4), 243–252 (1999). https://doi.org/10.1002/(SICI)1520-6378(199908)24:4%3c243::AID-COL5%3e3.0.CO;2-3
58. Kotera, H.: A spectral-based color vision deficiency model compatible with dichromacy and anomalous trichromacy. Color Imaging Conf. **23**(1), 127–132 (2015). https://doi.org/10.2352/CIC.2015.23.1.art00022

59. Green, P., Nussbaum, P.: Daltonization by spectral filtering. Electron. Imaging **32**, 1–5 (2020). https://doi.org/10.2352/ISSN.2470-1173.2020.15.COLOR-237
60. Grossenbacher, P.G., Lovelace, C.T.: Mechanisms of synesthesia: cognitive and physiological constraints. Trends Cogn. Sci. **5**(1), 36–41 (2001). https://doi.org/10.1016/S1364-6613(00)01571-0
61. Cytowic, R.E.: Synesthesia: A Union of the Senses. 2nd edn. The MIT Press, Cambridge (2002). https://doi.org/10.7551/mitpress/6590.001.0001
62. Dinopoulos, A.: Synesthesia: Or the Diary Writes July Blue. Parisianou Publications, Athens (2019). (in Greek)
63. Wang, X., Gao, Z., Qian, H., Xu, Y.: Jukepix: a cross-modality approach to transform paintings into music segments. In: 2018 IEEE International Conference on Robotics and Biomimetics (ROBIO), Kuala Lumpur, pp. 456–461. IEEE (2018)
64. Wreschner, E.E., Bolton, R., Butzer, K.W., et al.: Red ochre and human evolution: a case for discussion. Curr. Anthropol. **21**(5), 631–644 (1980)
65. De Lumley, H.: Les fouilles de Terra Amata à Nice: Premiers résultats. Bull. Mus. Anthropol. Prehist. Monaco **29**, 29–51 (1966)
66. Mészáros, G., Vertes, L.: A paint mine from the early Upper Palaeolithic age near Lovas (Hungary, County Veszprem). Acta Archaeol. **1–2**, 1–32 (1995)
67. Dart, R.A., Beaumont, P.: Evidence of iron ore mining in Southern Africa in the middle stone age. Curr. Anthropol. **10**(1), 127–128 (1969). https://doi.org/10.1086/201014
68. Tarlach, G.: What the ancient pigment ochre tells us about the human mind. Discover Magazine (2018)
69. Rosenqvist, T.C.: Seeing with color: psychophysics and the function of color vision. Synthese **202**(1), 20 (2023). https://doi.org/10.1007/s11229-023-04226-y
70. Matisse, H., Fourcade, D.: Écrits et propos sur l'art. Hermann, Paris (1972)
71. Rose-Greenland, F.: Color perception in sociology: materiality and authenticity at the Gods in Color Show. Sociol. Theory **34**(2), 81–105 (2016). https://doi.org/10.1177/0735275116648178
72. Pandiani, D.S.M., Pescarin, S.: Beyond static colors: an interactive participatory design perspective on color-centric experiences. Int. J. Conserv. Sci. **13**, 1691–1706 (2022). https://ijcs.ro/public/IJCS-22-125_Pandiani.pdf
73. Turner, V.: The Forest of Symbols: Aspects of Ndembu Ritual. Cornell University Press, Ithaca (2016)
74. Paoletti, J.B.: Pink and Blue: Telling the Boys from the Girls in America. Indiana University Press, Bloomington (2013)
75. Byrne, D.: Colors/Pink. Cabinet Magazine 11 (2003). http://cabinetmagazine.org/issues/11/pink.php

Colour Through Time

Irina Crina Anca Sandu[1(✉)], Cristiana Barandoni[2], Letizia Monico[3], Arthur Clay[4], Irina Mihaela Ciortan[5], Brenda Doherty[3], Roberta Iannaccone[2], Harmut Kutzke[6], Catlin Langford[7], Donata Magrini[2], Sofia Pescarin[2], Sivert Thue[1], Raquel Santos[8], and Giorgio Trumpy[5]

[1] MUNCH Museum, Oslo, Norway
{irina.sandu,sivert.thue}@munchmuseet.no
[2] CNR ISPC, Florence, Italy
{cristianabarandoni,robertaiannaccone,donata.magrini,sofia.pescarin}@cnr.it
[3] CNR SCITEC, Perugia, Italy
{letizia.monico,brenda.doherty}@cnr.it
[4] HSLU, Lucerne, Switzerland
arthur.clay@hslu.ch
[5] Colour Science Lab, NTNU, Gjøvik, Norway
{irina-mihaela.ciortan,giorgio.trumpy}@ntnu.no
[6] Department of Collection Care and Management, Cultural History Museum, Oslo, Norway
hartmut.kutzke@khm.uio.no
[7] V&A Museum, London, UK
l.burgio@vam.ac.uk
[8] Centre for Humanities, NOVA University of Lisbon, Lisbon, Portugal
raquel.st@protonmail.com

Abstract. This chapter explores the evolution of colour theories and their impact on the perception of art throughout history. It traces the development of colour theories, beginning with ancient philosophers like Democritus and Plato, through Isaac Newton's introduction of primary colours to modern theories, by scholars like Johann Wolfgang von Goethe and Albert Henry Munsell. It highlights how colour perception is subjective and influenced by cultural and social norms. It also examines how artists like Van Gogh, Monet, and Rothko used colour in their work to evoke emotional responses, communicate messages, and create depth. Additionally, it discusses the symbolic meanings of colours in various historical and cultural contexts, showcasing how artists have harnessed the power of colour to represent emotions, social status, and spiritual symbolism. In the Greek and Roman world, colour was not only associated with its aesthetic value but also had a great symbolic meaning, often as a representation of social status, divine power, and a means of art appreciation, such as the carvings and temple decorations, where bright colours were used to represent mythological and religious messages.

Keywords: Colour interpretation · Colour theories · Colour symbolism

S. Pescarin et al. (Eds.): *Chromatic Visions*, LNCS 15750, pp. 29–80, 2026.
https://doi.org/10.1007/978-3-032-07792-9_2

1 Colour Theories and the Perception of the Artistic Message

1.1 Basics of Colour Theory

One of the seven elements that are important for art objects made in different techniques and using different materials is colour, followed by value, texture, space, line, form, and shape[1]. The colour theories define colours thorough their main characteristics: hue, saturation (chroma), and brightness (lightness) [1]. The perception of colour is a subjective phenomenon and is also influenced by social and cultural norms. Scientific and theoretical investigations into colour go back thousands of years, when writings of philosophers such as Democritus, Plato and Aristotle emerged [2]. Other notable theories of colour have been developed through the centuries in the work of significant scholars, scientists, and artists, including: al-Kindi, Leonardo da Vinci, Robert Fludd, Isaac Newton, Johann Wolfgang von Goethe and others. Leonardo da Vinci was the first to talk about the various colour principles. Isaac Newton introduced a new theory of colour in 1666, in which he revealed the concept of primary colours. The entomologist Moses Harris was influenced by Newton's theory in creating the first colour wheel in 1766, which consisted of the primary colours. In the early 20th century, German painter Johannes Itten extended the previous version of the colour wheel by adding secondary and tertiary colours. He also came up with the idea of classifying warm and cool colours [3], while Johann Wolfgang von Goethe, in his "Theory of colours" (1810), had introduced the idea of colour temperature linked to the psychological impact of different colours on mood and emotion. The colour wheel invented by Newton speaks about three primary colours (blue, red, yellow), three secondary colours (green, orange, purple) and tertiary colours (red-orange; red-purple/violet; yellow-orange; yellow-green; Blue-green; Blue-purple/violet).

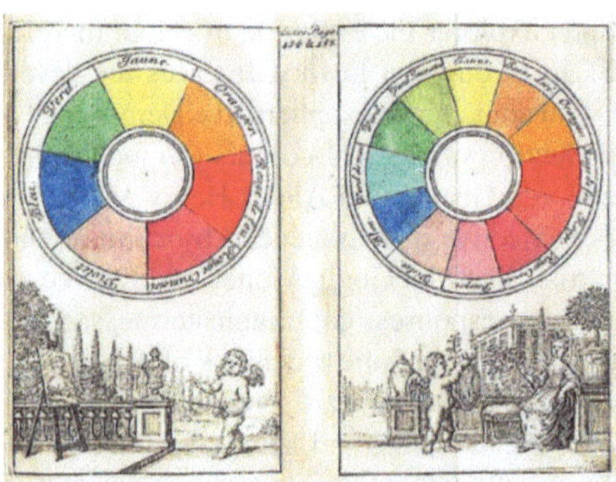

Fig. 1. Drawing thought to be Claude Boutet's seven-colour and twelve-colour circles (from Wikimedia – CC commons license) (Color figure online)

The basis for the colour wheel (Fig. 1) is rooted in Newton's experiments with prisms. Several colour schemes emerge from the use of different colours [4, 5]. While complementary and analogous are two of the well-known colour schemes, there are also

[1] https://artandthecities.com/en/2022/11/27/colour-theory.

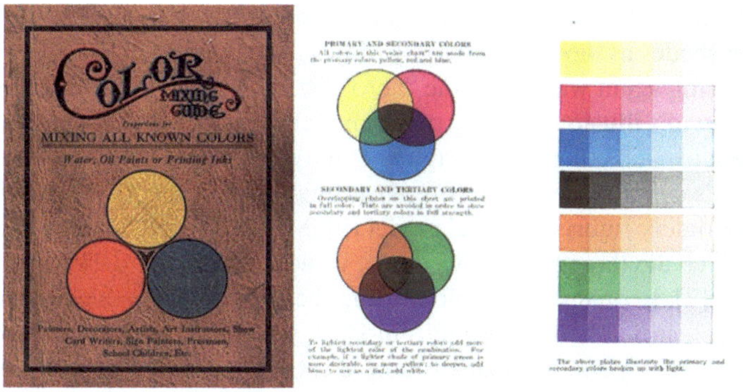

Fig. 2. Colour Mixing Guide: proportions for mixing known colours (Wikimedia – CC license) (Color figure online)

split-complementary and triadic, and others like square and tetradic [5]. Complementary colours are opposite or "across" one another. For example, red and green, blue and orange, and yellow and purple. This also applies to all the tertiary colours that are also opposite/across each other. This is referred to as a colour scheme and provides contrast in a visual composition due to their opposite placements (Fig. 2). Another popular scheme includes analogous colours, which are adjoining colours; just like opposites can be a pair, so too can colours that are side-by-side. For example, red, red-orange, red-purple/violet, or orange, yellow-orange, red-orange, etc. The split-complementary colour scheme applies to what is referred to as the "dominant" colour, which can be the primary or secondary colour and the two colours next to its opposite colour. If we use red as an example, the two colours that are on either side of red's complementary colour, which is green, would be blue-green and yellow-green, respectively. The triadic colour scheme applies colours that are an even distance from one another. There are usually three spaces between each colour. As an example, this could look like green, orange, and purple. Similarly, the combination of the primary colours will also make a triadic colour scheme. The German painter Philipp Otto Runge was the next to propose a colour wheel model. His 1807 model took Mayer's notion of three "pure" colours, plus black-and-white, and spread these ideas over and inside a 3D colour sphere. In 1839, his model gave way to Michel Eugène Chevreul's hemispherical system (Fig. 3).

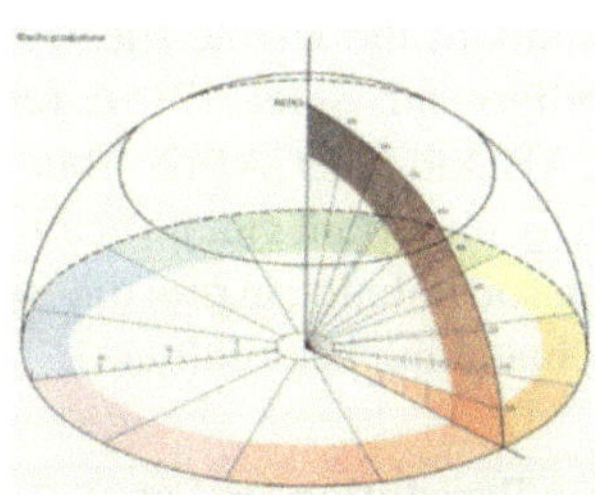

Fig. 3. Hemispherical system of colour wheel according Michel Eugene Chevreul

Chevreul arranged his 72 colours in a hemisphere, with similar proportional relationships between shades as those proposed by Mayer. The use of black and white as a lightening or darkening agent was called the "nero" factor. He also described a phenomenon known as Chevreul's illusion: the way two identical colours of different intensities, when placed adjacent to each other, seem brighter at the edge where they join. In 1900, Albert Henry Munsell's cylindrical system (Fig. 4), brought colour theory into the 20th century with an appropriately futuristic visual model [6]. Munsell opted for a three-dimensional cylinder, in which the three axes showed hue, value (lightness or darkness), and chroma (colour purity) [7]. In quantifying colour using these three values, this model described colours more scientifically than previous models, which themselves cracked the colour wheel concept wide open in favour of more ersatz shapes: Hermann von Helmholtz's cone in 1860, William Benson's tilted cube in 1868, and August Kirschmann's grandiloquent sounding "slanted double-cone" from 1895. A.H. Munsell and Newton shared the concept of likening colour notation to music notation. In his original colour wheel (1704), Newton included musical notes correlated with colour beginning with red and dividing the circle by the musical scale starting with D and ending with the octave of D.

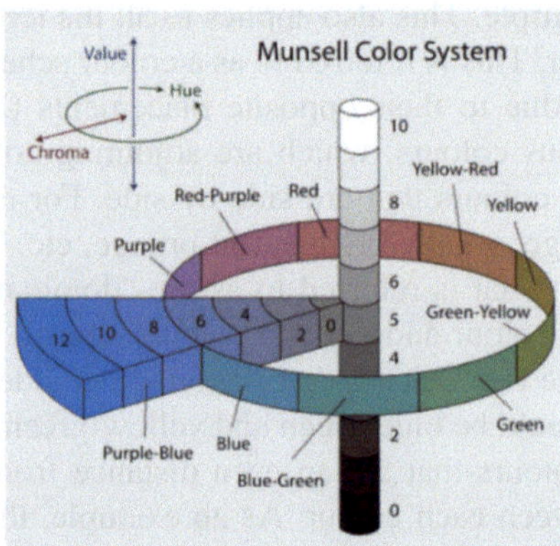

Fig. 4. Munsell colour system representation

It was no surprise that violet and purple colours are located next to red on the wheel, since they are considered non-spectral and mixtures of red and violet light. Munsell's model and subsequent books of colour follow a similar order as Sir Isaac Newton, who represented on a wheel the colours of the visible spectrum of light - ROY G BIV (red, orange, yellow, green, blue, indigo and violet) [7]. A few other models have emerged since Munsell – notably CIELAB and CIECAM2 – but Munsell's system is still used in different fields where colour is important (identification of skin and hair colours for forensic pathology, for matching soil colours, in prosthodontics for selecting tooth shades for dental restorations, and for matching beer colours).

1.2 Use of Colours and Their Perception in Art

Since the beginning of human history people have wanted to communicate and render the world using colour. The oldest pigment used for artistic purposes is **red ochre**, in the powdered form of iron oxide, which has been found in prehistorical cave paintings and burial sites around the world. Since then, naturally originated colours from minerals, stones, shellfish, clay, plants and insect have been prepared through a process of drying, grinding, and mixing with a binding medium to produce a more fluid substance that can be applied to walls, panels, or canvases in the form of malleable paint.

For the Western world, trade relations and colonial expansion introduced new and exotic pigments. Many of them were expensive (for example the lapis lazuli powders brought from Afghanistan or cochinille from Mexico) and so their use in a work of art could also reflect the wealth and power of those who had commissioned it. The industrial revolutions in manufacturing and transportation that characterised the birth of the modern and contemporary world left their mark on the invention of industrially produced pigments. Modern and contemporary artists, as Edvard Munch, were eager to experiment with new synthetic colours also displaying novel properties that would allow a longer elaboration of the paint on canvas or other supports. But these new paint formulations, introduced in the form of collapsible tubes to be easily sold, transported and used in an artist studio, also came with challenges due to stability and compatibility issues. Collapsible metal paint tubes were invented and patented by the American artist John Rand in 1841. By 1842 collapsible tin tubes in Rand's design were being commercially sold by paint suppliers such as Winsor and Newton [8]. In one letter, Edvard Munch wrote about which colourmen he purchased from his paint tubes. In this text, Munch emphasised the fact that he had used Winsor & Newton colours, as he regarded this brand as top quality, almost echoing an illustrated price list of Winsor & Newton's from 1895. Here, Winsor & Newton was stating that the pigments used in their tube colours were "the most brilliant and durable that can be manufactured," while the oils in the tubes were likewise "of the purest and most perfect quality" [8]. Ultimately even the colours themselves are not quite what they used to be, as the chemical transformation of the paint over time, and the altered environments and lighting in which they are seen, alter our perception of them. The relationship with colours is ever evolving, and our experience of works of art changes accordingly. Learning what a colour meant to the artist who used it transforms our understanding and perception of both colour and the artwork [9]. Artists like Piet Mondrian utilized primary colours in abstract paintings, such as Composition with Red, Blue, and Yellow (1930). An example of complementary use of colours in art is Vincent van Gogh's The Night Café (1888), where the red of the walls complements the green of the ceiling as shown in Fig. 5 [10]. Van Gogh wrote many letters to his brother Theo van Gogh, and often included details of his latest work.

In one of the letters from 1888, he describes this painting:

> *I have tried to express the terrible passions of humanity by means of red and green. The room is blood red and dark yellow with a green billiard table in the middle; there are four lemon-yellow lamps with a glow of orange and green. Everywhere there is a clash and contrast of the most alien reds and greens, in the Figures of little sleeping hooligans, in the empty dreary room, in violet and blue. The blood-red and the yellow-green of the billiard table, for instance, contrast with the soft tender Louis XV green of the counter, on which there is a rose nosegay. The white*

clothes of the landlord, watchful in a corner of that furnace, turn lemon-yellow, or pale luminous green[2].

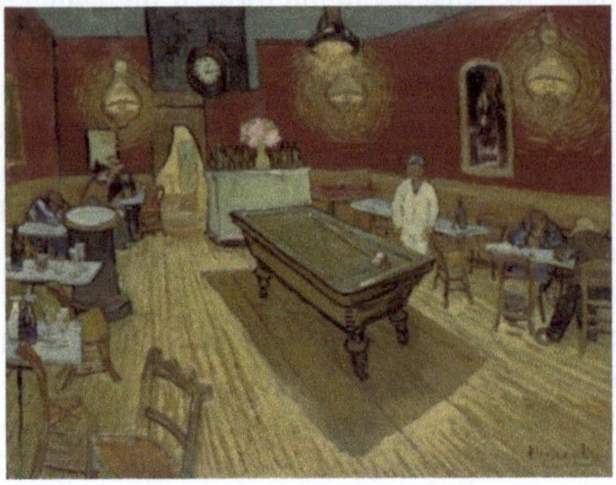

Fig. 5. Van Gogh's "Night café", 1888, Yale University Art Gallery, New Haven[3]

Another example of translation of these colour theories in art is from Mark Rothko's paintings like *Orange, Red, and Yellow* (1961), where we can see large areas of orange and a band of yellow at the top of the composition with a red background (Fig. 6). The red-orange centre of the painting tints the yellow above it with just a bit of green. The yellow above seems to tint the orange with blue. He believed that colours had the power to communicate and trigger emotional reactions. His use of warm colours like red and orange aimed to create a sense of passion and intensity, while cool colours like blues and purples might evoke tranquillity. The psychology behind Rothko's colour choices shows an artist that understood the fact that different colours could evoke different emotional responses in viewers. His paintings were often composed of stacked or floating rectangles of colour, creating a layered and immersive experience. The juxtaposition of these colours within the canvas created a dynamic interplay, inviting viewers to engage with their own emotions and memories.

Colour value refers to the level of light or darkness of a colour. It is determined when looking at a black and white image. This can also be viewed on what is referred to as a gradient or a grayscale, which will show how colours go from lighter to darker, namely from white to black, including the transitionary colours in between. A low-key colour value means it is darker and created when black is added to a colour; a high-key colour value means it is lighter and created when whiter is added to a colour. Colour value in visual arts is important because it is what creates the effect of depth or three-dimensionality [4]. The intensity in colour is also usually referred to as *saturation* or *chroma*. This is commonly described as the "brightness or dullness" of a colour. When a colour is unmixed it is known as being in its "pure" or more brilliant form [11]. Colour intensity can be applied when it is needed to create spatial depth or draw attention to a

2 https://www.webexhibits.org/vangogh/letter/18/533.htm.

3 https://www.vincentvangogh.org/the-night-cafe.jsp-, credits Wikipedia CC.

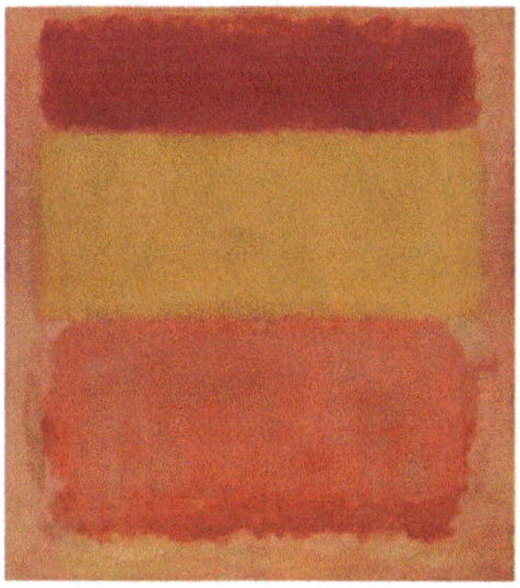

Fig. 6. Rothko's "Orange, Red and Yellow", 1961, private collection (https://www.markrothko.org/orange-red-yellow/, credits wikipedia CC.). (Color figure online)

focal point. An example of colour in art that utilises a high level of colour saturation can be seen in Ernst Ludwig Kirchner's *Seated Girl (Fränzi Fehrmann)* (1910). In Claude Monet's *Impression, Sunrise* (1872), there is a lower colour saturation, however, higher intensity is evident in the sun, which becomes the focal point of the composition (Fig. 7). This painting has been first exhibited at the exhibition of Impressionists in Paris, in April 1874 and it is considered to be the inspiration for the name of the Impressionist movement.

Colour temperature refers to some colours like reds or yellows that appear warmer and others like blue or green that appear cooler. As example of warm colours, we can consider the yellows in the Amsterdam version of "Sunflowers" (1889) by Vincent van Gogh's (Van Gogh Museum, Amsterdam), while for cold colours we can choose Claude Monet's "*Water Lilies*" (1906) at the Art Institute of Chicago (Fig. 8).

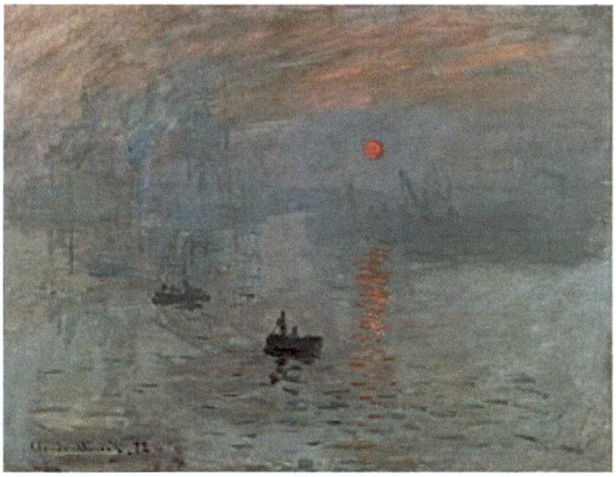

Fig. 7. "Impression, Sunrise" by Monet, 1872, AIC[4]

[4] En.wikipedia.org/wiki/impression,_sunrise.

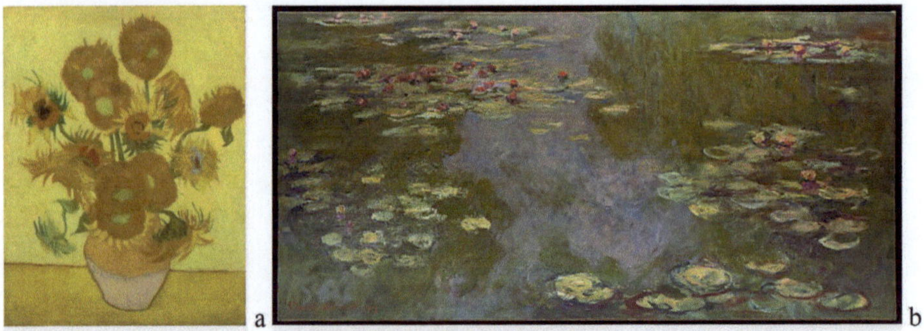

Fig. 8. a) "Sunflowers" by van Gogh, 1889, Van Gogh Museum Credits: Van Gogh Museum, Amsterdam (Vincent van Gogh Foundation); b) "Water lilies" by Monet, 1906, Metropolitan Museum[5]

1.3 The Symbolism of Colours in Painted Artworks

The use, relevance and symbolisms of colours in the history of art, however, is as broad and complex as human history itself. The exact same colour used by different artists can express divinity for one, wealth for another, or the artist's innermost feelings towards his subject matter. The raw pigments used in the manufacture of the paint give us important clues on economic and political contexts, encompassing the history of trade, exploration and exploitation of certain populations. The understanding of a colour is tied to a particular context (cultural, social, religious etc.), so for example we see reds of the past differently than those who saw them first [9].

Colour has always been recognized for its *symbolic power* [12, 13]. Our understanding and interpretation of colour symbolism has changed over time and varies from culture to country. The following overview gives some specific examples of colour-use and Mark Rothko meaning from the Western artistic tradition. It also points out some more general tendences of colours and their symbolic connotations.

Red through its association with fire and blood is used to represent danger, anger and violence but also passion, jealousy, love and sacredness. In Paul Gauguin's *Vision after the Sermon*, Jacob wrestles with the angel in a blood red field of spiritual battle, a metaphor for his internal struggle against the will of God. Other artists as Barnett Newman used red and brown/purple to symbolically represent characters as *Adam* and *Eve* (Fig. 9). He painted Eve, the Adam's companion, as a surface of red with a strip of brown on the right side, as shown in the figure. In fact, the combination of colours suggests a space of void in between the blood and the earth, the primordial sin and the essence of Eve's meaning in biblical context.

Yellow is the colour of the sun and power, being related to symbolism of life, energy, happiness, hope and wisdom. Vincent van Gogh's *Sunflowers* (Fig. 8a) is painted almost entirely with yellow and without any shadows, expressing the radiance of sunshine rather than giving us a detailed description of what the flowers look like. Van Gogh also uses

[5] A) https://www.vangoghmuseum.nl/en/collection/s0031V1962; b) https://www.metmuseum.org/art/collection/search/438008.

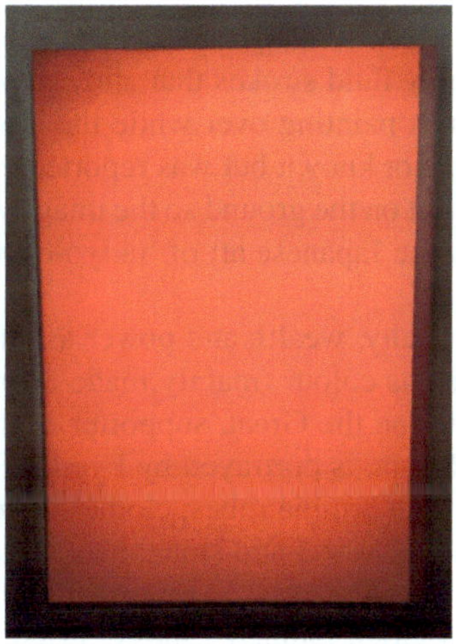

Fig. 9. Eve by Barnett Newman, 1950, Tate Britain (credits Sandu -MUNCH exhibit Shape of Freedom)

yellow as the symbol of hope, gratitude and friendship as the *Sunflowers* series was painted to welcome his friend Paul Gauguin to the Yellow House in Arles.

Orange symbol s creativity, change, energy, endurance and Autumnal energy. As a secondary colour it combines elements of the colours used to mix it: the creative passion of red with the energy and joy of yellow [4]. Rothko encouraged viewers to stand close to his large paintings so that they became spiritually immersed in the experience of colour. His painting *Orange and Yellow* is the door to an inferno of colour with a radiant energy that invites the spectator to open their emotions to "a spiritual kinship with primitive and archaic art".

Green is the colour of grass and trees and all that is associated with nature, health, cleanliness, abundance and growth. However, it is also used to represent more negative traits such as envy and inexperience. Cézanne's painting of *The Bridge at Maincy* (1879, oil on canvas, 1879–80, Musée d'Orsay, Paris) is a formal composition of horizontal, vertical and diagonal lines whose rigor is somewhat relieved by the curves of the bridge[6]. Paul Cézanne called his paintings, "constructions after nature", as he saw painting in abstract terms as the construction and arrangement of colour on a two-dimensional surface. The composition was simply a vehicle to assist with the realization of this surface structure of pattern and colour.

Blue is the coolest and most calming of all the colours, representing the sky, heavens but also deep waters of seas and oceans. In classical mythology, blue was the colour associated with the gods, Venus and Jupiter. In Christianity, it becomes the symbol of the Virgin Mary as Queen of Heaven. The calmness of blue is seldom more visible than in the Whistler's *Nocturne, Blue and Silver: Chelsea (*1871, oil on wood) representing

[6] https://www.pubhist.com/w13909.

a view at twilight from Battersea looking across to Chelsea [14][7]. All the nocturnes were painted with an extremely fluid strokes that allowed Whistler to build thin layers of luminous colour, sometimes painting over while the lower layer was still wet. The exact recipe of the mixture is not known but was reportedly so thin that the wet canvas sometimes had to be painted flat on the ground so the image would not slide off [15]. The style is strongly influence by the Japanese art of 'ukiyo-e' which translates as 'pictures of the floating world' [16].

Purple is the colour of royalty, wealth and power as only rich people could afford to wear clothes of this luxurious colour (mainly made from rare and expensive dyes). Catherine II, known as Catherine the Great, supporter of the development of the arts, literature and education in Russia, is portrayed by Fyodor Rokotov wearing a gown of the finest purple silk draped with ermine robes, clothes worthy of her noble status.

Brown is the colour of earth, wood and stone, evoking craftsmanship and the connection with earth and mud. It is also used to represent humility, dark environments or humble origins. Vincent Van Gogh painted "*Shoes*" (1888, oil on canvas) by using a darker palette of his earlier work to reflect the humility of the subject: a humble peasant in rugged earthy browns suggesting the hardship of their owner's life and to pay respect to the dignity of manual labour world' [4].

Black and its association with darkness, death, heavy emotions or feelings is used to represent evil, witchcraft, fear and mourning. *The Widow* (1921–22, woodcut) by Käthe Kollwitz is one of a series of prints from a portfolio called *Krieg* (War) which deals with the human tragedy of the first World War. This is a desolate image of a grief-stricken wife who is embracing the memory of her lost husband.

Grey is the natural colour of some metals and stone, but it also has some negative associations with the cloudy or foggy weather, boredom, decay, ashes, dust and old age. Grey is a mixture of black (death) and white (peace) and as such it is also associated with death and mourning. *Goat Skull, Bottle and Candle* (1952, oil on canvas) or *Black Jug and Skull* (1946, Tate) by Pablo Picasso are similar motives known also as modern *vanitas* still life - a traditional genre that addresses the idea of human mortality.

White and its association with light is used to represent peace, purity, cleanliness, innocence, hope and goodness. In 1915, the Russian artist Kazimir Malevich developed a geometric style of abstract art which he called *Suprematism* (*Suprematist Composition: White on White*, 1918, oil on canvas). He believed that pure abstract form had a spiritual power with an ability to open the mind to "the supremacy of pure feeling" and there was no purer colour for that purpose than white. Malevich believed White was the colour of infinity and signified a realm of higher feeling, a utopian world of pure form that was attainable only through nonobjective art.

Since prehistoric times humans started exploring the effect colour has on them, and although colour was used as an attempt to heal illness at first, it paved the way for what we recognize today as *colour psychology* (as part of *colour theory* also). In the 20th century, Carl Jung started to study how colours affected human emotions. He defined the four temperaments in terms of colours: Cool Blue, Earth Green, Sunshine Yellow and Fiery Red. Everyone has all the colour energies in their personality, to differing degree

[7] "Nocturne, Blue and Silver: Chelsea", 1871, Tate Britain: https://www.tate.org.uk/art/artworks/whistler-nocturne-blue-and-silver-chelsea-t01571.

[17] hoped to be able to use colours and art to help psychotherapy patients to heal from trauma. Along time colour psychology became a well-known field of study (Fig. 10).

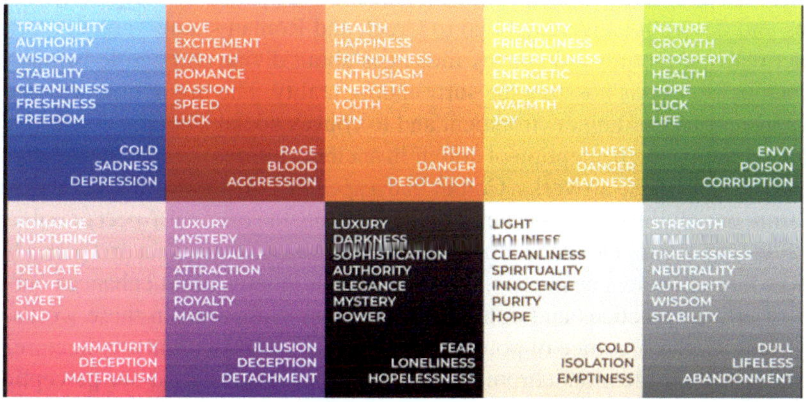

Fig. 10. Colour psychology: meaning behind each colour in the visible spectrum (credit: London Image Institute 2020). (Color figure online)

Many artists, filmmakers, surface designers, and visual artists of all kinds use insights from this field to create or enhance specific emotions in their artwork. In *The Pool of London* (1906, oil on canvas, Tate Britain) André Derain uses the contrast between warm and cold colours to express the noise and activity of this busy dockyard[8]. He creates the illusion of depth in the painting by using warmer colours in the foreground which gradually become cooler towards the background. This organized arrangement of colours in a landscape is called *Aerial Perspective* [4]. Derain was one of a group of artists known as *Les Fauves* (the wild beasts), according to a label given by a critic who was outraged by the bold colours in their art [4]. The artistic establishments of those times were offended by this wild use of colours as they respected control and restraint in the use of colour. The "Fauves" artists believed that colour had a direct link to our emotions, and they loved to use it at the highest pitch possible. The function of colour in their paintings was not to describe their subject matter, but to express the artist's feelings about it. Their ideas paved a path in the free use of colour for future generations of artists, giving them the freedom to explore colour as a subject in its own right.

2 Colours in Classical Sculptures

Colour has been a primary means of communication since the dawn of humanity. Polychromy (from the Greek *poly* = many and *chroma* = colour) was a regular characteristic of sculptures in most ancient societies, from Mesopotamia and Egypt to the Greeks, Etruscans and Romans. A debate about the practice of adding colour to statues and architectural elements emerged, among art historians, since the early 19th century. Greeks and Romans did not leave their sculptures and buildings in their original white marble

[8] https://en.wikipedia.org/wiki/The_Pool_of_London_(painting).

or stone state. Instead, they applied pigments to enhance and bring to life these artworks [18]. The use of colour in ancient art was widespread and had important cultural and symbolic significance. Discovering the original polychromy contributes to better understand the artistic intentions and cultural context of the artworks. Originally, colourful patterns and decorations were applied to Greek and Roman statues; colours were used to depict realistic features, to convey messages about divine and heroic beings, and to reproduce as closely as possible the surrounding reality. The use of colour was used to create a visually appealing environment and to convey social status.

Understanding the perception of colour in ancient cultures is essential to fully appreciate and interpret these artworks. Colour had a specific connotation and symbolism that varied from one civilisation to the next. By interpreting the original aspect of polychrome statuary, we can attempt to reconstruct the intended emotional impact and gain a broader understanding of ancient aesthetics. Studying colour in ancient art enhances our appreciation of artistic practices and cultural implications embedded in these works of art. Acknowledging the presence of polychromy in ancient statuary and architecture allows us to move beyond the monochromatic views that have dominated our perceptions for centuries. It adds a new layer of complexity and richness to our understanding of ancient cultures and their artistic expressions. Understanding and characterising the remains of pictorial materials on ancient stone artefacts is therefore a crucial issue and must be approached with care. Gathering as much information as possible about the original paint is indeed extremely important, as it offers a new concept that brings us closer to the original appearance of the sculpture. The lack of attention paid to this issue over the years has led to a significant misinterpretation of artworks and ancient artistic culture, not only from a perceptive point of view, but also from a semantic one, since colouring has been associated with a series of meanings and significances.

A correct reading is primarily made difficult by the small quantity and extent of surviving pigments, due to the vicissitudes the sculptures have undergone as the exposure to the elements or burial for over two thousand years; and even if the colour in some cases survived, it has been often extensively removed after hasty archaeological excavation or improper restorations, the latter specifically intended to reveal the Neoclassical white «pure form» of sculptures [19]. Physical remains are often incomplete, and even when traces remain, we cannot be sure that the hues and intensities remaining on a figure, reflect its original appearance.

In recent years, the scientific community has been engaged in an important debate on several research questions concerning the appearance of these coloured statues. A network of scholars working on this topic meets every two years at the "International Round Table On Polychromy In Ancient Sculpture And Architecture"[9]. The results of these debates are subjects to appreciation but also continued to generate debate, especially when the results are shown to the public in exhibitions such as "Gods in Colours[10]" or, more recently, "Chroma: ancient sculpture in colour[11]" at the MET in New York. The research questions that arise from the study and reconstruction of ancient polychromy are: *Which painting materials were used? What was the painting technique?*

[9] https://www.polychromyroundtable.com/

[10] https://www.liebieghaus.de/en/exhibitions/gods-color-golden-edition.

[11] https://www.metmuseum.org/exhibitions/chroma

Which was the sequence to apply colours? How were colours mixed to achieve particular effects? How many layers?

Preliminary evidence of original polychromy can be obtained through autoptic observation and archaeometry investigation. High-magnification images make it possible the mapping and locating on the surface of the tiny remnants of polychromy, even if they are no longer visible to the naked eye. Scientific campaigns and the integration of the results obtained through non-invasive and micro-invasive scientific analyses allow us to characterise raw materials, including pigments and sometimes binders. Archaeometry provides detailed and comprehensive instruments and methodologies to gather data about the materials, their composition and the state of conservation in polychrome stone artworks [19–21]. A variety of conventional micro-invasive techniques could be applied to comprehensively characterise the materials, such as, observations with optical and scanning electron microscope (SEM) of sample's surface or cross-sections [22]; mineralogical characterisation of materials with X-ray diffraction (XRD) [23]; compounds identification with Infrared spectroscopy (FTIR) and gas-chromatography-mass spectrometry (GC-MS) [24, 25]. Since sampling represents a permanent intervention into the original material, it is evident that non-invasive methodologies have to be preferred. Among these, imaging techniques to map the spatial distribution of colour's traces [26], UV-VIS-NIR spectroscopy [27], Raman spectroscopy [28], X-ray fluorescence (XRF) and MA-XRF [29, 30] to identify pigments and colourants, have proven to be a valid support to avoid sampling or to drastically reduce their number. Additionally, the acquisition of these type of measurements on the artwork can be carried out rapidly and *in situ*, facilitating the exchange between the participating scholars while the measurements are still in progress [26].

In addition to the scientific data, the reconstruction of the paint palette is also based on the analysis of primary and secondary sources that describe recipes, techniques and the preparation of pigments. As an example, Pliny the Elder's *Naturalis Historia*, written in 77 C.E. directly discusses the tradition and history of art in Greece and Rome, focusing on materials, artist histories, wall painting techniques, and sculpture [31]. The contribution of Pliny's monumental work as well as providing a commentary on the materials used during imperials time, also implies moral reflections on his contemporaries, examining the relationship between the viewer and the natural world; a point of particular interest in Pliny's discussion of coloured marbles is well highlighted in the discussion of the artistic role in the treatment of pigments, as we can read in book 35. The most recent critic has carefully evaluated the text, which is not free from incoherent digressions, but it is extremely important for the discussion of categories such as natural philosophy, the divinity of nature alongside with much more pragmatic and empirical discussions on pigments. He discusses these themes in great depth in this book, in which he frames the history of painting alongside naturalistic forays aimed at recovering the origin of some colours. In the same book we also find some well-documented insights into the use, or abuse, of metals in Greco-Roman architecture art. In parallel, of particular interest are his digressions on the theme of paintings used by ambitious politicians and conquerors as trophies. Pliny also did a thorough discussion on the treatment of pigments: thus, we know that for example red lead (*minium*), an extremely expensive red pigment, was considered sacred by the Romans who in the past used it only and exclusively on rituals, such as illuminating the face of the statue of Jupiter Optimus Maximus or adding it to ointments during the celebration of some recurrences.

What do we have today? What's remaining? Sometimes, raw materials themselves, discovered in original pots, provide the evidence of the use of selected pigments in a specific context.

What do we know about colour application techniques? Some scholars argue, for example, that the surface of marble was not intentionally roughened to enhance its capability to receive the colour [32], however this cannot be accepted as a general rule; on the other hand, we can state, with reasonable certainty that the presence of decorative geometric motifs resulted in a series of regular interval decorations, such as for example those discovered in borders or garments. Multispectral investigations have also revealed the presence of preparatory drawings or small incisions on the surface, a sort of *ante litteram* sinopia, indispensable in some circumstances to emphasise anatomical details. A particular care was due to human or god's figures whose eyes had to be intense and penetrating so that the *sculptures seemed alive.* The painting technique may have made use of a complex combination of pigments with highlights and shadows which are difficult to find in place and retrace. Moreover, while the reconstruction of the paint palettes used in antiquity can be considered satisfactory, mainly thanks to the historical sources, a little is known about ancient painting techniques and how colours were layered on statues, and how they might have been blended, or shaded. Moreover, while literary sources indicate the use of different binders (e.g., wax, egg), these are often hard to be analytically detected, but surely would have affected the final appearance of a work. Ancient statues were painted by artists using a complex mixture of pigments and binders and by an artistic technique and a personal sensitivity that is impossible to reproduce today. The colour palette showed that the choice was much wider than we can imagine today; some scholars even argue that no space was left without colour [33].

Hypotheses can be made, but without sufficient data, they are open to debate and discussion. For instance, studies suggest that painters did not apply colours in a uniform, flat manner. Instead, they aimed to achieve a *natural aspect* of their work. The lack of information about the original appearance of the works, including the amount of colour on the surface and the appearance of the final finishing, makes our understanding today very difficult, which is one of the reasons why today there is much discussion about the quality and finish of 1-to-1 reconstructions in resin or plaster. Even when based on physical evidence, it is close to impossible to recreate the ancient polychromy as it would originally have appeared [33] and this involves a substantial difficulty in understanding the meanings related to colours since - nonetheless important to stress - colours were part of lived life from the earliest history of mankind [34].

What can we do today to try and understand the use and the final appearance of those works of art? Despite, for almost a century, archaeological exploration of Greece and its borders has continuously delivered more and more evidence of a polychrome affecting all the plastic expressions of Greek art, friezes, pediments, metopes, isolated statues, groups, steles: it remains today to understand the reasons for such white resilience, against all the evidence. Jockey believes that the roots of this superiority of white could be traced back to antiquity itself, when the passion for the excellence of the form of the

Greek masterpieces of the 5th or 4th centuries, pushed during Neoclassicism to privilege the reproduction of sculpture leading the absence of colours as normality [35].

The last twenty years have seen museums and scholars closely engaged in experimenting with new reconstruction techniques, thus guaranteeing a remarkable experiential database. Experiments of this kind, however, were limited to the physical reconstruction of pictorial paintings proposing copies of originals on which the ancient colours have been repositioned (Fig. 11). This kind of reconstruction has been a real keystone in understanding the relationship between the ancient world and the visitor of museums, recently becoming the subject of exhibitions [36, 37].

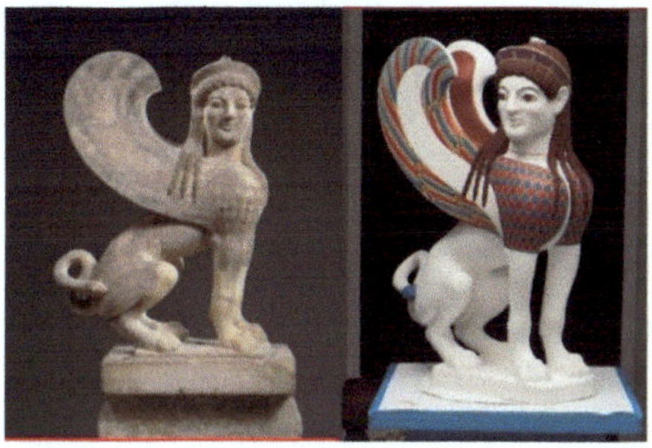

Fig. 11. Left: Detail of marble capital and finial in the form of a sphinx. Greek, Archaic period, ca. 530 B.C. H. with acroterion 142.6 cm. Munsey Fund, 1936, 1938 (11.185d, x); Right: Reconstruction of the sphinx based on the pigments identified[12]

However, there is not a general agreement on the 1-to-1 scale reproduction of these prototypes, both for ideological reasons and due to the significant approximation applied, although the colours are repositioned following scientific research. Recently, attempts to reconstruct polychromy, especially in the field of statuary, have increasingly resorted to sophisticated and complex applications, tested both on works, groups or statues in their original context, mostly using AR or Projection Mapping technologies. These experiments are of considerable importance since the pictorial framework of the sculpture dialogue closely with the environment within and for which it was set. Understanding the work within its architectural context is challenging today. The use of digital technologies offers the opportunity for an untried path to explore: to cite an example in 2012, Dimitrios Pandermalis, former Director of the Acropolis Museum, launched the 'Archaic Colours' initiative, regarding the polychromy of archaic sculpture seamlessly incorporated into the ongoing exhibition featuring archaic sculptures from the Acropolis. This addition introduces to the use of individual pigments, 1:1 physical reconstruction, and detailed explanatory texts. Among these reconstructions stands the 'Peplos Kore' (Fig. 12) which has been meticulously *brought to life:* this sculpture acts as a 3D canvas within an engaging online interactive digital experience available on the museum's website, complemented by a range of other interactive activities [38]. The project was carried out within the Eu CHESS project (ICT-2009.4.1). Therefore, a colour reconstruction is

[12] Images from: https://www.metmuseum.org/perspectives/articles/2022/8/new-research-greek-sphinx.

increasingly necessary, particularly if we move from the concept of a colour reconstruction to that of a 'potential appearance' or a proposal reconstruction, based on ancient sources study and archaeometry surveys. On the other hand, abandoning the idea of the whiteness of classical sculptures could lead to misinterpretation and, in some cases, support false and dangerous ideologies, such as the "white supremacist" propaganda that still utilises these artworks (Fig. 13).

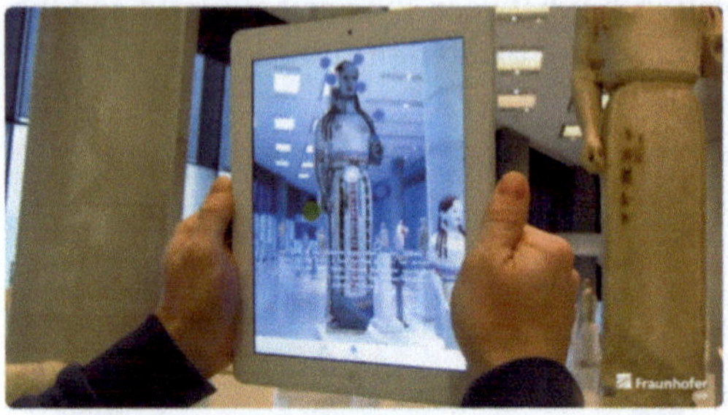

Fig. 12. Peplos Kore Project with AR to tell stories at Acropolis Museum (Credits: Fraunhofer)

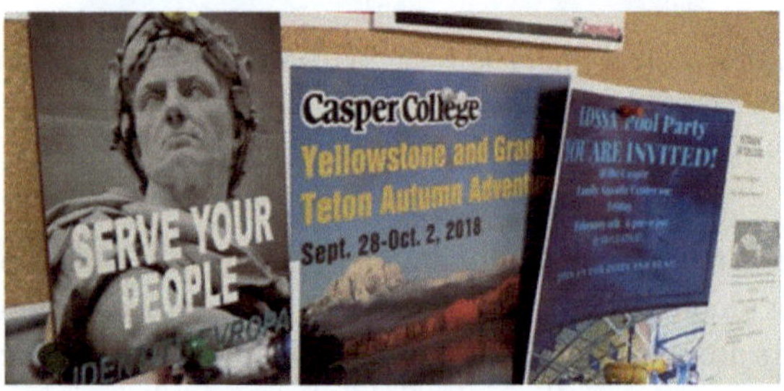

Fig. 13. A white supremacist flyer on a wall (source: Start Tribune journal online)

This example underlines the importance of colour. To date, however, there is still no common approach to understanding the most correct method, nor are there common requirements for the scientific reliability of these reconstructions. These are used to establish parameters for this experimentation. Combining reconstruction with scientific research data is an even more challenging task, since most of the scientific techniques carried out, give as their final output spectral data, which are difficult to integrate in three-dimensional digital reconstructions.

The statuary collections in archaeological museums all over the world are considered real mines of information on polychromy in the Greek and Roman contexts. As we know, traces of the original colours on sculptures have disappeared over time so when there are examples of preservation, it is worth examining in depth, since they are real rarities. Statues exhibited at the National Archaeological Museum of Naples (MANN), representing different iconographies of Venus of Pompeian provenience, still preserve evident coloured traces, and could be suitable for such inspections (Fig. 14, a-c). It is a series of statues depicting the iconography of Venus Anadyoneme, a portrait of the goddess who, as she comes out of the water, wringing out her hair after bathing by raising her arms. Among the numerous examples in the museum, two have been selected whose colours are perceptible even to the naked eye. This iconography, which dates to 4th century BC became widely spread from the mid-3rd century BC onwards. The sculpture from the House of Camillus is directly inspired by the iconography developed by Alexandrian Greek workshops: the iconography derives from a famous painting by Apelles purchased by Augustus in Cos and brought to Rome for the temple of the Divo Julius. The restoration of the painting commissioned by Vespasian could be one of the reasons why this iconography became widely spread in the Roman, in Pompeii where is often linked in the cult to Isis expressing religious syncretism. From this point onwards, numerous variations developed, characterized by postures, the presence or absence of clothing, the hairstyle; the myth is associated with the theme of birth and water source, ideal architectural setting within Pompeian homes as decorations for fountains or near aquatic environments, such as gardens or nymphaea.

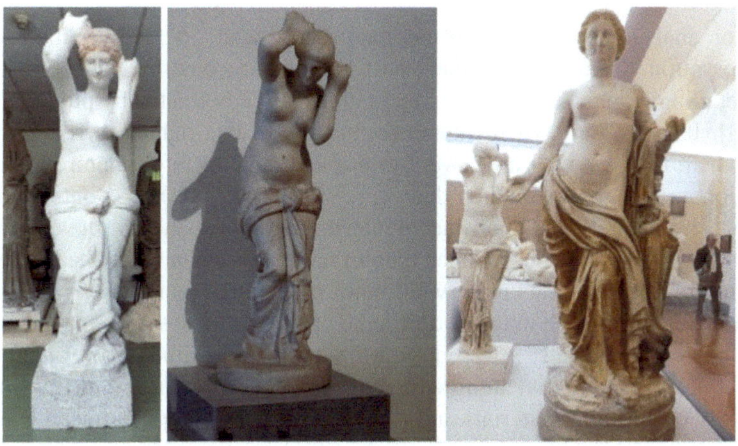

Fig. 14. Examples of Venuses at MANN museum. From left to right: a. Anadyomene, INV:6292; b. Anadyomene from the Temple of Isis, INV: 6298; c. Lovatelli Venus, INV 109608

The INV 6292 comes from the Casa del Camillo (VII 12, 22.23) in Pompeii and already at the time of its discovery in 1863 the excavation diaries report a note which documents the presence of "red" colour on the robe and on the band that holds her hair back. She is depicted in the typical gesture of squeezing her hair to remove excess water after bathing; her elegant hairstyle is enlivened by voluminous waves divided into strands, held together by a thin band. Partially naked, she has a large cloak wrapped around her hips, tied at the pubis with a large knot [38]. The second, INV 6298, comes from the Temple of Isis in Pompeii where it was found next to the wall of the south portico, east section. The small statuette was made in white marble from the Aegean islands and has her hips wrapped in a mantle. In literature, reference is made to traces of gilding on the hair, breasts and a necklace, while the cloak still partially retained a blue colour decoration [39]. Although the sculptural rendering is of inferior quality compared to INV 6292, this Venus is further evidence confirming the relationship with the cult of Isis, whose syncretism is well documented in Pompeii [40]. The last one INV. 109608, is the renowned Lovatelli Venus, found in Pompeii in 1873, in the lararium built on the east wall of the peristyle of the House of Diomedes (I 2, 17): it is one of the rarest examples of classical statuary dating back to the 1st century AD whose pictorial framework is substantially intact. The goddess rests her left arm on an archaic-style idol of Spes, placed on a black-painted rock. She wears a green chiton and a yellow peplos, with a modius adorning her head. Traces of colour, including yellow in the hair and on the himation, are evident across the surface. The idol, positioned on her left side, also displays various coloured traces. This model is reminiscent of a famous original from the 4th century, evoking the iconography of Aphrodite Urania. Additionally, it is hypothesized that the Roman copies derived from the model found in Greece and that the nudity of the torso was a Roman variant to better suit the taste of imperial era [41].

3 Colours in Roman Frescoes

This section analyse specifically Roman painting on walls, describing the problem of their conservation, discussing approaches to study and reconstruct the original colours, without detailing Roman wall painting techniques, already treated by publications such as [42, 43]. It is well known in literature that the events that occurred on frescoes, in particular in the Vesuvian area, were multiple and of different nature: damage due consumption and tear of time when they were still in situ, seismic events, burial under blankets of ash and lapilli and finally accidental damage due to their discovery. Excavated about 150 years ago and exposed since then to rain wash when not protected by temporary structures, wall paintings underwent both to chemical (acid gases) and biological (colonisation) attacks which may be considered one of the most serious problems of their conservation. What normally affected the wall paintings in situ was prevalently constituted by saline compound that along with magnesium and sodium sulphates, influenced the durability of building materials and caused discolouration phenomena compromising their aesthetic value [44]. It must also not be forgotten news from the chronicles of the excavations which report for example how in official circumstances, during state visits or parades, frescoes were wetted with solutions of water and ammonia to make the colours bright and vivid. Finally, a further aspect that must be taken into consideration is also the intervention of detachment of the panels from the walls for their transfer to museums where, further restorations carried out over time, have conditioned and still conditions the original aspects. Mass tourism is only the last plague which may have a negative impact on those very fragile witness of the past.

Walls of Roman buildings were painted using the "*fresco*" technique. The paint was applied to a layer of plaster, when this was wet. All details were added after this phase, painted on partially dry or on dry plaster. As described also in Pliny [31], a preliminary sketch of the final composition was drawn, the so-called "*sinopia*", drawn on a layer of rough plaster, called "*arriccio*". After this phase, a final layer of smooth plaster ("*intonachino*") was applied. The outline of the foreseen paint was engraved in short time, into this coat of plaster; at this point, the painters could begin the work, starting from the top part of the wall (Fig. 15). These artists were called "*pictor parietarius*" and their technical instruments included plumblines, punches and squares to be able to draw perfect architectures as well as precise details. All outlines were then painted. There was also the "*pictor imaginarius*" who would be specialised in painting the figurative scenes directly on the wall or in wooden panels. From the repetition of some pictures, scholars have understood that during Roman times there might have been "cartoons", circulating among the artists, to enable them to copy part of compositions or/and decorative motives and, eventually, adding their variations. Paintings could be also produced on marble surfaces and not only on walls and on wood.

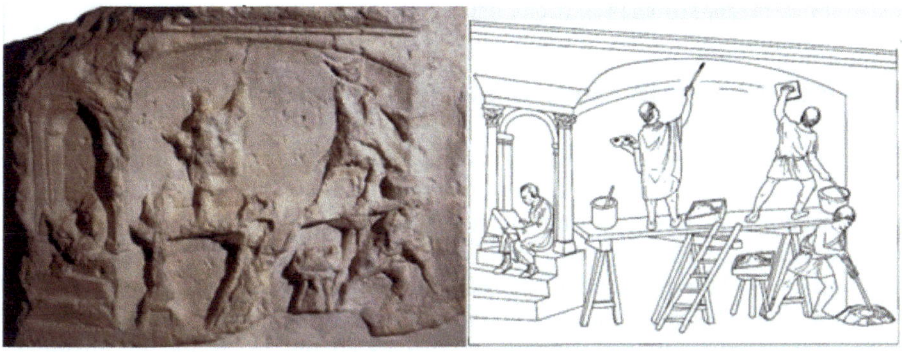

Fig. 15. Painters' workshop [43]

Regarding the type of colours, painters used, as Pliny explains, the pigment employed might have been of mineral, vegetal or animal origin. They were divided into two categories: the so-called "*florid*", the most expensive ones, that were brighter, and the "*plain*" less bright and expensive, and more widespread. Main colours used were:

- **Red**: obtained from hematite or from calcination of yellow ochre (Fig. 16). Among the reds, we know that also a lead containing pigment was used, known in antiquity as "*vermilion*", mercury sulphide, which tended to blacken from exposure to light alone or in combination with chlorine-containing compounds [45–47];
- **Green**: extracted from minerals containing celadonite or glauconite (the most precious was made from malachite);
- **White**: produced from calcium carbonate;
- **Black:** obtained from carbon made by burning pine resin or bark or vines but also from ivory and bones;
- **Blue**: made of a mixture of sand, potassium nitrate or sodium nitrate and copper. This mixture was used to obtain the most used type of blue, the so-called "*Egyptian Blue*", initially produced in Egypt at Alexandria and later introduced in Italy (Campania, Pozzuoli), spreading from here to Gaul (Fig. 16). The mixture was burnt in a kiln, in small balls shapes reduced to fine powder. The most precious of these blues was "*armenium*" (from azurite) and "*scythicum*" (known as *lapis lazuli*).

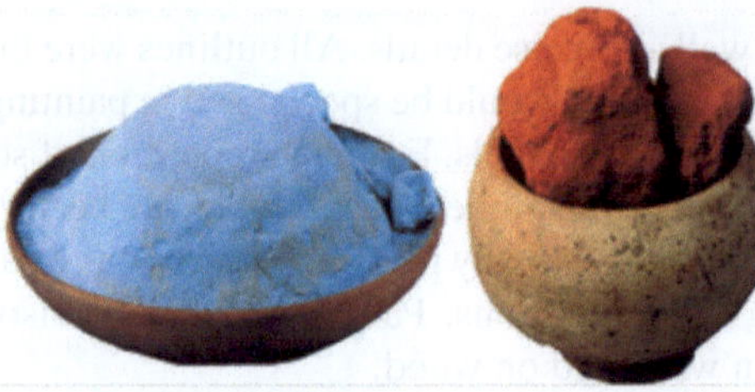

Fig. 16. Cups with Egyptian blue (Pompeii MANN inv. n. 117338) and red ochre (Pompeii MANN inv. n. 112265) (from [43]) (Color figure online)

4 Colour Change in Modern and Contemporary Paintings

Darkening, fading and yellowing, often accompanied by flaking, crumbling and chalking of the paint, affect tens of thousands of paintings and works in museums, collections and archaeological sites with dramatic threats for their understanding, preservation, and management of this heritage. Such irreversible colour changes are the result of transformations of the constituent materials (pigments, binding media, varnishes) of the work of art, thus depending on their chemical nature and the past conservation history of the work of art itself (e.g., storage and display conditions, restoration treatments, thefts or other vandalic acts). Among numerous examples, the darkening of vermilion red threatens a series of 15th-16th century paintings, including those by Pieter Paul Rubens [47], as well as Pompeian wall paintings [45, 48, 49]. In the latter, a change from yellow to red was also often observed in the areas where ochre pigments were used [50]. Similar colour changes of red and ochre pigments are also visible in other Pompeian wall paintings displayed the MANN museum today and documented as early as the 1980s (see Chap. 6).

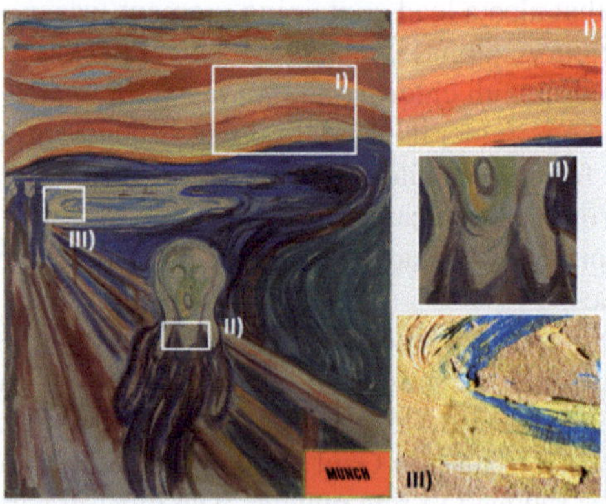

Fig. 17. Discoloured areas [I) and II)] and flaked off region [III)] cadmium yellow areas of The Scream (ca. 1910?) from MUNCH Museum, Oslo, NO [64]. (Color figure online)

Other yellows of low durability and high sensitivity to external factors (radiation, relative humidity etc.) are orpiment, chrome yellows and cadmium yellows. The discolouration of orpiment is present in the works of art of Medieval Italian masters as well as 17th century Flemish and Dutch painters, such as those authored by Cimabue [51], Pietro Lorenzetti [52], Daniël Seghers [53], Adriaen Coorte [54], and Jan Davidszoon de Heem [55]. The darkening of chromate-based yellows affects, instead, artworks by George Seurat [56], Paul Gauguin [57], and Vincent van Gogh [57–59], while the whitening of cadmium yellows threatens paintings by James Ensor [60], Henri Matisse [61], Joan Mirò [62] and Edvard Munch [63, 64] (Fig. 17).

Among blues, the darkening of French ultramarine, cerulean and Thénard's blues is also visible in masterpieces by Munch [65], while the fading of Prussian blue was observed in a series of 18th century Danish Golden age paintings (Statens Museum for Kunst, Copenhagen) [66]. Here a dramatic fading of the original blue colour is visible only in the areas exposed to light, while below the frame the colour is intact (Fig. 18).

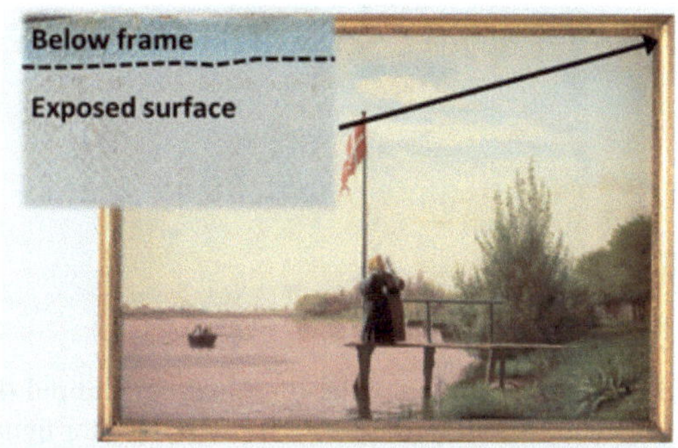

Fig. 18. Photograph of "A View of Lake Sortedam from Dosseringen Looking towards the Suburb Nørrebro outside Copenhagen" by Christen Købke (1838, oil on canvas, 53 x 71,5 cm, KMS359. Statens Museum for Kunst, open.smk.dk, public domain). Detail shows the blue sky in an exposed area and below the frame respectively [66]. (Color figure online)

Ultramarine blue paints are known for having poor drying properties in oil. Although the pigment is noted to be comparatively permanent, there have been reports of changes in colour due to instability of the pigment [67]. Recent research has linked the discolouration to the degradation of the binding medium, understood to be a result of a of pigment–binder separation [68]. For instance, the phenomenon of blackening of deep blue paint observed in *Old Man in Warnemünde (1907)* by Edvard Munch (Fig. 19) may be seen in relation to reports on similar phenomena, dating from the period of early manufacturing of synthetic ultramarine [69, 70]. As synthetic ultramarine became available on the market in 1907, it was noted that the paint had issues of blackening, at the time attributed to impurities or additions of iron, lime, magnesia and potash, or to the paint being mixed with white lead [71]. Aluminium and Sulfur speciation investigations by synchrotron radiation-based micro-X-ray fluorescence and micro-X-ray near edge structure spectroscopy show that the composition of the ultramarine pigment used in *Old Man in Warnemünde* has a uniform distribution across the surface, in both dark and brighter blue areas [65]. This could suggest that darkening is likely due to transformations of organic components rather than to the pigment itself. The presence of natural wax in the paint sample is interpreted as a probable additive from the manufacturers. Such additions have been identified in ready-made paint from the period [71, 72] and have also been indicated by FT-IR analysis in ATR mode of certain paint tubes in the Munch Museum's reference collection [73].

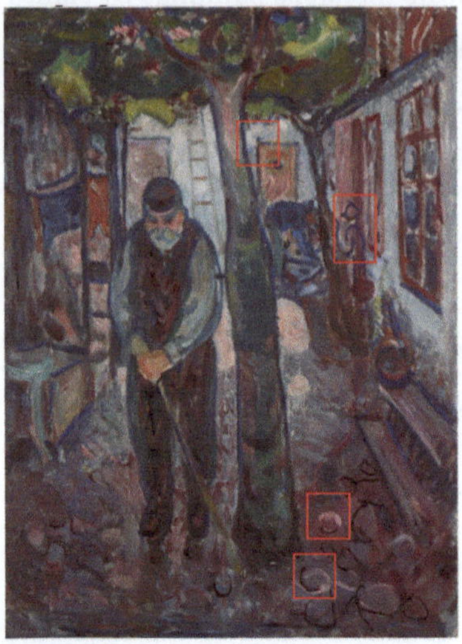

Fig. 19. Areas (marked in red squares) with blue (ultramarine) and red (lake) darkening in the painting Old Man in Warnemünde (1907) (Photo @MUNCH). (Color figure online)

Ongoing research on blue paint tubes from MUNCH's collection showed that the presence of metal soaps could be held accountable for the separation of binding medium and pigments particles, resulting in an enriched medium phase that leads to superficial darkening [70, 73]. These processes can be responsible of black grains observed on the surface of other Ultramarine and cobalt-based blue areas of few paintings from *Warnemünde* period (e.g. *Drowned boy, 1907–1908*) or other periods (*Death of the Bohemian, 1915–1920*). Additional coloured materials that suffer from chromatic changes are specific classes of red lakes (i.e., cochineal-, alizarin- and eosin-based lakes), as found in some works of art of the American watercolourist Winslow Homer [74], in a series of works on paper by Gauguin [75] and in several paintings by Munch [65, 69] and Van Gogh [76–78]. The degradation phenomena described in red lake paints in Van Gogh's *Glass with Yellow Roses*, and *Portrait of an Old Woman* [76], exhibits cracks and spotting, in addition to browning, like the red lake paints in Munch's *Old Man in Warnemünde.* In a red sample from *Old Man in Warnemünde* (Fig. 19), alizarin was identified as the colourant, with a possible tin/calcium-based and/or aluminium/calcium based-substrate, which might suggest a mixture of two types of red lake [65]. The deterioration, resulting in a brown discolouration of the red lake used by Van Gogh, related to the inclusion of starch.

One technique that allow the light induced colour change to be detected and monitored is micro-fadeometry tests (MFT), by suggesting a safe threshold dose of light for the studied object [79]. Currently there is no established protocol for MFT experiments. Work in this direction has been performed in H2020 project "Integrating Platforms for the European Research Infrastructure ON Heritage Science" (IPERION HS, GA n. 871034 [80]). Thus, the measurement typically consists in irradiating a selected point on an object with a specific white light with an intensity of the order of a few mW and simultaneous registration of colour changes caused by this illumination via colourimetric measurements. The light probe is focused on a small spot (typically less than 0.6 mm diameter); it follows that on one hand the fluence of light (expressed as power/area) at the measured spot is some order of magnitudes higher than equivalent conventional lighting systems, on the other hand the area of the paintings exposed to irradiation is strongly limited. A single micro-fading test takes up to 10 min and several measurements are usually performed for selected hues present on an object. To additionally ensure that minimal damage occurs, in some cases, the experiment is terminated once a colour change reaches units of just-noticeable-differences of human perception [81]. Moreover, the portability of the micro-fading allows it to be carried out for different types of artworks and in different conditions, such as Spanish Levantine rock art outdoor as shown by [82], prints, watercolours, and textile in museum environments [83]. In such context, further information regarding lightfastness of certain painting materials were successfully obtained by MFT (i.e. darkening of red areas containing vermillion) [84]. Nonetheless, micro-fading only assesses the photodegradation of the materials and does not indicate the possible chemical/structural changes that might have occurred due to other aging factors. A good example in this sense is that of the cadmium yellow pigment, which fades mainly because of exposure to moisture instead of light [64]. In this case, the micro-fading test captures a low extent of light-induced colour change, while in practice, the discolouration of certain types of formulation of cadmium yellows results from the exposure to humidity.

The micro-fading data may support the prediction of future colour change, which is straightforward from a mathematical perspective, as various interpolation and regression techniques can be applied. For instance, Morris [85] created a look-up table of colour changes for increasing light dosages, based on micro-fading tests carried out on paint mock-ups. Hendriks et al. [86] predicted the photodegradation of the red and yellow colours in Van Gogh's "*The Bedroom*", for a light dosage of up to 30 Mlux hr. As input, the authors used micro-fading data from aged mock-ups of red lakes and chrome yellow paints and based on these, altered the image of the painting in a commercial image editing software. In a similar way, Brokerhof et al. [87] rendered colour changes for a collection of Dutch city maps from the 17th century in an image editing software for various levels of fading and light exposure. Riutort-Mayol et al. [82] applied Gaussian process interpolation to extend the single-point micro-fading measurements to the two-dimensional surface of the artworks, and to forecast the photodegradation behaviour beyond the limited temporal dimension considered during the measurement. Ciortan et al. [81] adopted time-series analysis to forecast the individual change of colour coordinates in the CIE L*a*b* space, rather than the aggregate colour change, for a set of samples measured on "*A Japanese Lantern*" by Oda Krohg.

Recently, a novel approach that allows for simulating not only the colour, but also the spectral changes induced by light-exposure was proposed by Ciortan and co-workers [88, 89]. The method is based on a tensor decomposition model that, from a set of micro-fading measurements, disentangles in one shot the reflectance of pure pigments, their concentration, and their fading rate as a function of time. Such knowledge, combined with hyperspectral image analysis, was then exploited to render the past and future appearance of the artwork by modulating the amount of light exposure.

It is important to mention that all colour simulations (future and past) derived from micro-fading and more conventional accelerated photoaging experiments are questionable with respect to the light reciprocity principle. The light reciprocity principle states that exposing a specimen for short periods of time under high levels of light intensity is equivalent to long exposure under low illuminations levels [90]. Several studies [91, 92] have shown that light reciprocity principle breaks down when the difference in illumination between the two scenarios is higher than 3 orders of magnitude. This is even more relevant for micro-fading, for which light intensities and/or fluence values are further increased. Perhaps this is one of the reasons why simulations of past appearance of a painting, extrapolated from micro-fading tests, is especially controversial as often the full record of light conditions that the painting was exposed to (perhaps even before it became part of a museum collection) is unknown. Apart from the different levels of intensity, the light-induced change of a specimen in accelerated aging tests is influenced by the spectral distribution of the light source as earlier described [93]. As a rule of thumb, the change is proportional with the amount of overlap between the absorption spectrum of a material and the emission spectrum of the light. An exception to this rule is given by Prussian blue pigments and chrome yellow oil paints [94, 95]. The inverse behaviour of Prussian blue was discovered to be related to the amount of oxygen in the environment where the photoaging experiment occurs [96]. In the case of chrome yellow, the sensitivity of the paint towards darkening not only after the exposure to the blue component but also to the green one of the visible lights was ascribed to the presence of pentavalent chromium compounds, resulting from the thermal interaction of the original chromate-based pigment with the oil [95].

The reconstruction of how colours would have originally been perceived is crucial, since their appearance strongly influences our perception of the work of art. Guidelines are also needed to preserve them when exhibited (a control of environmental conditions and lighting is fundamental). In such context, through studies of micro-samples from artworks, laboratory paint mock-ups and direct non-invasive examinations of painted surfaces, a large quantity of data has been acquired in the last two decades from the above-mentioned case studies, thus helping to unveil how the constituent materials of the paint and environmental factors may affect the overall colour change phenomena. In the case of chrome yellows and cadmium yellows, for example, analytical investigations have contributed to help Museums to optimize the long-term conservation condition of some iconic paintings, such as the *Sunflowers* by Van Gogh (version of the Van Gogh Museum, Amsterdam) [97] and *The Scream* (ca. 1910) by Munch (version of the MUNCH museum, Oslo) [64]. In addition, experimental datasets have contributed to develop models for the virtual reconstruction of the original colour scheme of the artwork intended by the painter. In such context, several projects or research groups were and still are involved in activities related to the digital colour reconstruction.

A pioneering work on paintings by Seurat and Van Gogh was conducted by Berns, Pozzi and co-workers [98–100] by following a hybrid strategy that combines visual evaluation of the painting, pigments identification in micro-samples using different analytical methods (e.g., infrared and X-ray fluorescence spectroscopy; gas chromatography-mass spectrometry; scanning electron microscopy coupled with energy dispersive X-ray spectroscopy; Raman spectroscopy), single-point reflectance visible spectroscopy measurements acquired across the painting surface, definition of the original optical properties of the painting via comparison with data obtained from paint reconstructions, colour-managed RGB imaging and image adjustments with dedicated photo-editing software. Paint reconstructions of Van Gogh's palette was a significant component of REVIGO (Reassessing Vincent van Gogh's colours), a project devoted to study the colour changes in Van Gogh paintings and drawings and sponsored by The Netherlands Organization for Scientific Research. Key technologies for REVIGO were hyperspectral visible spectrum imaging and macro XRF scanning mapping methods, as well as machine-learning techniques [101–103]. The optical database resulting from the paint reconstructions has given an important contribution toward understanding Van Gogh's working method and improvements in colour reconstruction methodologies.

The application of machine learning techniques for processing and modelling large spectral imaging datasets has drastically expanded over the last 5 years, by permitting to perform advanced mapping of artistic painting materials and to identify pure and mixed pigments via application of spectral data obtained from artistic materials [104–108]. To date, however, there is still no standard method that allow for a reliable virtual colour restoration or colour change prediction of an object to be performed. Deep knowledge from the study of laboratory samples and paintings to understand how the paint composition and environmental factors influence the extent of colour changes is still missing and further experiments are required. The PERCEIVE project, by applying digital and machine learning-based methods to large experimental datasets from discoloured laboratory samples, Pompeian wall paintings and artworks by Munch, has developed a "Predictive System" that uses the analysis of colour change during the time, giving the possibility to "play forward the clock" and see how the artwork would look in a span of time if no specific preventive conservation measures are put in place.

5 Munch's Colours

"That a tree can be red or blue – that a face can be blue or green – they know it is wrong – From childhood they have known that leaves and grass are green, and that skin-colour is slightly red. – They can't understand that this is meant seriously – it must be humbug done in sloppiness – or with a deranged mind. Preferably the last".

(Munch, 1891–1892).

Modern and contemporary artists, like the Norwegian painter Edvard Munch (1863–1944), have altered the way we view colours. Munch wrote the quote above in a notebook around the time he made his first sketches of an early *Scream*-version. Today it works as a reminder of how much our perception of what is acceptable colour-use in an artwork has changed. Most of us willingly accept both a symbolic and an expressionistic approach to the use of colour.

Edvard Munch wrote "*I do not paint what I see but what I saw*". This statement suggests that Munch perceived his work as a product of the cumulative emotion of the mind's eye. "Intentionally and consciously between seeing something and realizing it in paint, he passes through a mental filter from which it later emerges transformed in the intensity of the remembered moment". He often uses colour not for naturalistic description but to convey the authenticity of feeling. The red colour of the sky in the Scream is correlative of the characters emotions: hopelessness and panic [109–111]. Munch had respect for scientific knowledge (he showed fascination with photography, radio, Xray and electricity,) but also valued in first place his own experience, considering his visual observations equally valid. The injury to his right eye in 1930 gave him different sensations that he started to express in his drawings and watercolours. In his in-depth study of his ocular disease, Munch linked the theme of *the Scream* to difficulties in perceiving colours and sounds. In one note the artist writes "What we hear are sound waves that are absorbed by the membrane of the ear. What we see are light waves that affect the membrane of the eye. Poetry is the perception of waves by the eye and the ear. Painting is what the brain perceives through the filter of the eye" [109]. Munch emphasises his suggestion with a reference to *the Scream*: "These light effects not only resulted in oscillation of the eye, they also resulted in oscillations of the ear – so that I actually heard a scream. I then painted the picture the Scream". Edvard Munch read popular manuals on the science behind colour and the powerful visual and physical effects that colour can have on us. These writings promoted that there are forces at work in the universe that impact the world around us but are imperceptible to the human eye—a notion that was very fashionable at the time, given the great public excitement over such recent scientific breakthroughs as the discovery of X-rays, electricity, magnetism, and the beginnings of quantum mechanics. These discoveries included new insights by physicists revealing how humans interpret colour, regarding the radiation spectrum and the way radiation interacts with matter and produces the effect of colours [109, 110]. For example, because red has a greater wavelength and is interpreted faster by the human brain, appears more striking. Red became a key- colour for the artist, being used in many masterworks from *Vampire* (1895) to the blood-red sky of *The Scream* (1893; 1910?). According to Jonathan Bober, curator of prints and drawings at the NGA, Munch was also familiar with popular spiritual theories propounded by Annie Besant and C.W. Leadbetter, whose influential book *Thought Forms* (1901) described individual colours as emitting different energies.

From an early-stage Munch also worked with new types of paint that were introduced on the market, experimenting with mixture of oil and tempera media, and even creating his own paints by mixing casein and glue or egg with powdery pigments purchased in several shops of painting materials in Oslo [112]. He was also fond of using crayons in addition to the more traditional painting materials, as many other artists of that time (Renoir and Pissarro preferred to use pastels due to the matt appearance of the medium, while Manet is said to have used pastels on canvases that were factory primed) [110–113].

In the *Sketches* for the monumental paintings made for Aula Magna at UiO he even used the unprimed canvas as an active part of his palette (and with time the canvas yellowed and darkened due to deposition of particulate from atmosphere, pollution and other factors). Several binding media have been identified such as: gelatine (from animal glue) based and linseed oil for *The Sun* (1910–1911), walnut oil and boiled linseed oil in *The Researchers* (1925–27). The matt appearance of the paint is due to the small amount of binding media (oil) in relationship to the amount of pigments and chalk [110–114]. Munch also used mixed techniques where oil brush strokes or even water-based colours (with organic components as glue and plant gums (gum Arabic or tragacanth) were overlapped or juxtaposed by lines of crayons (wax and oil-based drawing materials). Some of the sketches were painted on primed linen canvas, while others were executed on unprimed cotton canvases, not intended for painting purposes [112].

Munch was an experimentalist and not very much interested in the stability or long-term preservation of his works. Along the 60 years of artistic production, he evolved from a more realistic style of painting to a symbolistic and modern expressionistic style. He is also renowned for his unconventional, direct, and personal style. He used less saturated colours than those used by his contemporaries, but the low saturation colour enhances a different, heavier, more charged atmosphere. His main interest was to obtain a matt appearance for his paintings. He therefore avoided the use of varnishes or other glossy surface materials. Colours become protagonist in Munch's innovative experimentation, applied in thin and/or thick brushstroke, squeezed directly from the tube, blended with crayon lines and contours, pure, mixed or overlapped, sometimes even applied with fingers or other tools than brushes. This taste for experimentation lead to severe consequence regarding the stability of paint layers and integrity of the representation [113, 114]. Nowadays many works present porous, powdery, decohesed paint layers, colour change phenomena (fading and darkening), abrasions and cracks, paint layers losses and lacunae. All these degradation-forms lead to a loss of the readability of the image depicted, but also to many questions related to long-term survival of these artworks. In few cases, as *Madonna* and *The Scream,* the integrity of the object has also been affected by theft (2004) and other vandalic acts [115]. A big halo produced by a liquid agent left a deturpated mark in the left corner of the Scream painting, the cardboard also being heavily damaged in this area. The decade 1892–1902 represents the most important period of Munch's activity: in these years the artist defined his pictorial language, which is enriched with symbolistic elements and would influence Expressionism. The recurring themes are love, death and existential anguish, depicted by deformed images and garish colours. In this period his iconic *Scream* was also born, emblematic of the condition of modern man, prey to anxiety, inner loneliness and mal de vivre. That "great infinite scream" which according to Munch pervaded Nature, became the manifesto of his artwork, an expression of the lonely scream of all humanity [115, 116]. Munch wrote about the moment he saw the striking sunset depicted then in the *Scream* (also initially named the *Scream of nature*):

> *I was walking along the road with two friends – when the sun went down. The Sky suddenly turned blood-red I paused, leaned against the fence tired to death – above the blue-black fjord and city blood in flaming tongues hovered. My friends*

walked on and I stayed behind quaking with angst – and I felt as though a vast endless scream passed through nature.[13]

It is obvious that Munch used colour in this painting to convey emotion and drama, the colours becoming an exacerbation of feeling supporting his vision of the clouds as real blood. There is a powerful contrast between exaggerated reds and oranges used for the sunset in the background, and the dull blues, greens, purples and greys used for everything else. The use of colour contrast in this painting works two ways: to show the intensity of the "blood red" sunset in the background and the drama of the figure on the bridge. Whilst the painting appears relatively simple, there is a clever contrast between two pairs of complementary colours: red and green, and orange and blue. This adds a subtle level of complexity and some interesting colour relationships to the painting. The "blood" red in the sky appears to be the strongest colour, which contrasts against the very weak green for the land. The ghoulish figure was painted with sickly colours: dull yellows, blues, and purples. Dark and light accents painted over the top hint at the "screaming" expression of the character. No attention was given by Munch to paint this figure with any sense of realism. It is likely a creation of Munch's own tormented mind and anguish [116]. Munch's approach to the union of senses and the belief that one might taste a colour or smell, a musical note, results in the visual depiction of sound and emotion. The visual music and visual poetry are here combined in the specific aesthetics of the artist, resulting from the curved lines and the strong colours used to accentuate the oppressive general state of the character and the artwork itself. As such, *The Scream* represents a key work for the Symbolist movement as well as an important inspiration for the Expressionist movement of the early 20th century. As Munch wrote it in his notebook "it is not the chair which is to be painted but what the human being has felt in relation to it". Through this artwork, Munch sought to express internal emotions through external forms and thereby provide a visual image for a universal human experience. Pettersen's late research [116] shows that there are several theories about how and why the *Scream* motif developed further, and she brings up new aspects that can point towards the dating as being 1893 and not 1910?, in one version in the MUNCH-collection. She studies not only the stylistic features and the sources of the time of creation and Munch's own writings, but also suggests that further chemical analysis of *The Scream* compared with other paintings (i.e. *Anxiety, Despair, Hands*) (Fig. 20) could bring new clues regarding the dating year. The first version of the Scream is considered to be what is known as *Despair*, dated 1892 (Fig. 20a) and its schematic representation of the sky and main characters vaguely resembles the motif in the later versions (form National Museum and also MUNCH, Fig. 20b and c). The prose poems related to *the Scream* are also contradictories and it is difficult sometimes to understand which version is mentioned.

In 1893, Munch rented two exhibition rooms on the second floor at 19 Unter den Linden in Berlin. Here *The Scream* was exhibited for the first time. In the catalogue, six of the 25 paintings were listed under the heading *Die Liebe*. *The Scream* was one of them, entitled *Verzweiflung* (*Despair*). A couple of newspaper reviews mention the painting. The first one was published in Berliner Tageblatt: "but the last picture: *Despair* has a caricature figure in which nothing human is to be found; the background, a white

[13] https://www.emunch.no/TRANS_HYBRIDMM_T2367.xhtml.

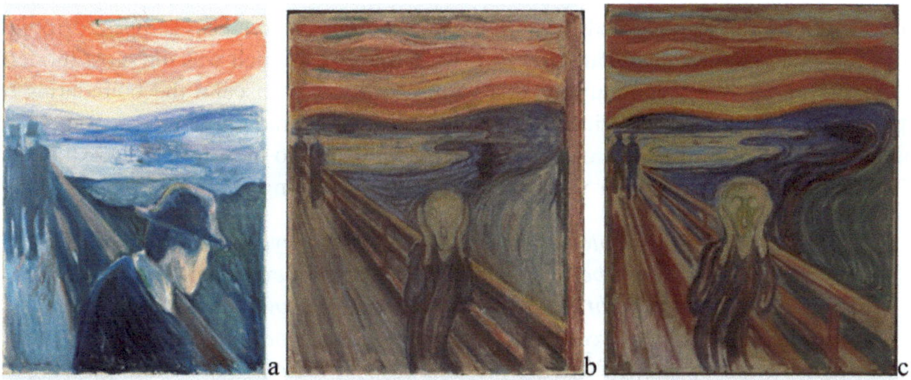

Fig. 20. a) "Sick Mood at Sunset. Despair", 1892, oil on canvas, 92 x 67 cm, signed lower left "E. Munch 1892", Thielska Galleriet Stockholm, Photo: Tord Lund, ©Thielska Galleriet. b) "The Scream", 1893, tempera and crayon on cardboard, 91 x 73.5 cm, signed lower left "E Munch 1893", Nasjonalmuseet Oslo, Photo: Børre Høstland, ©Nasjonalmuseet; c) "The Scream", 1910(?), tempera and oil on cardboard, 83.5 x 66 cm, Munchmuseet Oslo.

mixture of the brightest red, green, and yellow brushstrokes: What that means, if it isn't himself – heaven knows!" [116]. The second review was published in the Norwegian Morgenbladet (1893): "The last picture *Despair* is a pure caricature. The background is a strange egg yellow, bright red and green sky, and the figure, a small man with long hair and convulsively open mouth, impossible to understand". Both these reviews mention three colours: red, green and yellow. There is a large area of green in the background to the right, under the red and yellow sky, in MUNCH's version of *The Scream* (Fig. 20c). In NaM's version, there is very little green colour to be seen. Only a few green lines are drawn in crayon on the right, and some green brushstrokes run along part of the painted red vertical area (Fig. 20). In NaM's version, there is very little green colour to be seen. Only a few green lines are drawn in crayon on the right, and some green brushstrokes run along part of the painted red vertical area (Fig. 20). One wonders which of the two versions is described above – is it really the National Museum's version? It is difficult to believe that its colours could have been described as red, green and yellow. The colour descriptions in these reviews undoubtedly suit MUNCH's version better.

In Stanislaw Przybyszewski's article *Psychischer Naturalismus* [117], the colours of *The Scream* are described as follows:

Finally, the last picture: Despair. On a bridge, or something similar, it doesn't really matter what it represents – stands a mythical animal with its mouth wide open. [...] Each pain a blood red stain; each long-awaited scream of pain a belt of blue, green, yellow stains; unbalanced, brutal next to each other, like the boiling elements of developing worlds in wild creative fervours.

Przybyszewski mentions not only red, yellow and green, but also the blue colour, which is present in both paintings [117].

The paintings of Munch have a provocative character due to the use of certain materials and developments in his experimental techniques of preparing and applying the paint.

His subject matter was marked by unmotivated blotches, scratches, scrapes, dripping paint and areas of unpainted canvas. From 1886 onward, Munch was the most modern of contemporary Norwegian artists, due to his ability to transcend the boundaries of common convention and practice, even regarding the most radical of expressions to be found in Realism and Impressionism [117]. In connection to the Christiania exhibition of 1892 Munch's use of colour was described in the following manner [116, 117]:

> *hideous, pale colours, as blotted and thin as the first layer of paint on a house wall, anemic colours like tuberculosis patients that have never known nourishment of blood or daylight or – at their worst – piercing yellow patches, hectic reds, and screaming blotches spread aimlessly and with no apparent meaning over the canvas...On top of that, he smears the canvas with lemony yellow and glowing reds, as if he had daubed it with bloody fingers – and calls it evening mood!*

In the manifest *Cosmos. The New Sublimity in Art* written in 1893 by Carl Gustaf Uddgren and Max Dauthendey, Munch is deemed the contemporary artist who had made the most progress in reproducing "the intimacy of colour": "For him, the main objective is to conserve the most sublime feelings of pleasure or displeasure that colours trigger in his brain. It is irrelevant to him what scene or event he depicts". Munch conveys a "main mood" through his use of colour!" [116]. Przybyszewski describes the Munch's style in *Psychic Naturalism* [118] saying about the shaping of landscape in *Despair*: "His landscape is the absolute correlation of such naked feeling; every single vibration of nerves, which have been directly exposed to the utmost ecstasy of pain, is instantly converted into an appropriate sensation of colour". Przybyszewski contends that Munch's painting introduces something new in comparison to contemporary symbolism by using colour to coupe spiritual life with pictorial expression. Colour is Munch's most essential tool and in colours he sees "mourning and screams rumination and wilting" [116].

6 Colours in Textiles

Throughout human history, colour has played an important role also in historical textiles. To trace the origins of textile dyeing, one must delve into the earliest history of mankind. Initially, clothing was fashioned from animal hides, but with the development of spinning and weaving, the art of dyeing textile fibres came closer to realization. The techniques of spinning, weaving, and dyeing likely emerged concurrently among various cultures, and the knowledge of these practices spread from one civilization to another [119]. It was through this dissemination of knowledge that the art of dyeing textiles evolved and flourished, leaving a rich legacy of vibrant colours that continue to captivate us in the modern age. The availability of certain pigments and dyes also affected the rarity and cost of particular colours, making them exclusive to the privileged classes such as the elite, royalty or religious orders [120, 121].

Based on origin, natural colourants can be classified as organic and inorganic, inorganic are pigments while organic colourants can be further grouped as dyes and pigments. Historical dyers since antiquity would expertly use natural resources such as parts of plants (leaves, root, bark, wood, seed, flower), minerals (ochres, carbonates), insects (cochineal, lac dye) and their mixtures to create various hues.

Colours often held deep cultural meanings and symbolism where different cultures associated specific colours with values and traditions. Colours were much more than just aesthetic choices; they were expressions of identity, beliefs, and social norms of their time. The diverse palette of colours used in these creations conveyed rich narratives about the people and societies that embraced them, offering a captivating window into the rich heritage of human history and culture. Tapestries and dress collections bore witness to the profound cultural meanings and symbolism attached to colours. These vivid expressions of identity, beliefs, and social norms showcased the diverse values and traditions of different cultures. Beyond being mere aesthetic choices, the colours woven into these tapestries and draped across garments conveyed intricate narratives about the people and societies that cherished them.

Studying historical textiles and use of colours offers valuable insights into the mindset and values of people from different periods and societies. The importance of colour in textile collections cannot be overemphasized, as it plays a central role in shaping the overall impact and significance of these collections. Colour is one of the primary factors that catch the viewer's eye by creating an immediate visual impact. The use of vibrant and well-coordinated colours enhances the attractiveness of a collection, making it more engaging and memorable. Colours can be used to convey specific themes, moods, or concepts in textile collections. Whether it's a historical period, cultural representation, or a specific artistic vision, colours help to communicate and reinforce the intended message. Different colours can also accentuate various design elements in textiles, such as patterns, embroidery, and textures. The right colour choices can draw attention to these details and enhance the overall design. For historical collections, understanding the original colours and their significance is crucial for preservation and restoration purposes. Accurately reproducing faded colours can provide a more authentic representation of historical garments.

The problem of fading in dress collections is a significant concern for museums, collectors, and those responsible for maintaining and displaying historical garments and textiles. It is a complex issue that requires careful attention and proactive conservation strategies to ensure the long-term preservation of historical textiles and garments for future generations. The fading of colours refers to the gradual loss or alteration over time, resulting in a less vibrant and often discoloured appearance. Several factors can contribute to fading such as exposure to light, particularly ultraviolet radiation (UV) from natural and artificial sources where prolonged exposure can break down the chemical bonds in dyes and pigments, leading to degradation and colour loss. Fluctuations of environmental conditions, namely, temperature, humidity, and air pollution can also affect fading and deterioration processes. The lightfastness and stability of dyes and pigments, and the use of certain textiles can significantly impact their susceptibility to fading due to their inherent chemical and physical properties. Considerations regarding previous exposure to harsh environmental conditions or improper storage and handling practices can exacerbate fading and accelerate colour loss. Nowadays museums are increasingly sensitive to the issues of fading and implement tailored strategies such as light control and environmental monitoring in display, storage and exhibition spaces as well as garment rotation and use of conservation grade materials for storage to help minimize fading [122].

Analysing and reconstructing original colours in textiles can be a challenging process. Various approaches and techniques are used through multidisciplinary collaborations between scientists, historians, and conservators to uncover and reproduce the original colours of historical textiles [123–125]. Studies of historical documents that depict similar textiles or garments from the same period can offer valuable clues about the original colours. Through visual examinations of the textile, areas of preserved pigments or dyes can be observed, and any patterns or motifs that may provide clues about the original colours may be identified. The fine details of the textile's surface and structure, including the distribution of colourants and any degradation or fading patterns can be revealed on microscopic examinations [125]. The analysis of dyes in fragile materials such as textiles, requires a continuous development of new analytical approaches that reduce the physical impact of analysis on the material under study whilst ensuring that maximum information is gained. Emphasis is given to the need for reliable, efficient, and ideally minimally or non-invasive analytical approaches for historical dye analysis [126]. In recent years, capturing images of a textile under different wavelengths of light, such as ultraviolet, visible, and infrared with multispectral and hyperspectral imaging techniques have been shown to reveal hidden features and faded colours not visible to the naked eye [127]. Non-invasive portable spectroscopic techniques (UV-Visible, fibre optic reflectance (FORS) Infrared and Raman spectroscopy) and colourimetry can help analyse the electronic and molecular composition and colour quantification of original textiles in-situ and assist in identifying the presence of specific dyes or pigments [128]. They can collectively provide insights into the original colourants used leading to a deeper understanding of fading mechanisms. Non-invasive methods are often used to guide object sampling should ulterior destructive chromatographic (HPLC) or spectrometric analyses (MS) be required for identification of complex dyes mixtures and associated degradation products [129]. Within PERCEIVE project, we construct experiments with historical recipes and traditional extraction and dyeing methods to recreate the colours observed in textiles by preparing a set of mock samples. This process involves using natural materials and conventional dyeing techniques often with the use of mordants to match the original colours as closely as possible. Artificial ageing of such ad-hoc laboratory samples can provide information on fading and degradation with close monitoring of alterations arising from modifications to the physical and chemical properties of the dyes and textiles. It is essential to note that reconstructing original colours is not always feasible, especially if a textile has undergone significant fading or deterioration over time. Moreover, employing such analytical approaches can help uncover valuable information about historical textiles and contribute to a deeper understanding of our cultural heritage. The work to be achieved is to supplement the laboratory and on-site methods with colour reconstruction techniques to restore and enhance digital colour information, to fill in missing or damaged areas in an image by estimating the pixel values based on the surrounding regions. Such techniques are particularly useful in chosen cases where colour data has been lost or degraded and aims to improve the overall quality and realism of images by filling. Colour reconstruction techniques could include interpolation, *demosaicing* or colourization by estimation of missing colour values based on surrounding known colour values; reconstruction of full-colour images from images captured with colour filter arrays; or prediction of plausible colour information based on the input grayscale data. Such methods do not always produce the most accurate results, especially in regions of complex colour variations but can be assessed accordingly [130, 131].

Artificial intelligence (AI) can be also utilized in combination for restoring faded textiles to their original or enhanced appearance. AI can assist in the analysis of multi-spectral imaging data for faded textiles to provide additional insights into the original colours, patterns and textures of the textiles. AI-based techniques can also help in reviving faded textiles or historical tapestries that suffer from colour degradation due to aforementioned ageing, exposure to light, and environmental factors [132]. AI can be applied to faded textiles through image restoration techniques, often based on deep learning models to reconstruct missing or degraded colours by learning from large datasets of well-preserved textile images such as those from specific experimental mock sets constructed in the laboratory. AI algorithms can estimate the original colours of faded textiles by analysing the remaining colour information and comparing it with known colour patterns from similar textiles or historical references as those case studies to be analysed in-situ. Deep learning models can be trained to predict plausible colour values for each pixel based on surrounding context and patterns where AI can be used to recognize and reconstruct faded patterns in textiles. By training machine learning models on patterns from well-preserved textiles, the AI can identify and recreate the original patterns in faded textiles. By artificially simulating fading and aging effects on digital textile images, in comparison with real laboratory studies AI can create larger datasets, which can be used to train models for colour reconstruction and restoration [133]. This entire process from the laboratory to in-situ analysis in PERCEIVE can permit for targeted colour reconstruction and restoration of the faded areas of the chosen textile case studies, minimizing the risk of applying corrections to the original areas. Combining AI with traditional analytical conservation methods can lead to more effective and responsible restoration practices for faded textiles whilst enhancing the experience of museum visitors [134].

7 Colours in Historical Coloured Photographic and Film Collections

Preservation of Cultural heritage, including motion pictures and photographs, is a critical endeavour that safeguards the visual history of societies. Both motion pictures and photographs share common challenges in terms of deterioration, technological obsolescence, and environmental factors. However, they also possess distinct characteristics that necessitate specific preservation strategies.

7.1 Dye Fading

A significant challenge in the preservation of colour photographic and film heritage is the fading of the image dyes, which compromises the original appearance and commonly results in a pink/magenta cast deriving from the dominating residual dyes [135]. Chemically, the breakage of intermolecular bonds in the dye molecules produces fading

[136], which visually manifests as the loss of contrast and reduction of the overall density of the image [137]. Moreover, as the stability characteristics of different dyes are usually not the same, the fading is uneven, resulting in colour casts (Dye Fade[14]). This irreversible form of deterioration is attributed to the instability of the image dyes used in photographic and cinematographic materials, particularly those produced between the 1940s and 1980s, and their susceptibility to storage conditions [138].

The gravity of this impermanence has been highlighted in audio–visual preservation communities since the 1970s [139, 140]. Over the years, commercial photographic and film material manufacturers have endeavoured to improve the stability of their colour stocks [138, 141]. Simultaneously, chemists, imaging and conservation scientists, in collaboration with the global archival community, have initiated multidisciplinary research focusing on the characteristics of chromogenic dyes, their subsequent aging [142–144] and possible measures for the preservation and restoration of affected photographic and film heritage [145]. Chemical restoration of dye fading is challenging as it involves targeting the physical locations of broken colour dye molecule bonds (Image Forming Materials[15]). However, digital restoration has been used alternatively to reconstruct the colour appearance of historical photographic and film materials [143, 144]. This process usually starts with a conventional RGB scan of the faded, historic materials and typically employs the available tools of colour grading on the raw-scans to adjust colour balance and saturation [145]. In film production, colour grading is deemed as a creative tool, empowering filmmakers to set the final aesthetics of the image to be screened. However, as a matter of fact, these colour grading tools are also employed for the digital restoration of faded materials, using the same creative approach to compensate for changes in colour balances and recover, as best as possible, the scene colour and grading choices and ultimately a supposed original colour. In the absence of external colour references (such as fragments of a non-faded positive print, input from the filmmakers, or, in the case of animation, original artworks), the determination of the supposed original colour becomes a subjective process, drawing on factors such as memory colour, knowledge of or assumptions about contemporary aesthetics, the presumed appearance of a colour process at a given time in film history, artistic implications drawn from a filmmaker's oeuvre, or even less defined local preferences such as the "British Technicolour look". Given the above limitations of a conventional digital restoration workflow, the capability of image processing to numerically reverse the fading and simulate the original colour appearance in a more objective manner has been investigated over last decades [145]. However, the feasibility of an objective and accurate colour reconstruction depends on the amount of residual colour present in the film, the quality of the digital image capture and the efficacy of the image processing techniques.

Objectives. Physically-based, quantitative approaches are adopted to infer the original appearance based on evidence of residual original materials. This is obtained in conjunction with AI architectures that generalizes colour/image reconstruction. Restoration methods relying on generative methods (e.g., inpainting, colourization) will be

[14] Dye Fade. Visual Decay Guide. Filmcare.org. Available online: https://www.filmcare.org/vd_dyefade.php (accessed on 21 March 2023).

[15] https://www.nfsa.gov.au/preservation/guide/handbook/image-forming-materials%22 /t %22_blank.

employed, as long as full honesty towards the process and the level of speculation are disclosed together with the results.

7.2 Autochrome

The Autochrome Lumière was the first commercially successful colour photography process. While colour photography processes were being developed soon after the announcement of photography in 1839, many of these early iterations could not be mass manufactured and were not commercially viable or accessible. Patented in 1903 and released publicly in 1907, the process was developed by the French Lumière brothers. The process comprises a glass plate, on top of which sits a colour filter made of microscopic starch granules dyed red, green and blue, and a panchromatic gelatine silver emulsion (Fig. 21). In the early twentieth century, the autochrome was lauded for its faithful and natural production of colour [146]. Today, autochromes are held in a range of museum collections internationally, including the Victoria and Albert Museum, London (V&A). Autochromes pose a number of conservation considerations. Notably, autochromes are highly light sensitive and the colours are prone to fading. In addition, autochromes can be damaged by a range of external factors, including humidity and moisture, which can cause greening, discolouration and silvering. For these reasons, many international public institutions do not display their autochrome collections and instead exhibit facsimile copies. This practice highlights the importance of developing non-invasive tools and methods that enable new ways of public access and understanding. Exposure to light can increase the appearance of dye fading in autochromes, a common form of degradation that affects the plates [147].

Our goal is to develop a virtual restoration approach that adopts spectral imaging techniques [148, 149] to separate the information corresponding to the two overlapping elements of the autochrome plates. Once the optical properties of the colour dyes have been separated from the photographic image, their original concentration can be selectively restored before being recombined. The virtual restoration can be presented in a tactile experience with a set of layers representing the photographic image and the colour dyes, where the latter can be replaced from one representing the current state to one representing the restored version.

7.3 Kodacolour

Since the birth of cinema in the mid of 1890s, inventors and researchers set out to apply techniques of colour photography to motion pictures [150]. However, decades had to pass before a practical solution to record colour information on motion pictures was developed. In the 1920s the industrial exploitation of 'autonomous colours' such as tinting and toning was at its climax [151], but at the same time a series of two- and three-colour photographic processes were trying to improve their results and finally enter the market. Colour processes based on temporal synthesis reunite colour separations on the screen by displaying them in the same rapid succession as they have been recorded. The primary colours were added to the black and white separations either by filtering the projected light with a spinning filter wheel, as in Kinemacolour [152], or by tinting the

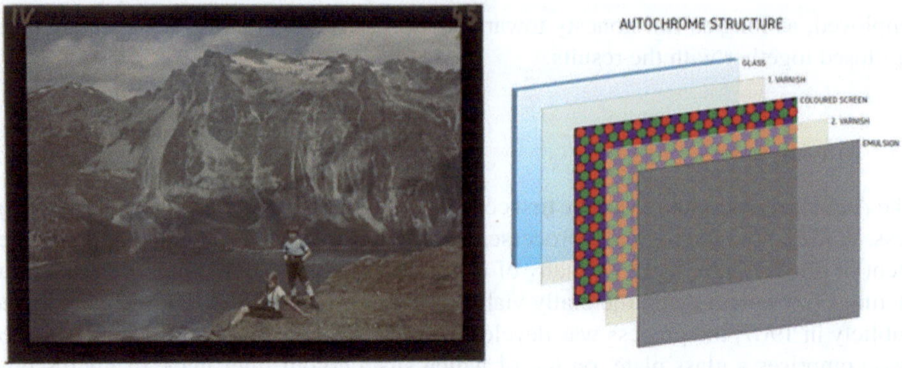

Fig. 21. Left: F. A. Paneth, Sils: Eva and Heinz at Muotathal in the Engadine, 1927; an example of Autochrome from the V&A Museum's collection, London. Right: The layered structure of the Autochrome (Nejc Urankar 2020).

individual film frames, as in Friese-Greene [153]. Since the colour separation images were taken at slightly different times, the reproduction suffered from pronounced colour fringes around moving objects. The lenticular film process, on the other hand, produced three colour separations taken at the same time, and recorded them in a single film frame. Kodak was the only company that successfully produced lenticular film at an industrial scale. The film was marketed from 1928 under the name Kodacolour. However, most of the inventions behind the process have been made before the company acquired the rights to exploit the technique [154]. The fundamental ideas behind the lenticular process for the photographic reproduction of colours can be found in the work done by Liesegang before the turn of the century [155] and by Lippmann a few years later [156]. Liesegang envisioned a 'pixelated' spatial encoding of colours with a perforated screen, while Lippmann described in detail the structure of a lenticular film that would allow a better representation of reality. In the early 1900s Berthon combined the two concepts and started endeavouring the application of the lenticular colour process to moving pictures [157]. In 1909 he patented a set-up including a tripartite red-green-blue filter as part of the imaging lens combined with a lenticular structure in front of the light sensitive panchromatic black-and-white film. The realization of this optical design had some major challenges. For instance, the lenticular structure engraved on the celluloid base defines the resolution of the image, so the lenticules had to be minuscule. For the technical implementation on an industrial scale, Berton found the assistance of Albert Keller-Dorian, who was the director of a company with expertise in engraving techniques [158]. Berton intended to produce a film with minuscule lenticules either in a honeycomb-like shape (Fig. 22) or as a linear array [157]. The Keller-Dorian-Berthon process was patented in 1915 [159] but the first World War delayed the development. The cooperation continued until the death of Keller-Dorian in 1924, resulting in no more than some short experimental shots.

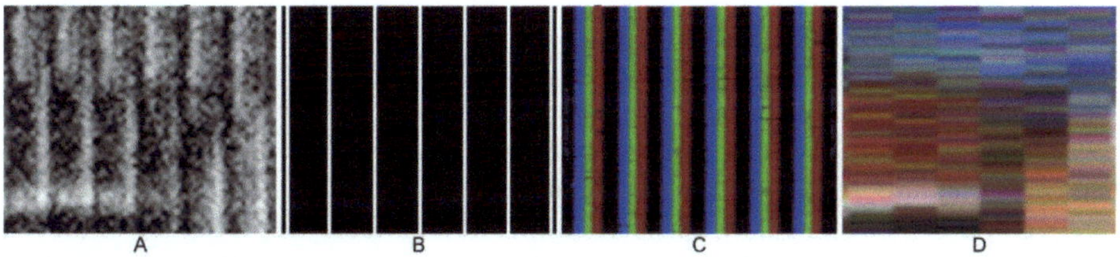

Fig. 22. The digital colour reconstruction process for a tiny fraction of a lenticular image – A: The input black-and-white image.

In 1926 a short film was successfully produced [160]. However, the scarce success of the financial investment forced Berthon to sell the patents to Kodak, who instead managed to develop the Kodacolour film product in a comparably short time. The result was a 16mm reversal film with a linear vertical lenticular screen, aimed at amateurs and released in 1928. AGFA was working on a very similar product, called Agfacolour, which was ready to hit the market in 1932 [161]. However, no trace of any actual footage shot by amateurs could be found by the authors so far. The process was likely abandoned before its release.

A widespread market success was necessary for a novel colour technique to survive for a sufficient time and have a significant impact on the landscape of cinema. If a new colour process allowed to use existent hardware for film recording and projection, it had more chances of success [158], while additional costs of specific equipment—that could become obsolete in a short period of time—would likely dissuade exhibitors and amateurs alike to adopt novel processes. Also, the possibility of producing copies for distribution was an essential requisite for the success of a film technique [162]. Kodacolour did not have these requirements, and it also had limitations in (i) image brightness—typical of all additive colour processes—(ii) image detail and (iii) colour gamut. After some success in amateur filmmaking, these limitations determined a quite sudden decline when a superior product was introduced by the same company. The introduction of chromogenic colours in the form of Kodachrome was a game changer for the film industry, and all other existing colour processes on the amateur market rapidly disappeared. A wide colour gamut could be reproduced by Kodachrome with subtractive synthesis, producing images with high level of detail using standard recording or projection equipment. Therefore, lenticular film rapidly lost appeal and Kodacolour disappeared from the market towards the end of the 1930s. A considerable amount of surviving lenticular films is today hidden. As a matter of fact, lenticular film appears like a 'normal' black-and-white film at a first quick look, so many lenticular films have most likely been misclassified by collectors who will need further training to identify them correctly. It is therefore important to revive awareness of the lenticular colour processes thus making these precious historical colour movies available again to a public and securing them for posterity.

8 Colours in Digital Artworks

This section investigates the influence of colour in Augmented Reality (AR) digital born artworks, with a specific focus on transparency and on the interplay between transparency in an AR "sculpture", or 3d object, and its impact on other elements within the artwork. Moreover, we will examine how transparency in AR art affects the perception of the real world, when viewed through the transparent elements of a sculpture. By analysing these interactions, the research aims to unravel the transformative effects of transparency in AR art on colour aesthetics and the overall visual experience for viewers, both within the artwork and in the context of the real world. We will start by examining a masterpiece by Leonardo da Vinci, which, despite being a painting, employs transparency in a remarkable manner. Building upon this observation, the text presents a comparison between the mixing of pigment and light, aiming to establish a connection between the traditional use of transparency in painting and its contemporary application in digital art. Through this exploration, we will highlight the distinctions and similarities between analogue and digital colour usage in artistic expression. With the concept of transparency and its broad applications in colour clearly outlined, the text proceeds with the presentation of a diverse types of digital artworks. Through the use of AR, these artworks skilfully merge virtual elements with reality, all made possible through the intentional incorporation of transparency.

8.1 Use of Transparency

During the Renaissance period, artists like Leonardo da Vinci used paints composed of natural pigments sourced from minerals, plants, and insects. These pigments, including earth-based ochre, Siena, and umber, as well as mineral-based lead white and vermilion, provided a diverse palette of colours. Additionally, colours like indigo from plants and carmine from insects offered further options. Artists ground these pigments into fine powders and mixed them with binders such as egg tempera or linseed oil to create paint. This allowed the pigments to adhere to surfaces while influencing drying times and finishes. Leonardo da Vinci painted the Mona Lisa using a technique called sfumato, which comes from the Italian word "sfumare," meaning "to tone down" or "to evaporate like smoke." This technique involves the subtle blending of colours and tones to create smooth transitions in light and shadow, resulting in seamless transitions between different areas of the painting, particularly notable in the sitter's enigmatic smile and the landscape background. To achieve this effect, Da Vinci applied thin layers of oil paint to the canvas, allowing him to create a sense of depth and three-dimensionality in the portrait. Additionally, he utilised his understanding of light and shadow to create a sense of luminosity and realism in the painting, particularly evident in the famous enigmatic smile of the Mona Lisa (Fig. 23).The distinction between mixing pigments and mixing light lies in the fundamental principles of colour interaction and the mediums through which they operate. Mixing pigments, known as subtractive colour mixing, involves combining materials like paints or inks, wherein the pigments absorb certain wavelengths of light and reflect others, resulting in the perception of colour. In contrast, mixing light, termed additive colour mixing, occurs when different coloured lights are combined, with each colour adding to the overall brightness and contributing to the final perceived colour. While mixing pigments subtracts wavelengths of light to create colours, mixing light adds wavelengths together, fundamentally differing in their approach to generating colour perceptions. Transparency serves as a pivotal element in both Leonardo da Vinci's Mona Lisa and contemporary Augmented Reality (AR) artworks. In Da Vinci's masterpiece, transparent layers of oil paint allow for subtle blending, depth, and the interplay of light, enhancing realism and captivating the viewer. Similarly, in AR artworks, transparency facilitates the seamless integration of virtual elements into the real world, creating immersive experiences with depth and interactivity. Whether in the Renaissance painting or modern digital art, transparency acts as a conduit for artistic expression, enriching the viewer's engagement and perception of space, dimension, and narrative.

8.2 Colour Perception in Traditional vs. Digital Art

The perception of colour remains however consistent across digital art and traditional art. Our visual system processes colours in a similar manner whether viewing a digital artwork on a screen or a traditional artwork on a tangible surface. Light containing various wavelengths interacts with the objects or surfaces in the artwork, and the reflected or transmitted light enters our eyes. Photoreceptor cells (cones) in our eyes detect different wavelengths and transmit signals to the brain, which processes this information to create the perception of colour (Ch. 1).

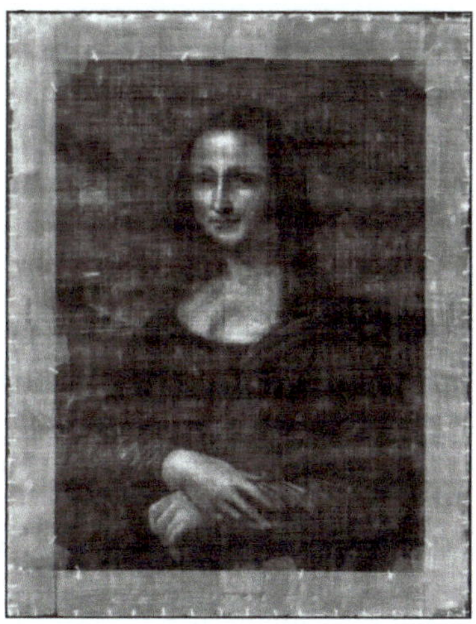

Fig. 23. "Mona Lisa" by Leonardo da Vinci. X-Radiography. A 2005 X-ray procedure on the 'Earlier Mona Lisa', revealed the sequence in which the painting was executed: the figure, the columns, the sky, and then the landscape [ref Mona Lisa Foundation 2012].

Therefore, the underlying principles of colour perception persistently apply across diverse mediums, despite variations in the methods of pigment mixing and coloured light blending. The difference between colour in painting and digital art lies in their creation, manipulation, presentation, and reproducibility. Traditional painting involves manual application of physical pigments using brushes, while digital art employs digital tools and software. Traditional artists directly control colour application with techniques like layering and blending, while digital art allows for non-destructive adjustments. Traditional paintings are unique physical artworks, while digital art can be easily reproduced and shared online. Both mediums offer distinct advantages and require different skill sets for colour expression (Fig. 24).

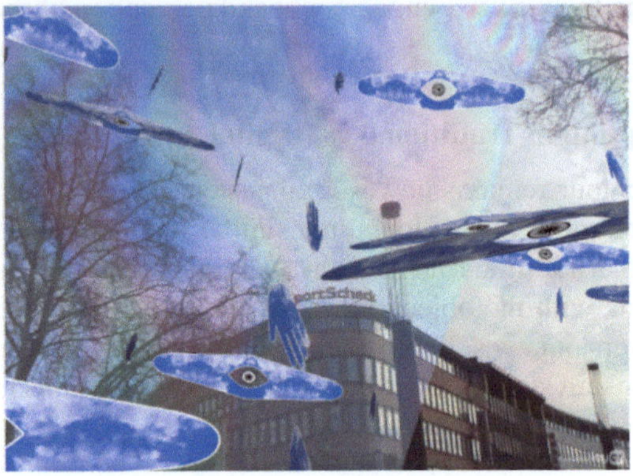

Fig. 24. Screen shot from the AR artwork The Coming of a New Dimension, Arthur Clay, 2015. Copyright held by Virtuale Switzerland.

8.3 Transparency in Digital Art

Transparency in digital art significantly impacts colour perception and visual aesthetics. Transparent elements enable the blending of colours, smooth transitions, depth, and dimension within the artwork which in turn also convey illumination, glow, and overlay effects that add complexity and interest. Additionally, transparency can alter colour intensity, softening or enhancing certain elements. In contrast to traditional art, transparency in digital arts, such as Augmented Reality (AR), takes on a dynamic role, seamlessly blending virtual elements with the real world. AR artists can create interactive and immersive experiences, where virtual objects appear to exist within the viewer's environment, challenging spatial perceptions and engaging audiences in real-time interactions. While traditional arts excel in static visual expression, AR opens new dimensions of creativity, enabling accessibility, interactivity, and spatial manipulation in the realm of digital art.

8.4 Case Studies

The diverse selection of AR artworks chosen for this study showcases the versatility and sophisticated use of transparency as a pivotal artistic element within the genre. Each artwork demonstrates unique and distinct approaches to employing transparency, ranging from skilfully mimicking glass textures to creating intricate patterns through the strategic overlapping of virtual objects. The study explores how these varied applications of transparency contribute to the overall aesthetic and immersive qualities of the AR artworks, providing valuable insights into the creative potential and expressive power of transparency in this evolving art form (Fig. 25).

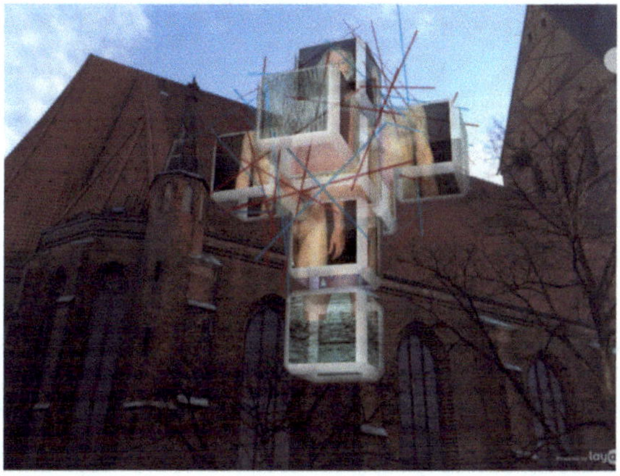

Fig. 25. A screenshot from the work 'Ecco Homo' by Arthur Clay with Ingo Lie, 2015. Copyright held by Virtuale Switzerland.

In "Ecce Homo," the artist employs "Transparency Dimensionalization" to intertwine colours and textures, skilfully creating an illusion of shifting and transformation. By partitioning the artist's self-portrait within a three-dimensional cross, the artwork engages observers from different angles, evoking a sense of dynamic interaction. In "Gaza 1:1 - The Name," "Transparency Layering" is deftly utilized to seamlessly integrate hand-written names onto a Moravian star. This technique evokes a profound sense of remembrance and contemplation against the vast expanse of the celestial realm, honouring the lives lost during Operation Protective Edge in Gaza. "The Coming of a New Dimension" explores the passage of time and space, symbolized by a time capsule element. Through meticulous gradation of translucency using "Transparency Layering," the artwork vividly captures the temporal metamorphosis, creating a sense of perpetual flux between the earthly and cosmic domains. In "Dante's Inferno," "Environmental Masking" is employed to isolate a virtual flood scene, skilfully captivating viewers within the tumultuous and haunting landscapes of the infernal realm, devoid of external distractions. "Crystal Coffin" exemplifies "Simulated Realism" through the virtuosic use of 90% translucent glass, imbuing the virtual pavilion with veritable authenticity. This evocative element embodies potent symbolism within the context of traditional

Chinese architecture, prompting contemplation on the intertwined dynamics of history, modernity, and ideological paradigms (Fig. 26).

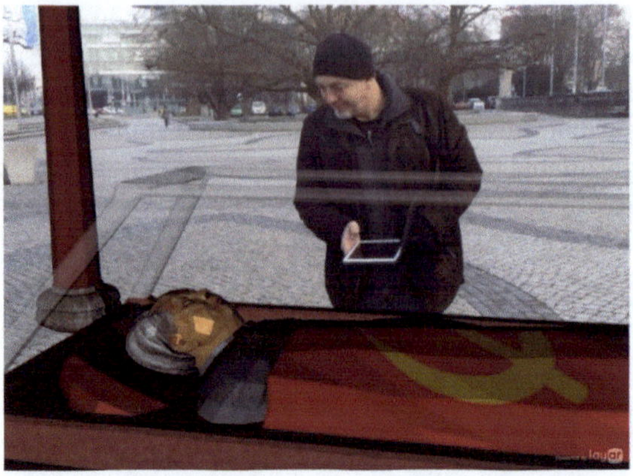

Fig. 26. A screenshot from 'Crystal Coffin' by Lilly & Honglei, 2015. Copyright held by Virtuale Switzerland.

The above cited AR artworks showcase the versatility and sophisticated use of transparency as a pivotal artistic element. Each artwork employs transparency in unique ways, ranging from mimicking glass textures to creating intricate patterns through overlapping virtual objects. These varied applications of transparency contribute to the overall aesthetic and immersive qualities of the AR artworks, revealing the creative potential and expressive power of transparency in this evolving art form. The comprehensive analysis of diverse AR artworks presented above underscores the significance of transparency as a transformative artistic means, as the utilization of transparency in these works transcends conventional artistic boundaries which result in captivating visual encounters that engage and enchant viewers by influencing the perception of analogue colour in real space with the blending in of coloured light.

Acknowledgments. This study was funded by the PERCEIVE project that received funding from the European Union's Horizon research and innovation programme under grant agreement No. 101061157.

Authors Contributions. Irina Crina Anca Sandu contributed to Sect. 1; Cristiana Barandoni, Donata Magrini, Roberta Iannaccone contributed to Sect. 2; Cristiana Barandoni, Sofia Pescarin contributed to Sect. 3; Irina Crina Anca Sandu, Sivert Thue, Letizia Monico, Irina Ciortan contributed to Sect. 4; Irina Crina Anca Sandu and Sivert Thue contributed to Sect. 5; Lucia Burgio, Brenda Doherty, Raquel Santos and Harmut Kutzke contributed to Sect. 6; Giorgio Trumpy (NTNU), Catlin Langford contributed to Sect. 7; Arthur Clay contributed to Sect. 8.

Disclosure of Interests. The authors have no competing interests to declare that are relevant to the content of this article.

References

1. Art and the Cities: Colour Theory: How to classify colours and their meanings in art. Art and the Cities. https://artandthecities.com/en/2022/11/27/colour-theory. Accessed Aug 2023
2. Brown, D.H., Macpherson, F. (eds.): The Routledge Handbook of Philosophy of Colour, 1st edn. Routledge, Taylor & Francis Group, London, New York (2023)
3. Gage, J.: Colour and Meaning: Art, Science and Symbolism. Repr. Thames & Hudson, London (2001)
4. Arty Factory: Colour Theory. Arty Factory. https://www.artyfactory.com/color_theory/color_theory.htm#google_vignette. Accessed Aug 2023
5. Colour Matters: Explore Colours. Colour Matters. https://colourmatters.com. Accessed Aug 2023
6. Ohta, N., Robertson, A.R.: Colourimetry: Fundamentals and Applications, 1st edn. Wiley (2005). https://doi.org/10.1002/0470094745
7. Munsell Colour Company: Munsell Colour Notation & Colour Test; Dimensions of Colour | Munsell Colour System. Munsell Colour Company. Accessed Sept 2024
8. Sandu, I.C.A., Sandbakken, E.G., Ferrer, J., Syversen, T.: Colours, brands, labels and other fancy stuff in the original art historical materials collection at MUNCH. In: 1st MUNCH2022 International Conference, vol. 13(1), pp. 1653–1664 (2022)
9. Art and the Cities: Art and the Cities Blog di arte e viaggi. Art and the Cities. https://artandthecities.com. Accessed 2023
10. Lant, A.: Vincent Van Gogh. In: Shestack, A. (ed.) Yale University Art Gallery: Selections, pp. 68–69. Yale University Art Gallery, New Haven (1983)
11. Palmer, S.E.: Colour, consciousness, and the isomorphism constraint. Behav. Brain Sci. **22**(6), 923–943 (1999). https://doi.org/10.1017/S0140525X99002216
12. Birren, F.: Colour Psychology and Colour Therapy: A Factual Study of the Influence of Colour on Human Life. Kessinger Publishing, Whitefish, MT (2006)
13. Feisner, E.A., Reed, R. (eds.): Colour symbolism. In: Colour Studies, 3rd edn., pp. 184–205. Bloomsbury, New York (2016). https://doi.org/10.5040/9781501303364.ch-014
14. Siwert, J.: Whistler's Nocturnes and the Aesthetic Subject. 0775–0775 (1995)
15. Mayer, L., Myers, G.: American Painters on Technique: 1860–1945. J. Paul Getty Museum, Los Angeles (2013)
16. Ono, A.: Japonisme in Britain: Whistler, Menpes, Henry, Hornel, and Nineteenth-Century Japan. Routledge Curzon, London (2003)
17. Hogan, S.: Healing Arts: The History of Art Therapy. Kingsley, London (2001)
18. Liverani, P., Santamaria, U.: Diversamente bianco: La policromia della scultura romana [Atti del convegno internazionale, Università di Firenze, 15 novembre 2010]. Edizioni Quasar, Roma (2014)
19. Ostergaard, J.S.: The polychromy of antique sculpture: a challenge to Western ideals? In: Brinkmann, V., Primavesi, O., Hollen, M. (eds.) Circumlitio: The Polychromy of Antique and Medieval Sculpture, pp. 78–105 (2010)
20. Abbe, M.: Recent research on the painting and gilding of Roman marble statuary at Aphrodisias. In: Proceedings of the Johann David Passavant Colloquium: Circumlitio. The Polychromy of Antique and Medieval Sculpture, München, Germany, 10–12 December (2008)
21. Magrini, D., Bracci, S., Bartolozzi, G., Iannaccone, R., Lenzi, S., Liverani, P.: Revealing Mithras' colour with the ICVBC mobile lab in the museum. Heritage **2**(3), 2160–2170 (2019). https://doi.org/10.3390/heritage2030130
22. Škapin, A.S., Ropret, P., Bukovec, P.: Determination of pigments in colour layers on walls of selected historical buildings using optical and scanning electron microscopy. Mater. Charact. **58**(11–12), 1138–1147 (2007). https://doi.org/10.1016/j.matchar.2007.05.013

23. Artioli, G.: Science for the cultural heritage: the contribution of X-ray diffraction. Rend. Fis. Acc. Lincei **24**(S1), 55–62 (2013). https://doi.org/10.1007/s12210-012-0207-z
24. Bonaduce, I., Blaensdorf, C., Dietemann, P., Colombini, M.P.: The binding media of the polychromy of Qin Shihuang's Terracotta Army. J. Cult. Herit. **9**(1), 103–108 (2008). https://doi.org/10.1016/j.culher.2007.08.002
25. Lluveras-Tenorio, A., et al.: An insight into Gandharan art: materials and techniques of polychrome decoration. Heritage **5**(1), 488–508 (2022). https://doi.org/10.3390/heritage5010028
26. Hain, M.: Multispectral analysis of cultural heritage artefacts. Meas. Sci. Rev. **3**, 9–12 (2003)
27. Bacci, M., et al.: Fiber optics reflectance spectroscopy in the entire VIS-IR range: a powerful tool for the non-invasive characterization of paintings. In: MRS Proceedings, vol. 852 (2004). https://doi.org/10.1557/PROC-852-OO2.4
28. Cosano, D., Mateos, L.D., Jiménez-Sanchidrián, C., Ruiz, J.R.: Identification by Raman microspectroscopy of pigments in seated statues found in the Torreparedones Roman archaeological site (Baena, Spain). Microchem. J. **130**, 191–197 (2017). https://doi.org/10.1016/j.microc.2016.09.003
29. Karidas, A.G., Brecoulaki, H., Bourgeois, B., Jockey, P.: In situ XRF analysis of raw pigments and traces of polychromy on marble sculpture surfaces. In: Non-destructive Examination of Cultural Objects, pp. 42–62. Adv. X-ray Anal. (2006)
30. Smilgies, D.-M., Powers, J.A., Bilderback, D.H., Thorne, R.E.: Dual-detector X-ray fluorescence imaging of ancient artifacts with surface relief. J. Synchrotron Rad. **19**(4), 547–550 (2012). https://doi.org/10.1107/S0909049512021164
31. Brinkmann, V.: La ricerca sulla policromia antica. In: Bankel, H., Liverani, P. (eds.) I colori del bianco: policromia nella scultura antica, pp. 55–72. De Luca, Roma (2004)
32. Kiilerich, B.: Towards a 'Polychrome History' of Greek and Roman sculpture. J. Art Historiogr. **15**, 1–18 (2016)
33. Brøns, C.: Ancient colours: perspectives and methodological challenges. In: The Value of Colour, pp. 311–337 (2019)
34. Jockey, P.: Le mythe de la Grèce blanche. Belin, Paris (2015)
35. Brinkmann, V., Ostergaard, J.S., Collareta, M., Potts, A.: The Colour of Life: Polychromy in Sculpture from Antiquity to the Present. Getty Publications, Los Angeles (2008)
36. Brinkmann, V.: Gods in Colour: Painted Sculpture of Classical Antiquity. Stiftung Archäologie, Munich (2007)
37. Sargent, M.L.: Research on ancient sculptural polychromy with focus on a 2nd century CE marble Amazon. In: Tracking Colour: The Polychromy of Greek and Roman Sculpture in the Ny Carlsberg Glyptotek, Preliminary Report, pp. 45–65 (2010)
38. Carrella, A.: Marmora Pompeiana nel Museo Archeologico Nazionale di Napoli: Gli arredi scultorei delle case pompeiane. Studi della Soprintendenza Archeologica di Pompei, pp. 55–67 (2008)
39. Bragantini, I.: Il culto di Iside e l'egittomania antica in Campania. In: De Caro, S. (ed.) Egittomania: Iside e il mistero, pp. 159–167. Electa, Milano (2006)
40. Anti, C.: Nuove repliche della Venere che si toglie il sandalo. Riv. Archeol. **45**, 12–25 (1927)
41. Ling, R.: Roman Painting. Cambridge University Press, Cambridge (1991)
42. Salvadori, M., Sbrolli, C.: Wall paintings through the ages: the Roman period—Republic and early Empire. Archaeol. Anthropol. Sci. **13**(11), 187 (2021)
43. Grifa, C., Barba, S., Fiorillo, F., et al.: The domus of Octavius Quartio in Pompeii: damage diagnosis of the masonries and frescoed surfaces. Int. J. Conserv. Sci. **7**, 885–900 (2016)
44. Cotte, M., Susini, J., Metrich, N., et al.: Blackening of Pompeian cinnabar paintings: X-ray microspectroscopy analysis. Anal. Chem. **78**(21), 7484–7492 (2006). https://doi.org/10.1021/ac0612224

45. Radepont, M., Coquinot, Y., Janssens, K., et al.: Thermodynamic and experimental study of the degradation of the red pigment mercury sulfide. J. Anal. At. Spectrom. **30**(3), 599–612 (2015). https://doi.org/10.1039/C4JA00372A
46. Elert, K., Pérez Mendoza, M., Cardell, C.: Direct evidence for metallic mercury causing photo-induced darkening of red cinnabar tempera paints. Commun. Chem. **4**(1), 174 (2021). https://doi.org/10.1038/s42004-021-00610-2
47. Anaf, W., Janssens, K., De Wael, K.: Formation of metallic mercury during photodegradation/photodarkening of α-HgS: electrochemical evidence. Angew. Chem. Int. Ed. **52**(48), 12568–12571 (2013). https://doi.org/10.1002/anie.201303977
48. Pérez-Diez, S., Pitarch Martí, A., Giakoumaki, A., et al.: When red turns black: influence of the 79 AD volcanic eruption and burial environment on the blackening/darkening of Pompeian cinnabar. Anal. Chem. **93**(48), 15870–15877 (2021)
49. Gliozzo, E.: Pigments—Mercury-based red (cinnabar-vermilion) and white (calomel) and their degradation products. Archaeol. Anthropol. Sci. **13**(11), 210 (2021)
50. Marcaida, I., Maguregui, M., Morillas, H., et al.: Raman imaging to quantify the thermal transformation degree of Pompeian yellow ochre caused by the 79 AD Mount Vesuvius eruption. Anal. Bioanal. Chem. **411**(28), 7585–7593 (2019). https://doi.org/10.1007/s00216-019-02175-5
51. Monico, L., Prati, S., Sciutto, G., et al.: Development of a multi-method analytical approach based on synchrotron radiation X-ray micro-analytical techniques and vibrational microspectroscopy methods to unveil the causes and mechanism of darkening of "fake-gilded" decorations in a Cimabue painting. J. Anal. At. Spectrom. **37**(1), 114–129 (2022). https://doi.org/10.1039/D1JA00271F
52. Howard, H., Najorka, J., Schofield, P.F., et al.: Degradation of fourteenth-century mordant gilding layers: synchrotron-based microfocus XRF, XRD, and XANES analyses of two paintings by Pietro Lorenzetti. Stud. Conserv. **69**(3), 193–208 (2024). https://doi.org/10.1080/00393630.2023.2201094
53. Vermeulen, M., Nuyts, G., Sanyova, J., et al.: Visualization of As(III) and As(V) distributions in degraded paint micro-samples from Baroque- and Rococo-era paintings. J. Anal. At. Spectrom. **31**(9), 1913–1921 (2016). https://doi.org/10.1039/C6JA00134C
54. Keune, K., Mass, J., Meirer, F., et al.: Tracking the transformation and transport of arsenic sulfide pigments in paints: synchrotron-based X-ray micro-analyses. J. Anal. At. Spectrom. **30**(3), 813–827 (2015). https://doi.org/10.1039/C4JA00424H
55. Broers, F.T.H., Janssens, K., Nelson, W., et al.: Two pathways for the degradation of orpiment pigment (As_2S_3) found in paintings. J. Am. Chem. Soc. **145**(16), 8847–8859 (2023). https://doi.org/10.1021/jacs.2c12271
56. Casadio, F., Xie, S., Rukes, S.C., et al.: Electron energy loss spectroscopy elucidates the elusive darkening of zinc potassium chromate in Georges Seurat's A Sunday on La Grande Jatte—1884. Anal. Bioanal. Chem. **399**(9), 2909–2920 (2011). https://doi.org/10.1007/s00216-010-4264-9
57. Monico, L., Janssens, K., Vanmeert, F., et al.: Degradation process of lead chromate in paintings by Vincent van Gogh studied by means of spectromicroscopic methods. Part 5. Effects of nonoriginal surface coatings into the nature and distribution of chromium and sulfur species in chrome yellow paints. Anal. Chem. **86**(21), 10804–10811 (2014). https://doi.org/10.1021/ac502841g
58. Monico, L., Van der Snickt, G., Janssens, K., et al.: Degradation process of lead chromate in paintings by Vincent van Gogh studied by means of synchrotron X-ray spectromicroscopy and related methods. 2. Original paint layer samples. Anal. Chem. **83**(4), 1224–1231 (2011). https://doi.org/10.1021/ac1025122

59. Monico, L., Janssens, K., Hendriks, E., et al.: Evidence for degradation of the chrome yellows in Van Gogh's *Sunflowers*: a study using noninvasive in situ methods and synchrotron-radiation-based X-ray techniques. Angew. Chem. Int. Ed. **54**(47), 13923–13927 (2015). https://doi.org/10.1002/anie.201505840
60. Van der Snickt, G., Dik, J., Cotte, M., et al.: Characterization of a degraded cadmium yellow (CdS) pigment in an oil painting by means of synchrotron radiation based X-ray techniques. Anal. Chem. **81**(7), 2600–2610 (2009). https://doi.org/10.1021/ac802518z
61. Pouyet, E., Cotte, M., Fayard, B., et al.: 2D X-ray and FTIR micro-analysis of the degradation of cadmium yellow pigment in paintings of Henri Matisse. Appl. Phys. A **121**, 967–980 (2015). https://doi.org/10.1007/s00339-015-9239-4
62. Gomez Lobon, M., Ghirardello, M., Juncosa Darder, E., et al.: A study of cadmium yellow paints from Joan Miró's paintings and studio materials preserved at the Fundació Miró Mallorca. Herit. Sci. **11**, 145 (2023). https://doi.org/10.1186/s40494-023-00987-4
63. Singer, B., Aslaksby, T.E., Topalova-Casadiego, B., Tveit, E.S.: Investigation of materials used by Edvard Munch. Stud. Conserv. **55**(4), 274–292 (2010). https://doi.org/10.1179/sic.2010.55.4.274
64. Monico, L., Cartechini, L., Rosi, F., et al.: Probing the chemistry of CdS paints in The Scream by in situ noninvasive spectroscopies and synchrotron radiation x-ray techniques. Sci. Adv. **6**(20), eaay3514 (2020). https://doi.org/10.1126/sciadv.aay3514
65. Ferrer, J.S., Sandu, I.C.A., Syversen, T., Cardoso, A.M., Candeias, A., Borca, C.N.: Investigating colour changes in red and blue paints – a preliminary study of art materials and techniques in Edvard Munch's *Old Man in Warnemünde* (1907). In: van den Berg, K., et al. (eds.) Conservation of Modern Oil Paintings, pp. 209–218. Springer, Cham (2019). https://doi.org/10.1007/978-3-030-19254-9_15
66. Fitenborg, T., Buti, D., Vila, A., et al.: Colour changes due to the fading of Prussian blue in Danish golden age paintings. In: Clarricoates, R., Dowding, H., Gent, A. (eds.) Colour Change in Paintings, pp. 27–38. Archetype Publications, London (2016)
67. Cato, E., Borca, C.N., Huthwelker, T., Ferreira, E.S.: Aluminium X-ray absorption near-edge spectroscopy analysis of discoloured ultramarine blue in 20th century oil paintings. Microchem. J. **126**, 18–24 (2016). https://doi.org/10.1016/j.microc.2015.11.021
68. Wijnberg, L.: Do we see what we know or do we know what we see? Conservation of oil paintings in the Stedelijk Museum Amsterdam. In: van den Berg, K., et al. (eds.) Issues in Contemporary Oil Paint, pp. 21–32. Springer, Cham (2014). https://doi.org/10.1021/acs.chemmater.3c01470
69. Ferrer, J.S., Hull, A., Syversen, T.: Investigating surface phenomena in paints and paintings by Edvard Munch (1863–1944) in the context of the artist's painting materials. In: Clarricoates, R., Dowding, H., Gent, A. (eds.) Colour Change in Paintings, pp. 126–130. Archetype, London (2016)
70. Boracchi, B.J., Strand Ferrer, J., Izzo, F.C., et al.: A multi-analytical study on 20 synthetic ultramarine blue and cobalt blue oil paint samples from tubes belonging to Edvard Munch's Studio. In: Bridging the Gap: Synergies Between Art History and Conservation, Proceedings (Postprints) of the Conference Organized by the National Museum in Oslo, November 2023, pp. 273–274 (2024)
71. Carlyle, L.: The Artist's Assistant: Oil Painting Instruction Manuals and Handbooks in Britain 1800–1900; with Reference to Selected Eighteenth-Century Sources, 1st edn. Archetype Publications, London (2001)
72. Van De Laar, M., Burnstock, A.: "With Paint from Claus & Fritz": a study of an Amsterdam painting materials firm (1841–1931). J. Am. Inst. Conserv. **36**(1), 1 (1997). https://doi.org/10.2307/3180080I

73. Strand Ferrer, J., Boracchi, B.J., Izzo, F.C., et al.: A multi-analytical investigation on darkening of blue pigments in Munch's paintings. In: InART24 Conference, Oslo, Oral Communication, 4–7 June 2024
74. Berrie, B., Casadio, F., Dahm, K., et al.: A vibrant surface: investigating colour, texture, and transparency in Winslow homer's watercolours. In: Sgamellotti, A., Brunetti, B.G., Miliani, C. (eds.) Science and Art, pp. 404–428. Royal Society of Chemistry, Cambridge (2020)
75. Daher, C., Sutherland, K., Stratis, H., et al.: Paul Gauguin's Noa prints: multi-analytical characterization of the printmaking techniques and materials. Microchem. J. **138**, 348–359 (2018)
76. Burnstock, A., Lanfear, I., Van Den Berg, K.J., et al.: Comparison of the fading and surface deterioration of red lake pigments in six paintings by Vincent Van Gogh with artificially aged paint reconstructions. In: Verger, I. (ed.) ICOM 14th Triennial Meeting The Hague, 12–16 September 2005, Committee for Conservation: Preprints, pp. 459–466. James and James, London (2005)
77. Hendriks, E., et al.: A comparative study of Vincent van Gogh's Bedroom Series. In: Studying Old Master Paintings: Technology and Practice. The National Gallery Technical Bulletin 30th Anniversary Conference Post prints, Archetype, pp. 237–243 (2011)
78. Centeno, S.A., Hale, C., Carò, F., et al.: Van Gogh's *Irises* and *Roses*: the contribution of chemical analyses and imaging to the assessment of color changes in the red lake pigments. Herit. Sci. **5**, 18 (2017). https://doi.org/10.1186/s40494-017-0131-8
79. Whitmore, P.M., Pan, X., Bailie, C.: Predicting the fading of objects: identification of fugitive colorants through direct nondestructive lightfastness measurements. J. Am. Inst. Conserv. **38**(3), 395–409 (1999). https://doi.org/10.2307/3179999
80. Keune, K., Monico, L., Duivenvoorden, J., et al.: IPERION HS D.5.3 Protocol to monitor objects and environment for preventive conservation and concept of digital platform infrastructure for preventive conservation (1.0). Zenodo (2023). https://doi.org/10.5281/zenodo.10141823
81. Ciortan, I.M., Poulsson, T.G., George, S., Hardeberg, J.Y.: Predicting pigment color degradation with time series models. In: Final Program and Proceedings, vol. 30, no. 1, pp. 250–257 (2022). https://doi.org/10.2352/cic.2022.30.1.44
82. Riutort-Mayol, G., Gómez-Rubio, V., Lerma, J.L., del Hoyo-Meléndez, J.M.: Correlated functional models with derivative information for modeling microfading spectrometry data on rock art paintings. Mathematics **8**(12), 2141 (2020). https://doi.org/10.3390/math8122141
83. Beltran, V.L.: Advancing Microfading Tester Practice: A Report from an Experts Meeting Organized by the Getty Conservation Institute, March 13–15, 2018. Getty Conservation Institute, Los Angeles (2019). https://hdl.handle.net/10020/gci_pubs/advancing_mft_practice
84. Chan, E., Aambø, M., Łojewski, T., Grimstad, I., Ufnalewska-Godzimirska, M., Sandu, I.C.A.: Microfade lightfastness testing on the scream versions in the Munch museum collection. Int. J. Conserv. Sci. **13**, 1623–1634 (2022)
85. Morris, H.R.: "Virtual Fading" of art objects: simulating the future fading of artifacts by visualizing micro-fading test results. J. Am. Inst. Conserv. **46**(3), 215–228 (2007). https://doi.org/10.1179/019713607806112279
86. Hendriks, E., Brokerhof, A.W., van den Meiracker, K.: Valuing Van Gogh's colours: from past to future. In: ICOM Committee for Conservation 18th Triennial Meeting Copenhagen Denmark 4–8 September 2017, Pulido & Nunes, ICOM Committee for Conservation (2017)
87. Brokerhof, A.W., Kuiper, P.C., Scholten, S.: Spread or sacrifice: dilemma for lighting policies. Stud. Conserv. **63**(sup1), 28–34 (2018). https://doi.org/10.1080/00393630.2018.1504439

88. Ciortan, I.M., Poulsson, T.G., George, S., et al.: Tensor decomposition for painting analysis. Part 1: pigment characterization. Herit. Sci. **11**, 76 (2023). https://doi.org/10.1186/s40494-023-00910-x
89. Ciortan, I.M., Poulsson, T.G., George, S., Hardeberg, J.Y.: Tensor decomposition for painting analysis. Part 2: spatio-temporal simulation. Herit. Sci. **11**(1), 84 (2023). https://doi.org/10.1186/s40494-023-00913-8
90. Saunders, D.: Museum Lighting: A Guide for Conservators and Curators. Getty Publications, Los Angeles (2020)
91. Del Hoyo-Melendéz, J.M., Mecklenburg, M.F.: An investigation of the reciprocity principle of light exposures using microfading spectrometry. Spectrosc. Lett. **44**(1), 52–62 (2011). https://doi.org/10.1080/00387010903508572
92. Liang, H., Lange, R., Lucian, A., Hyndes, P., Townsend, J.H., Hackney, S.: Development of portable microfading spectrometers for measurement of light sensitivity of materials. In: International Council of Museums, Committee for Conservation (ICOM-CC) Triennial Conference, Lisbon, Portugal (2011). https://irep.ntu.ac.uk/id/eprint/17689/
93. Saunders, D., Kirby, J.: Wavelength-dependent fading of artists' pigments. Stud. Conserv. **39**(sup2), 190–194 (1994). https://doi.org/10.1179/sic.1994.39.Supplement-2.190
94. Lerwill, A., Brookes, A., Townsend, J.H., Hackney, S., Liang, H.: Micro-fading spectrometry: investigating the wavelength specificity of fading. Appl. Phys. A **118**(2), 457–463 (2015). https://doi.org/10.1007/s00339-014-8645-3
95. Monico, L., Janssens, K., Cotte, M., et al.: Synchrotron-based X-ray spectromicroscopy and electron paramagnetic resonance spectroscopy to investigate the redox properties of lead chromate pigments under the effect of visible light. J. Anal. At. Spectrom. **30**(7), 1500–1510 (2015). https://doi.org/10.1039/C5JA00091B
96. Lerwill, A., Townsend, J.H., Thomas, J., Hackney, S., Caspers, C., Liang, H.: Photochemical colour change for traditional watercolour pigments in low oxygen levels. Stud. Conserv. **60**(1), 15–32 (2015). https://doi.org/10.1179/2047058413Y.0000000108
97. Hendriks, E., Vellekoop, M.: Van Gogh's Sunflowers Illuminated: Art Meets Science. Amsterdam University Press and Van Gogh Museum (2019). https://doi.org/10.5117/9789463725323
98. Berns, R.S., et al.: Rejuvenating the colour palette of Georges Seurat's A Sunday on La Grande Jatte—1884: a simulation. Colour Res. Appl. **31**(4), 278–293 (2006). https://doi.org/10.1002/col.20223
99. Berns, R.S.: Digital colour reconstructions of cultural heritage using colour-managed imaging and small-aperture spectrophotometry. Colour Res. Appl. **44**(4), 531–546 (2019). https://doi.org/10.1002/col.22371
100. Pozzi, F., Basso, E., Centeno, S.A., Smieska, L.M., Shibayama, N., Berns, R., et al.: Altered identity: fleeting colours and obscured surfaces in Van Gogh's Landscapes in Paris, Arles, and Saint-Rémy. Herit. Sci. **9**(1), 1–19 (2021)
101. Geldof, M., Proaño Gaibor, A.N., Ligterink, F., Hendriks, E., Kirchner, E.: Reconstructing Van Gogh's palette to determine the optical characteristics of his paints. Herit. Sci. **6**(1), 17 (2018). https://doi.org/10.1186/s40494-018-0181-6
102. Kirchner, E., Van Der Lans, I., Ligterink, F., Hendriks, E., Delaney, J.: Digitally reconstructing Van Gogh's Field with Irises near Arles, Part 1: Varnish. Colour Res. Appl. **43**(2), 150–157 (2018). https://doi.org/10.1002/col.22162
103. Kirchner, E., Geldof, M., Hendriks, E., Gaibor, A.N.P., Janssens, K., Delaney, J., et al.: Recreating Van Gogh's original colours on museum displays. Electronic Imaging **31**, 1–6 (2019)
104. Kleynhans, T., Schmidt Patterson, C.M., Dooley, K.A., Messinger, D.W., Delaney, J.K.: An alternative approach to mapping pigments in paintings with hyperspectral reflectance image

cubes using artificial intelligence. Herit. Sci. **8**(1), 84 (2020). https://doi.org/10.1186/s40494-020-00427-7
105. Chen, A., Jesus, R., Vilarigues, M.: Identification and visualization of pure and mixed paint pigments in heritage artwork using machine learning algorithms. SN Comput. Sci. **4**(2), 115 (2022)
106. Vermeulen, M., et al.: XRFast: a new software package for processing of MA-XRF datasets using machine learning. J. Anal. At. Spectrom. **37**(10), 2130–2143 (2022). https://doi.org/10.1039/D2JA00114D
107. Xu, B.J., et al.: Can deep learning assist automatic identification of layered pigments from XRF data? J. Anal. At. Spectrom. **37**(12), 2672–2682 (2022). https://doi.org/10.1039/D2JA00246A
108. Liu, L., et al.: Neural networks for hyperspectral imaging of historical paintings: a practical review. Sensors **23**(5), 2419 (2023). https://doi.org/10.3390/s23052419
109. Shine, K., Berman, P.: Edvard Munch the Modern Life of the Soul, The Museum of Modern art, NY, exhibition catalogue (2006)
110. Munch, E., Ydstie, I., Munch-museet (Oslo, Norway) (eds.): The Scream: Munch Museum; Vigmostad Bjørke, Oslo: Bergen (2008)
111. Plahter, U., Topalova-Casadiego, B.: The Scream by Edvard Munch: painting techniques and coloring materials. In: Studying Old Masters Paintings, The National Gallery Technical Bulletin, Ed. M. Spring, pp. 244–252 (2008)
112. Sandbakken, E.G., Tveit, E.S.: Edvard Munch's monumental sketches (1909–1916) for the Aula of Oslo University, Norway: conservation issues and treatments. Stud. Conserv. **57**(sup1), S258–S267 (2012). https://doi.org/10.1179/2047058412Y.0000000030
113. Stein, M.: The Sun and the Researchers. On the conservation of two of the drafts for the University Aula competition, in Munch's laboratory. The Path to the Aula, Munch Museet catalogue, 2011–2012
114. Frøysaker, T. (ed.): Public Paintings by Edvard Munch and His Contemporaries: Change and Conservation Challenges. Archetype Publications, London 139 (2015)
115. Neri, D. (2021). https://www.barnebys.com/blog/edvard-munch-the-colours-of-the-soul
116. Pettersen, P.: Edvard Munch - The Scream. IJCS Special Issue 2022 (2022). https://ijcs.ro/public/IJCS-22-104_Pettersen.pdf
117. Buchhart, D., Munch, E., Fondation Beyeler, Kunsthalle Würth (eds.): Edvard Munch: signs of modern art; [Fondation Beyeler, Riehen/Basel, March 18 - July 15, 2007, Kunsthalle Würth, Schwäbisch Hall, August 4 - end of December, 2007] (2007)
118. Przybyszewski, S.: Psychischer Naturalismus. Neue Deutsche Rundschau (Freie Bühne), V. Jahrgang, III. und IV. Quartal (Berlin: S. Fischer Verlag, 1894), 153–154 (1894)
119. De Graaff, J.H.H., Roelofs, W.G.T., Van Bommel, M.R.: The Colourful Past: Origins, Chemistry and Identification of Natural Dyestuffs (2004)
120. Cardon, D.: Natural Dyes: Sources, Tradition, Technology and Science. Archetype Publications, London (2007)
121. Mussak, R.A.M., Bechtold, T.: Natural colorants in textile dyeing. In: Bechtold, T., Mussak, R. (eds.) Handbook of Natural Colorants, pp. 315–338. Wiley, Chichester (2009)
122. Robinson, J., Pardoe, T.: An Illustrated Guide to the Care of Costume and Textile Collections (2000)
123. Colombini, P., Andreotti, A., Baraldi, C., Degano, I., Lucejko, J.J.: Colour fading in textiles: a model study on the decomposition of natural dyes. Microchem. J. **85**, 174–182 (2007)
124. Joosten, I., van Bommel, M., Keijzer, R.D., et al.: Micro analysis on Hallstatt textiles: colour and condition. Microchim. Acta **155**, 169–174 (2006). https://doi.org/10.1007/s00604-006-0537-6

125. Doherty, B., et al.: Identifying Brazilwood's marker component, Urolithin C, in historical textiles by surface-enhanced Raman spectroscopy. Heritage **4**, 1415–1428 (2021). https://doi.org/10.3390/heritage4030078
126. Zippel, E.: Analysis of textile colorants and coloured textiles. Rev. Prog. Col. **34**, 1–11 (2004)
127. Gulmini, M., Idone, A., Diana, E., Gastaldi, D., Vaudan, D., Aceto, M.: Identification of dyestuffs in historical textiles: strong and weak points of a non-invasive approach. Dyes Pigm. **98**(1), 136–145 (2013)
128. Rezic, I.: Historical Textiles and Their Characterization. Cambridge Scholars Publishing, Newcastle upon Tyne (2022)
129. Dyer, J., Tamburini, D., O'Connell, E.R., Harrison, A.R.W.: A multispectral imaging approach integrated into the study of Late Antique textiles from Egypt. PLoS ONE **13**(10), e0204699 (2018). https://doi.org/10.1371/journal.pone.0204699
130. Degano, I., Ribechini, E., Modugno, F., Colombini, M.P.: Analytical methods for the characterization of organic dyes in artworks and in historical textiles. Appl. Spectrosc. Rev. **44**(5), 363–410 (2009)
131. Park, C., Kang, M.G.: Color restoration of RGBN multispectral filter array sensor images based on spectral decomposition. Sensors **16**(5), 719 (2016)
132. Lukac, R., Plataniotis, K.N.: Color Image Processing: Methods and Applications. CRC Press, Boca Raton (2018)
133. Karthikeyan, B., Sztandera, L.M.: Analysis of tactile perceptions of textile materials using artificial intelligence techniques: Part 1: forward engineering. Int. J. Cloth. Sci. Technol. **22**(2/3), 187–201 (2010)
134. Pereira, F., Carvalho, V., Vasconcelos, R., Soares, F.: A review in the use of artificial intelligence in textile industry. In: Machado, J., Soares, F., Trojanowska, J., Yildirim, S. (eds.) Innovations in Mechatronics Engineering, icieng 2021. LNME, pp. 377–392. Springer, Cham (2022). https://doi.org/10.1007/978-3-030-79168-1_34
135. Chatterjee, S., Trumpy, G., Ruedel, U.: Digital unfading of chromogenic film informed by its spectral densities. Heritage **6**, 3418–3428 (2023). https://doi.org/10.3390/heritage6040181
136. Reilly, J.M.: Storage Guide for Colour Photographic Materials, pp. 9–11. Image Permanence Institute, Rochester, NY, USA (1998)
137. Chambah, M., Rizzi, A.: Perception-based digital motion picture restoration and quality evolution. In: Parraman, C., Rizzi, A. (eds.) Colour Coded, pp. 95–109. Society of Dyers and Colourists, West Yorkshire, UK (2010)
138. Wilhelm, H.G.: The Permanence and Care of Colour Photographs: Traditional and Digital Colour Prints, Colour Negatives, Slides, and Motion Pictures. The Center for the Image Grinnell, Grinnell, USA, 1st edn. (Following SPRINGER: Wilhelm, H., Brower, C.: The Permanence and Care of Colour Photographs: Traditional and Digital Colour Prints, Colour Negatives, Slides, and Motion Pictures. 1st edn. Preservation Publ. Co., Grinnell, Iowa (1993))
139. Adelstein, P.Z., Graham, C.L., West, L.E.: Preservation of motion-picture colour films having permanent value. J. SMPTE **79**(11), 1011–1018 (1970). https://doi.org/10.5594/J13565
140. Miyagawa, T., Shirai, Y.: Fading of colour photographic dyes I. J. Imaging Sci., 29, 216–218. Heritage **6**, 3427 (2023)
141. Pietro, G.D.: Chapter 4: Examples of using advanced analytical techniques to investigate the degradation of photographic materials. In: Physical Techniques in the Study of Art, Archaeology and Cultural Heritage, pp. 155–198. Elsevier (2007)
142. Kurosaki, T., Itamura, S., Rokutanda, S., Teramoto, T., Miyagawa, T.: Fading of colour photographic dyes III: isolation and identification of reductive fading products of 1-phenyl-3-methyl-4-aza-(p-diethylaminophenyl)-2-pyrazoline-4,5-dione by bisulphite. J. Photogr. Sci. **36**, 79–82 (1988). https://doi.org/10.1080/00223638.1988.11736971

143. Rizzi, A., Gatti, L., Kránicz, B., Berolo, A.J.: Mixed perceptual and physical-chemical approach for the restoration of faded positive films. In: Proceedings of the 4th European Conference on Colour in Graphics, Imaging, and Vision, Society for Imaging Science and Technology, Barcelona, Spain, 8–13 June 2008 (2008)
144. Liu, L., et al.: Digital restoration of colour cinematic films using imaging spectroscopy and machine learning. Sci. Rep. **12**, 21982 (2022)
145. Flueckiger, B., Pfluger, D., Trumpy, G., Croci, S., Aydın, T., Smolic, A.: Investigation of Film Material–Scanner Interaction; Version 1.1, pp. 1–88. University of Zurich: Zurich, Switzerland (2018)
146. Pénichon, S.: Twentieth-Century Colour Photographs: Identification and Care. Getty Conservation Institute, Los Angeles (2013)
147. Lavédrine, B., Gandolfo, J.-P., Capderou, C., Guinée, R., McElhone, J.P.: The Lumière Autochrome: History, technology, and Preservation. Getty Conservation Institute, Los Angeles (2013)
148. Barker, G., Hubička, J., Jacobs, M., et al.: Finlay, Thames, Dufay, and Paget colour screen process collections: using digital registration of viewing screens to reveal original colour. In: Colour Photography and Film: Sharing Knowledge of Analysis, Preservation, and Conservation of Analogue and Digital Materials (2023)
149. Cucci, C., Casini, A., Stefani, L., Cattaneo, B., Picollo, M.: A novel transmittance vis–NIR hyper-spectral imaging scanner for analysis of photographic negatives: a potential tool for photography conservation. Sensors **23**(7), 3562 (2023). https://doi.org/10.3390/s23073562
150. Pfluger, D., Garmsen, L., Trumpy, G.: Modern methods for the visualization of lenticular film colours. Colour Cult. Sci. J. **14**(1), 64–72 (2022)
151. Yumibe, J.: Moving Colour: Early Film, Mass Culture, Modernism. Rutgers University Press, New Brunswick, NJ (2012)
152. Kindem, G.: The demise of Kinemacolour: technological, legal, economic, and aesthetic problems in early colour cinema history. Cine. J. **20**(2), 3–14 (1981). https://doi.org/10.2307/1224830
153. Bedding, T.: Moving pictures in natural colours. Mov. Pict. World **4**(30–31), 230–231 (1909)
154. Capstaff, J.G., Seymour, M.W.: The Kodacolour process for amateur colour cinematography. Trans. Soc. Motion Pict. Eng. **12**, 940–947 (1928)
155. Ahriman, X.Y.: Von der zukünftigen Photographie: Ein neues Prinzip der Farbenphotographie. Photographisches Archiv **37**, 249–251 (1896)
156. Lippmann, G.: Épreuves réversibles: Photographies intégrales. Comptes rendus hebdomadaires des séances de l'Académie des Sciences **146**, 446–451 (1908)
157. Berthon, R.: A suggested new method of colour photography direct at one exposure. Br. J. Photogr. **57**, 421–423 (1910)
158. Jacquet-Loew, J.: La cinématographie en couleurs Keller-Dorian. Bulletin de la Société industrielle de Mulhouse **89**, 421–426 (1923)
159. Keller-Dorian, A., Berthon, R.: Patent E.P. 24,698. London (1914? 1915)
160. Ede, F.: Un épisode de l'histoire de la couleur au cinéma: le procédé Keller-Dorian et les films lenticulaires. 1895, 71, 187–202 (2013). https://doi.org/10.4000/1895.4787
161. Eggert, J.: Kurzer Überblick über den Stand der Farbenkinematographie. Bericht über den VIII. Internationalen Kongress für wissenschaftliche und angewandte Photographie, Dresden 1931. Leipzig: J. A. Barth, 214–221 (1932)
162. Mitchell, R.A.: Is lenticulated colour-film practical? International Projectionist, XXVI(10), 5–9, 29–30 (1951)

Authenticity in the Design of Digital Heritage Applications

Sofia Pescarin[1(✉)], Ivana Cerato[1], Irina-Mihaela Ciortan[2], Giuseppe Città[3], Arthur Clay[4], Samuele Spotti[1], Sophia Sotiropoulou[5], and Giorgio Trumpy[2]

[1] CNR ISPC, Florence, Italy
{sofia.pescarin,ivana.cerato}@cnr.it
[2] NTNU ColourLab, Gjøvik, Norway
{irina-mihaela.ciortan,giorgio.trumpy}@ntnu.no
[3] CNR ITD, Palermo, Italy
giuseppe.citta@cnr.it
[4] HSLU, Lucerne, Switzerland
arthur.clay@hslu.ch
[5] FORTH IESL, Heraklion, Greece
sophiaso@eap.gr

Abstract. This chapter explores the concept of authentic experiences and how it could be applied to design new engaging applications and user experiences dedicated to coloured collections. The authors analyse how the concept was used in different domains. They then explore users' behaviours and their understanding of this concept by designing a Cultural Probe Kit and involving a group of participants into an experiment. They then summarise the results and propose a theoretical framework in which authenticity is made of three interconnected dimensions: the 'Self' (representing cognitive, emotional and sensory engagement characteristics of individual users), the 'Others' (dealing with social relationships) and the 'World' (as physical or digital environment). As a conclusion, the chapter presents examples of how such framework can be practically applied to Interactive Media projects.

Keywords: Authenticity · Hybridity · XR applications

In the last three years we have been exploring the concept of authenticity and specifically of an "authentic experience", with the goal of better design interactive applications in the field of Digital Heritage. During the PERCEIVE project, we have further studied and detailed the concept, developing an overall conceptual framework to support designers' work.

Authenticity encompasses various dimensions beyond mere realism. In [1] and [3] we have described how "authenticity" is made of three dimensions: the "Self", the "Others" and the "World". An authentic experience, in fact, deeply involves, first, the user (the Self) on cognitive, emotional, and sensory levels. This involvement transforms impersonal virtual elements into something personal and intimately connected to the user. The role of the self is crucial, intertwined with its relationships with others (representing their social dimension) and the environment. Such prominent position of the perceiver led

S. Pescarin et al. (Eds.): *Chromatic Visions*, LNCS 15750, pp. 81–127, 2026.
https://doi.org/10.1007/978-3-032-07792-9_3

us to initially represent the concept of authenticity as a triangle, with the Self positioned at the apex. Each dimension is defined by multiple "components" and each component can be further detailed, identifying several different "elements". The organised structure of dimensions, components and elements was used to build our theoretical framework that is described here, represented in Fig. 1, Fig. 2, Fig. 3.

In this chapter, we synthetically describe previous studies on "authenticity" in various fields. We have included also an overview of the origin of the concept, to be able to re-organise our knowledge in line with our scope. After this chapter, we have described a survey we have carried out to further explore qualitatively the behaviours and thoughts of potential users about authenticity in the field of Cultural Heritage (3.2). We have then described the "authenticity framework" (3.3) and we have included a detail on some of its components (3.5 and 3.6). Furthermore, we have included examples of digital applications that can be considered as representative (3.7). We have concluded this overview, demonstrating how a framework can be used to identify design strategies that can be adopted to develop more authentic User Experiences (UX) in the field of Digital Heritage (specifically directed to the representation of coloured collections) (3.4) and also to set the ground for an "authenticity index" useful in evaluation (3.8).

1 Authenticity: A Multi-disciplinary State of the Art Analysis

In philosophy and psychology, individuals have historically been examined in light of their interactions with others and with the physical environment. This concept has been explored across various disciplines, its roots extending back to ancient Greek philosophers. Derived from the Greek word *authentikós* (*autós: self*), it has been a subject of philosophical discourse, notably within Western philosophy, where there has been a longstanding focus on the individual. This focus often delineates between a "private" and a "public" self, shaping its relationship with society. Questions such as "*Who am I*?" and "*What is my authentic self*?" Is at the bottom of the concept of self-judgment [2]. Prior to Hegel, sincerity, honesty, and moral values were considered more relevant than authenticity, linked to the potential of knowing oneself and acting accordingly, honest and truthful by society. However, the concept of authenticity also encompasses being true to oneself for personal fulfilment, independent from societal expectations, in a continuous search for a balance between identity and authenticity. This highlights the significance of recent psychological and social science research on introverted and extroverted traits [4]. Extroverts have been found to more readily perceive authenticity, while, on the other hand, studies have also shown that introverts, when prompted to act as if they were extroverted, were equally capable of perceiving authentic experiences [5]. Building on this idea, [6] Wilt et al. have defined a "Subjective Authenticity" that emerges when one behaves in accordance with their genuine self-concept. This definition highlights the importance, also in UX design, to let individuals express themselves, discover personal meaning, and engage in reflection; only through this self-expression or self-reflection it is possible to nurture a perception of authenticity. Moreover, there are other components that have been recently studied as being relevant. It is the case of "concentration" and "sense of direction". Grandey et al. [7], for instance, has studied how authenticity emerges only when distractions are limited, while Pescarin and Martinez Pandiani [8] have highlighted that this process impacts also knowledge acquisition and engagement in serious games. Also, disorientation may produce a negative effect on the perception of authenticity such as any situation that can create a discomfort in the participants of an experience [9] [10]. There are other aspects, connected to the personal disposition or to the context a person live in, that are also relevant and may modify significantly the sense of authenticity, including: personal expectations, emerging emotions and sensations, the previous knowledge one own and the desire to acquire new knowledge and to remember it (the "memorability" quoted by [11]), the possibility to act and to make a choice [12] and also the challenges one may experience, since they can be interpreted as opportunities to grow [13]. All these aspects relate to authenticity and to its "**Self**" dimension. Although many studies have been conducted to explore them, we prefer here to quote Karl Jasper, a psychologist and philosopher who probably better than others was able to summarise the complexity of this dimension already in 1919. In his "Psychology of WorldViews" [14], he wrote that authenticity emerges when it resonates with an individual's deepest self, enduring, evolving, and evolving alongside the person. Jasper therefore, beyond the concepts described in the first part of this chapter, had already introduced the notions of depth (the opposite of superificiality), time, and evolution. Following Jasper, an authentic experience is therefore something perceived as deeply personal and meaningful, unique to everyone, subject to change over time, and capable of being nurtured through introspective engagement.

Although pivotal to any experience, the Self is continuously in relation to other people (its social or "**Others**" dimension). Therefore, one important aspects to consider is the interconnecting components between the Self and the Others, by mean of formal or informal communication, expressed and explicit (low-context) or internal and implicit (high-context), spoken with natural language or with symbolic gestures. Authenticity, in fact, has been defined as a social psychological phenomenon, structured around social practices [15] such as activities shared with others (challenges, competition, cooperation), ruled by social norms. This phenomenon is based on action. According to Coupland [16], authenticity has a performative aspect ("*authentication*"), intended as an active process that involves action, verbal exposition and dialogue. Dialogue is therefore a crucial component in authenticity, not only when other people are around, but also when they are not (inner dialogue) or when they are potentially there (potential dialogue), such as when there is the desire to exchange experiences and thoughts, after an experience. The value of verbal exchange emerges also in what has been defined as the "mediated dialogue" approach by McKenney et al. [17], that can be facilitated by community building, development of personal connections, exploration of alternative perspectives and by a summarising and reflective moment at the end of the experience (https://www.sitesofconscience.org/). This approach was adopted also to design interactive VR experiences, as in the case of the Emotive EU project [18], where social gestures have been developed to strengthen embodiment and social dimension, reinforced by solicited dialogue among users, triggered by provocative questions and "emotive" tone of voice.

The last pillar of authenticity, is the external world, intended here both as the physical world, where the experience takes place, but also as the digital environment with digital objects, either reproduction of physical ones or as reconstruction or completely digitally born virtual elements. The "**World**" dimension is particularly relevant in this book, where the materiality of the artworks is at the bottom of any digital representation.

The component of "truthfulness", that requires a process of verification [2], is often used also in tourism studies, where users often report to search for something "faithful to an original" or "of unquestioned origin or authorship". Authenticity is among the most important expectations for tourists [11] and it is often mentioned in relation with realism, accuracy and objectivity [19], with realism being strictly connected to a physical context, its characteristics and properties. However, according to Vico [20], this faithful representation may not be authentic or a true reproduction of the artwork: a digital object represents the original not only in terms of appearance, but also in terms of context, meaning and use. Meticulous documentation of the digitisation and interpretation process are therefore crucial to maintaining a sense of authenticity in digital replicas through validation and interpretation processes. In Computer Science and specifically in the development of VR and XR experiences, authenticity is frequently associated with the sense of presence, interaction and immersivity that produce an "embodiment" in a digital space, but also with the graphical realism of the virtual environments [21]. Beyond the realism, there are other two aspects that emerge as crucial in serious games studies for the perception of authenticity: randomness and unexpected elements [22]. There is a last aspect that is emerging in recent studies of design and architecture, that is the concept of "atmosphere", a dynamic subjective phenomenon, perceptual, sensory, and emotive [23]. Although it is a subjective element and it should be included in the "Self" dimensions, this concept has a direct relation with the "World", and to the effort designers and developers put in the creation of a "mood" to be applied either to a single digital scenario or to an entire UX. It can be therefore interpreted as a quality of a place that can "move" people, making them feel emotions or triggering sensations or also memories. In tourism studies, Zimmermann [24] stated that developing an atmosphere of the past can satisfy users' needs of authenticity, while it also can produce "place attachment" in cultural tourism [25].

2 Qualitative Analysis on Behaviours and Perception of Authenticity

In this section we describe an experimental survey carried out to further extend, explore and test the initial findings obtained with the analysis of previous works. With this experiment we aimed at identifying and understanding how perceptions of authenticity vary across different contexts, focusing specifically on the cultural one. We have therefore adopted a specific qualitative method that required the design and development of Cultural Probe Kits (CPK). Described by Gaver et al. [26], CPKs are tools for self-reporting (i.e. diaries or activity workbooks), that allow participants to observe, think about, and describe their experiences, in situations where it is not possible to undertake direct observations, or when it is to avoid interference of researchers or in situation where there are privacy issues and time limits. They are distributed to participants, facilitating the documentation of particular sensations, emotions, or experiences [27].

In 2022, we have specifically designed a CPK for this purpose [3]. Although acquired inputs have been only qualitative, this exploration proved to be extremely useful in the integration and refinement of the concept of authenticity, detailing components and elements. During that experiment we have involved a uniform group of participants, consisting of potential users of virtual museums [28].

In the CPK we included a diary and an activity workbook. The activities were meant to explore multiple aspects around the following research questions:

- how do people perceive an authentic experience in a museum setting?
- how do people perceive authenticity in daily life?
- are there behaviours common to many users connected to authenticity?
- how do people perceive authenticity in a digital scenario?

Our objectives included the following: exploring the relationship between perceived authenticity and emotions; identifying the primary emotion associated with authentic experiences (among Fear, Happiness, Sadness, Disgust, Anger, and Surprise); determining the predominant senses and corresponding stimuli involved (among sight/visuals, hearing/sounds, touch, smell, and taste); investigating the correlation between authenticity, trustworthiness, and reliability; gaining insights into individuals' perceptions of genuine digital experiences; examining the relationship between the realism of digital scenarios and the perception of authenticity; and comparing the stimuli and triggers encountered in everyday life with those experienced in digital contexts.

In line with our initial research question, we structured our methodology to include a series of tasks for participants during their museum visit. These tasks involved capturing two distinct photos: one representing authenticity and the other depicting inauthenticity, accompanied by explanatory comments. Additionally, participants completed a writing task in which they reported a personal story describing a significant, authentic life event, supported by a relevant photograph. This narrative approach aimed to delve deeper into participants' thoughts and emotions surrounding their perception of authenticity. To address the third question, participants were encouraged to keep a diary for eight days, documenting their experiences to track the frequency of specific events over time. Finally, we administered a series of online questions tailored to explore the final research question. The findings revealed several key insights:

- *Authenticity Perception in Museum Settings.* Participants captured and commented on photos representing authentic and inauthentic experiences at the Palazzo Poggi Museum. Most participants felt that the absence of specific elements would alter their experience, highlighting the importance of these elements. Factors influencing perceptions of authenticity included expectations (45%) and, most notably, knowledge (75%). Participants also reported that experiencing the event alone or with others could have affected their perception of its authenticity.
- *Authenticity in Daily Life.* Participants shared personal stories and emotions related to authentic experiences in their lives. The narratives depicted diverse emotions and sensory perceptions contributing to authenticity, such as unique, authentic, and most notably, emotionally resonant moments including peace, love, sorrow, anxiety, oddity, and honesty.
- *Repeated/Common Behaviours and Thoughts.* Diary entries reflected similar themes to those identified in the museum setting, highlighting the significance of emotions and new experiences (engagement in something previously untried).
- *Authenticity in Digital Experiences.* Participants acknowledged a connection between emotions and authenticity in digital experiences. Reported emotions and feelings

included: Sadness; Sense of Loss; Spontaneity; Happiness and Joy; Pity; Human Pity; Human connection and empathy; Deep emotions, Feeling of being involved; Love and passion; Relief; Tiredness. It was also reported that factors such as reliability of the institution and immersive elements influenced authenticity perception. Participants identified various authentic digital experiences, including virtual environments and exhibitions.

Metaphors. Moreover, during the experiment we have collected a number of interesting comments and potential metaphors that could be used in the design of an authentic experience. Participants reported in their story, that authenticity is like:

- ... Entering somebody's **house**
- ... Something that is **coming to life**
- ... a **travel** with friends, where you experience many different elements
- ... an **earthquake** that shake you and made you feel many emotions and act
- ... a **night walk with a good friend** where the silence makes you focus on senses and sounds and thoughts and everything become intimate
- ... a party o**r a ceremony** in a natural place, where there are people and much more and you have time to talk and exchange with strangers
- .... a memorable **activity with friends in the nature** with a lot of exchanges, laughs, emotions, feelings
- ... a physical **travel back in time** where everything seems real and surprising

Results. As a result, we have identified the following main results:

1. Emotions triggering authenticity may come from Immersive Sounds, Hybridity (Real-Digital interconnection) and Interactivity. Emotions such as happiness and surprise were closely associated with authenticity, while fear and disgust received minimal recognition. Sight and hearing were the most cited senses involved in perceiving authenticity.
2. Involvement (sense of presence) is fundamental in the perception of authenticity and it can be built by an Involving Atmosphere and Involving Narrative
3. Authenticity may provoke a generated interest (beyond the experience)
4. Emotional impact and unpredictability were identified as crucial indicators of authenticity in digital and real-life experiences.

Overall, the study highlights the multifaceted nature of authenticity perception and its implications for design and user experience.

3 Authenticity Conceptual Framework

The Self encounters, identifies with, and collides with the Other (who is also a Self) within a temporal and spatial environment (the World), through a system of interactions to achieve an authentic experience [3]. Authenticity, therefore, begins with the Self; it touches the innermost self and endures, develops, and evolves with it. We have confirmed that authenticity is characterised by three interconnected dimensions, with most components being unique to one of them (i.e., others–dialogue, world–atmosphere). These three dimensions are: the Self, the Other, and the World.

From our investigations, it has become evident that grasping the essence of authenticity within hybrid and XR cultural experiences demands a thorough and multidimensional examination. This segment is dedicated to delving into the intricate nature of authenticity, elucidating its emergence through the convergence of theoretical frameworks and practical observations. Through this method, our aim is to delineate a conceptual model capable of effectively guiding both the crafting and evaluation of genuine experiences. The complexity inherent in authenticity reveals itself through myriad facets, reflecting the richness and diversity of its dimensions and constituents. These discernible traits transcend mere summation, weaving together in distinct configurations that yield deeply personal and profound encounters. Our inquiry commences with the premise that authenticity defies static definitions or narrow criteria; rather, it manifests through a mosaic of qualitative and quantitative elements, interacting in intricate and often unpredictable ways. Within this context, literature and empirical findings assume pivotal roles in unveiling authenticity's multifaceted nature. While literature furnishes us with theoretical underpinnings and fundamental concepts, empirical insights gleaned from fieldwork and qualitative analyses enrich this foundation with firsthand, contextual revelations. The synergy between these approaches offers fresh perspectives, suggesting that each authentic encounter arises from a unique amalgamation of interconnected components, inherently imbued with personal and subjective perspectives. Nonetheless, cognizant of this complexity, our aim was to distil a broad and concise definition of an "authentic experience," with the specific intent of aiding the design of interactive experiences. Addressing this complexity necessitates a comprehensive approach that acknowledges and integrates the myriad facets and interrelations of authenticity's constituents. This entails meticulous consideration of how these elements coalesce to engender experiences that resonate with authenticity. It is within this intricate tapestry of elements that the true essence of authenticity resides, rendering each experience distinctive and irreplicable.

In the following section we detail dimensions, components and elements.

3.1 The Self

The dimension of "**Self**" pertains to the users as the primary focal point of the digital or hybrid experience. It comprises 8 primary components, each encompassing specific elements (although we have outlined 34 elements here, this list is not exhaustive) (Fig. 1).

1. Disposition (personal)
2. Context (personal)
3. Cognition
4. Perception
5. Emotion
6. Embodiment (personal)
7. Familiarity (personal)
8. Time

1. **Personal Disposition** denotes an individual's inclination to perceive an experience as authentic, influenced by personality traits or contextual, cultural, and social factors. Emphasizing this component prompts visitors, for example, to share personal anecdotes or reflections, fostering individual inclinations and emotions within a safe and enjoyable environment. This encompasses shared elements and action-oriented design strategies that elucidate how designers can leverage these concepts effectively:

1. A Attention / Distraction	"***Limit Distractions***" i.e. the design includes visual (eye-catching elements), interactive or narrative elements that help limiting distractions
1. B Curiosity	***"Include Curiosity triggers***" i.e. the designer includes positive feedback
1. C Extroversion	***"Encourage extroverted behaviours"*** i.e. the designer makes users acting extroverted
1. D Identification: Identity	***"Develop identification"*** i.e. the designer adds initial identification tasks to personalise the experience.
1. E Sense of direction	***"Support Sense of Direction"*** i.e. the designer provides users with maps

2. **Personal Context** refers to the context in which a person lives and acts; a digital experience designed to give relevance to this component can include elements that bridge the context of the application with the context of the visitors. It includes:

2. A Challenge	***"Create challenges"*** i.e. the design includes challenges or tasks that make users struggle to solve them.
2. B Choice	***"Offer Choices"*** i.e. designers include mechanics where users can choose which direction to take
2. C Expectations	***"Address expectations"***
2. D Meaningfulness	***"Engage users meaningfully"*** i.e. designers engage users with metaphors through which they can find themselves motivated or with mechanisms that make the self-reflecting of the personal meaning
2. E Value (acting honestly and truthfully, in accordance with the true self)	***"Adopt valuable concepts"*** i.e. designers include narrative elements that can be taken as valuable for the users or may use "justice" as an abstract concept around which to develop the experience.

3. **Cognition** refers to personal cognitive characteristics.

3. A Comprehension	"***Make content understandable***"
3B. Knowledge	"***Generate knowledge***" i.e. users are given stories that increase their knowledge.
3C. Memory and recall	"***Ease recall***" i.e. the narrative is hierarchically organised to facilitate the recalling of facts [8]
3. D Reflection	"***Provoke Reflections***" i.e. designers include provocative questions that stimulate reactions, soliciting reflection.
3. E Self-monitoring	"***Stimulate self-awareness***" i.e. specific hints or requests are included to facilitate self-reflection and awareness about themself, the body, their position with the space and the others.

4. **Perception** refers to the different senses people use during an experience and that can fully stimulate sensory responses through the body.

4. A Hear	***"Add auditory stimuli"***
4. E Taste	***"Make users taste"***
4. B Touch	***"Develop tangible interaction"***
4. C Sight	***"Use visual elements"***
4. D Smell	***"Include sources of smells"***

5. **Emotions** are mental states triggered by neurophysiological responses, which can be variously connected to ideas, feelings, behavioural reactions, and a level of pleasure or discomfort [29]. Experiences that focus on emotions may use for instance narrative or dramatic storytelling elements that stimulate emotional resonance,

5. A Anger	***"Provoke Anger"*** i.e. the experience, through the narrative or the setting, moves the users towards a state of anger.
5. B Disdain	***"Provoke Disdain"***
5. C Fear	***"Create a status of anxiety and fear"***
5. D Joy	***"Solicit Joy"***
5. E Sadness	***"Provoke sadness"***
5. F Surprise	***"Create Surprise"***

6. **Embodiment** is the sense of one's own body (feeling), requiring that the person is aware of where they are (self-location), what they do (agency) [30]. Through embodiment, the sense of presence is strengthen, activating interaction and immersivity.

6. A Interaction	***"Include interaction mechanics"***
6. B Immersivity	***"Immerse users physically and mentally"*** i.e. the design includes visual, auditory, narrative elements that build the perception of being immersed in a place
6. C Sense of presence	***"Create a sense of presence" i.e.*** designers include a mix of different elements that increase the sense of being "there"

7. **Familiarity.** The experience makes the visitors feel comfortable.

7. A Accessibility	***"Make the experience accessible"***
7. B Comfort	***"Create comfort"***
7. C Usability	***"Make the experience comprehensible"***
7. D Ergonomics	***"Adapt elements of the experience to users"***
7. E Security	***"Let users feel secure"***

8. **Time** is the personal evolution one can experience

8. A Evolution (personal)	***"Create a personal evolution"*** i.e. narrative accompany the user towards an increasing deeper understanding

3.2 The Others

The self has a fundamental **social** dimension, we have named "Others", following Davidson's epistemology, about the three types of knowledge (of the self, of other minds, and of the external world [31]). The "Others" signify encounters, identifications, and collisions between selves within a temporal and spatial environment (the World), facilitated by interactions. Humans, as social beings, thrive on establishing and maintaining connections with others, leading to the development of diverse communication methods and interaction patterns. An individual's authenticity is thereby shaped by the social context in which they exist. The proposed framework suggests that authenticity can be enhanced also when soliciting this social dimension, increasing the sense of belonging through group affiliations, crafting collaborative experiences that foster exchange dynamics, incorporating guides or moderators, or designing hybrid experiences for co-presence to augment social embodiment. This dimension encompasses 7 distinct components, each characterized by various elements (we present here a not exhaustive list of 22 elements) (Fig. 2):

- Language
- Social practices
- Social norms

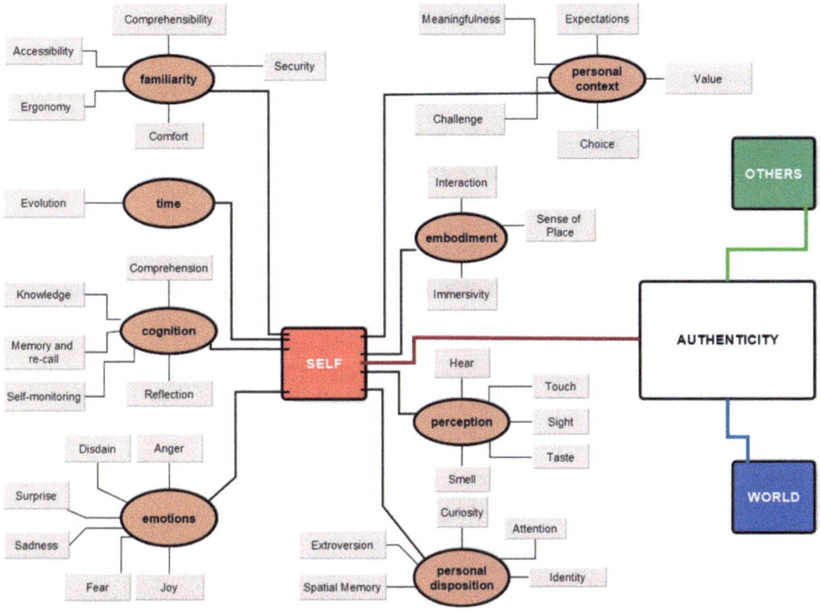

Fig. 1. - The "authenticity" conceptual framework with the dimension of the "Self", its components and elements.

- Unpredictability (social)
- Embodiment (social)
- Familiarity (social)
- Time
- **9. Language** refers to the human capability to exchange with others using natural language. It may involve and include:

9. A Exchange	***"Stimulate exchange"*** i.e. the designer assigns to a participant the role of a guide with the goal of stimulating the exchange among the other participants.
9. B Perspective	***"Explore alternative perspectives"*** i.e. the design promotes perspective giving or taking, improving empathy
9. C Provocation	***"Ask provocative questions"*** i.e. the experience includes provocative questions that trigger exchange
9. D Storytelling	***"Develop stories"*** i.e. curiosity is triggered by narrative.

10. **Social Practices** refers to the activities involving more than one participant of an experience:

10. A Challenges	***"Create team challenges"***
10. B Competition	***"Develop competitive mechanics"***
10. C Cooperation	***"Include cooperative mechanics"*** i.e. co-creation tasks
10. D Interaction (with o.)	***"Include social interaction"***
10. E Relationships	***"Build relationships"*** i.e. the mechanics help establish a connection among participants

11. **Social Norms** refers to elements that may impact the relation or exchange between the self and the others, such as constraints, rewards and penalties:

11. A Constraints	***"Get users to respect constraints"***
11. B Rewards	***"Add proportional rewards"***
11. C Roles	***"Assign roles"*** i.e. the experience includes the assignment of roles to participants with the goal of improving relation
11. D Penalties	***"Include penalties"***

12. **Unpredictability (social)** refers to unexpected behaviours that might happen:

12. A Casualty	***"Enable casual influences"***
12. B Influences (group)	***"Promote positive group influences"influences"***
12. C Unexpected	***"Leave unexpected social behaviours happening"***

13. **Embodiment** (social) is the sense of one's own body in relation with other people. It can be triggered by meaningful and symbolic interaction gestures (such as "high five") thus strengthening the sense of belonging to a group.

13. A Gestures (Social)	***"Develop social gestures"***
13. B Tone of voice	***"Define suitable tone of voice"***

14. **Familiarity** (social) as feeling of being part of the group as a family member:

14. A Communion	***"Facilitate community building"***
14. B Connection	***"Develop connections"***
14. C Sense of Belonging	***"Strengthen / develop sense of belonging"*** i.e. during the experience participants become aware to belong to the same group (social cohesion [32]).
14. D Hospitality	***"Provide welcoming social atmosphere"***

15. **Time** refers to the progress of the relationship within a group of people.

15. A Evolution (social)	***"Develop increasingly social relations"***

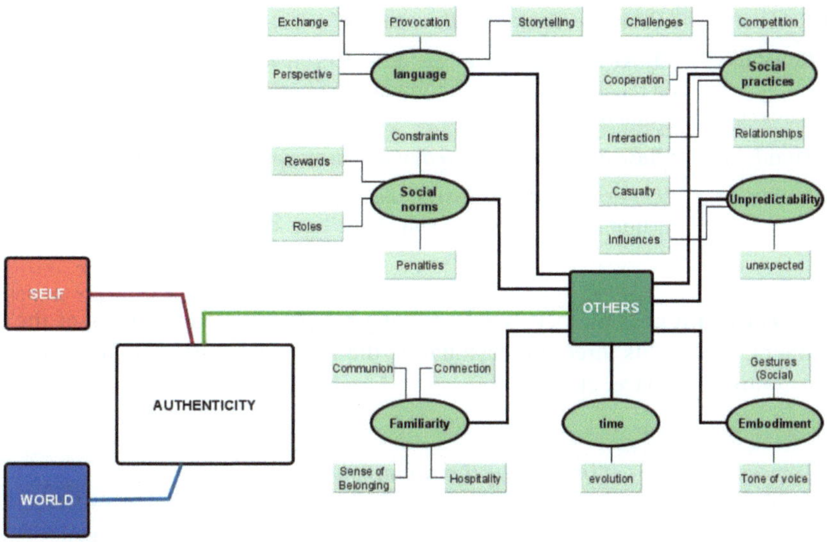

Fig. 2. The authenticity concept framework with a detail of the "Others"

3.3 The World

The last dimension, the "**World**", refers to an "authentic" perception of the environment. It might refer to the physical setting where the experience takes place or /and also to the digital environment (the scene and the objects), especially in XR applications. To strengthen this dimension, designers can use 3d realistic environments, make participants aware of their reliability, validation and interpretation, strengthen embodiment, develop an atmosphere (using effects of light, material, sounds), or include unexpected elements. It includes 7 components (Fig. 3):

- Verification
- Physical Context
- Atmosphere
- Unpredictability (environmental)
- Embodiment (environmental)
- Familiarity (environmental)
- Time

- With **Verification** we intend the realism, reliability, validation and accuracy of the environment (the space and objects) where the experience takes place. An authentic experience would not only immerse users in realistic scenarios, but also give them the possibility to be aware of their quality and validity. Nevertheless, reliability and truthfulness do not necessarily imply the realism.

16. A Realism	***"Create realistic digital scenarios"***
16. B Reliability	***"Make assets reliable"*** i.e. designers create digital reconstructions based on reliable sources of information and provide ways for participants to be aware of them.
16. C Truthfulness	***"Base digital reconstruction on true/scientific information"***
16. D Validation	***"Let expert validate assets"*** i.e. digital assets are validated by experts or institutions whose name and "fame" are known

17. **Physical Context** is the perceived and tangible physical components of the environment, such as its three dimensionality, the material aspect, the laws of physics that control the space, etc.

17. A Dimension (spatial)	***"Include the three dimensions"***
17. B Physics	***"Apply physical laws"*** i.e. the scenario of an application is modelled to be credible, as ruled by physical laws
17.C Visual Aspect	***"Assign visual properties to objects (material, colours)*** " i.e. elements are rendered to make users perceive their real colours and material characteristics.

18. **Atmosphere** refers to the development of an appropriate mood, made of a visual style, sounds, lighting and other environmental settings that contribute to enrich feelings and emotions in the visitors.

18. A Light	***"Add appropriate lighting"*** to obtain specific psychological effects
18. B Sound	***"Create a soundscape"***
18. C Synesthetic	***"Mix visual, audio and tangible effects"***
18. D Visual Style	***"Assign a coherent visual style"*** i.e. the scene is rendered in a coherent way with the goal, the desired mood and narrative

19. **Unpredictability** (environmental) refers to random side events that happen during the experience, unpredictable for people.

19. A Originality (novelty)	***"Use original / novel assets"***
19. B Peculiarity	***"Highlight peculiar aspects"***
19. C Randomness	***"Develop unexpected events"*** i.e. the system is programmed to make events happen randomly or procedurally
19. D Uniqueness	***"Underline uniqueness"*** i.e. participants are shown the uniqueness of objects or environments

20. **Embodiment** (environmental) refers to the immersion in the environment, enabled by real time feedback, interaction, perception of immersivity and the sense of being there.

20. A Feedback	"***Provide users with appropriate feedback***"
20. B Immersivity	"***Include high level of immersivity***" i.e. stereoscopic visualisation (high level of immersivity), spatial audio, engaging narrative.
20. C Interaction	"***Let users interact with environment***" i.e. users can interact with the world using tangible objects provided with sensors

21. **Familiarity** refers to the sense of comfort towards the environment (i.e. in the environment there are "markers" as meaning making elements, or maps).

21. A Comfort	"***Create a comfortable environment***"
21. B Reference (spatial)	"***Include spatial markers***" (as arrows, panels, lighted paths…).
21. B Safety	"***Develop a safe space***" i.e. no dangerous elements are included
21. D Sense of place	"***Strengthen sense of place***" i.e. familiar elements are inserted, soliciting past memories.

22. **Time** refers to the recreation of time flow during an experience:

22. A Natural conditions	"***Include natural conditions***" i.e. weather or seasonal variation
22. B Time flow	"***Make users perceive time flow***" i.e. a countdown or a clock or the movement of the shadows together with the time of the day

Among the above-mentioned components and elements, specific projects could be interested to focus on a subset of them. For instance, the PERCEIVE project, whose main goal is the development of new methods to study, reconstruct, preserve and communicate the colours of historical objects and artworks, has been mostly interested to aspects referred either to the increase of awareness and participation of the citizens toward the fragility of colour, or to the realism and reliability of the colour reconstruction and rendering:

SELF – *Personal Disposition* – Curiosity
SELF – *Personal Context* – Meaningfulness
SELF – *Cognition* – Comprehension, Knowledge, Reflection
SELF – *Perception*- Sight, Touch, Hear
SELF – *Emotions* – Joy, Surprise
SELF – *Embodiment* – Interaction, Immersivity, Sense of Presence
SELF – *Familiarity* – Accessibility, Usability
OTHERS – *Language* – Exchange, Perspective, Provocation, Storytelling
OTHERS – *Social Practices* - Cooperation and co-creation
WORLD – *Verification*, Realism, Reliability, Truthfulness, Validation
WORLD – *Physical Context* - Dimension, Physics, Visual Aspect
WORLD – *Embodiment (environmental)* - Feedback, Immersivity, Interaction
WORLD – *Familiarity* - Sense of place

Developing hybrid or XR experiences just focusing on these sub-set of elements, doesn't limit the potential authenticity of the result, for the user, but create a different distribution of them, ending into a full authentic engagement.

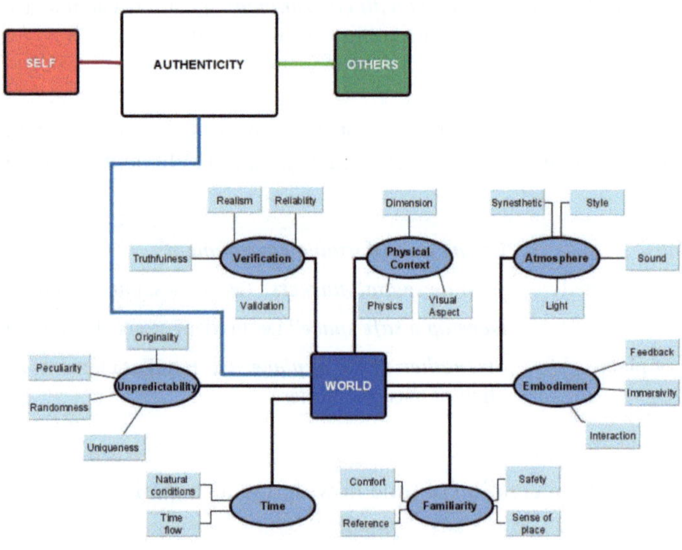

Fig. 3. Authenticity concept framework with a detail of the "World" dimensions.

In the next sections, we have better detailed two of these important components: "verification" (Sect. 4) and physical context (Sect. 5).

4 Towards a Tool for the Design of Authentic Experiences

How designers can develop ideas in line with the authenticity framework? To answer to this question, we have developed a tool that starts from the elements and actionable design hints described above and supports two stages of design: "identification of audience's goals" and "idea generation".

Within the European project GIFT (https://gifting.digital/), a deck of cards known as "Visitor Toolbox" has been redesigned and made available for museum professionals and designers working for cultural institutions [33]. This tool organised and supported co-creation activities following the five stages of design: 1. Context definition, 2. Institutional goals, 3. Design brief, 4. Ideation and storyboarding, 5. Disruption.

Unfortunately, although extremely useful, it doesn't enable to design taking into consideration cognitive and emotional goals that the designers and institutions could take as reference in their project. Therefore, within PERCEIVE project we have first of all created an extension for the Visitor Toolbox, adding specific cards to a new stage (Audience goals) and to the ideation phase. After a testing phase, we have then decided to develop a new deck of cards, initially as a paper boardgame design experience, and finally as a collaborative webapp, also enabling hybrid co-design sessions. This tool will be described in Volume 2. We include here the description of the different included elements for the design stages.

Design Stage: Audience Goals. During this design stage, the team focuses on the authenticity components and select which of them are a priority for the project and for the cultural institutions. Here we have included only general indications of the potential goals.

- Self. STIMULATE SENSORY RESPONSES
- Self. STIMULATE EMOTIONAL RESPONSES
- Self. STRENGTHEN PERSONAL EMBODIMENT
- Self. DEVELOP SENSE OF FAMILIARITY
- Self. DEVELOP PERSONAL EVOLUTION
- Others. IMPROVE LANGUAGE EXCHANGES
- Others. FOSTER SOCIAL PRACTICES AMONG VISITORS
- Others. FOSTER SOCIAL NORMS AMONG VISITORS
- Others. INCLUDE SOCIAL UNPREDICTABILITY
- Others. IMPROVE SOCIAL EMBODIMENT
- Others. DEVELOP SOCIAL RELATIONS INCREASINGLY
- Others. IMPROVE GROUP FAMILIARITY
- World. USE PLAUSIBLE AND REALISTIC ELEMENTS
- World. PHYSICALLY SIMULATE THE ENVIRONMENT
- World. INCLUDE INTERACTIONS AND FEEDBACK BETWEEN VISITORS AND ENVIRONMENT
- World. CREATE APPROPRIATE MOOD (VISUAL STYLE, SOUNDS, LIGHTING)
- World. LET VISITORS TIMEFLOW
- World. MAKE VISITORS FEELING FAMILIAR WITH THE ENVIRONMENT
- World. INCLUDE ENVIRONMENT UNPREDICTABILITY

Stage 5: Ideation. In this stage the team selects three "ideation" cards that are used as hints for idea generation; we have therefore shaped the cards as to include suggestions that help designers to focus on triggers, mechanics and strategies dedicated to authenticity, also suggesting specific potential actions that the team may adopt:

Self. PERSONAL DISPOSITION *Limit Distractions* *Include curiosity triggers* *Encourage extroverted behaviour* *Develop identification* *Support Sense of Direction*	Self. PERSONAL CONTEXT *Create challenges* *Offer Choices* *Address expectations* *Engage users meaningfully* *Adopt valuable concepts*
Self. COGNITION *Make content understandable* *Generate knowledge* *Ease recall* *Provoke Reflections* *Stimulate self-awareness*	Self. PERCEPTION *Add auditory stimuli* *Make users taste* *Develop tangible interaction* *Use visual elements* *Include sources of smell*
Self. EMOTIONS *Provoke Anger* *Provoke Disdain* *Create a status of anxiety - fear* *Solicit Joy* *Provoke sadness* *Create Surprise* Self. PERSONAL EMBODIMENT *Include interaction mechanics* *Immerse physically and/or mentally* *Create a sense of presence*	Self. PERSONAL FAMILIARITY *Create comfort* *Make the experience comprehensible* *Adapt elements of the experience* *Make the experience accessible* *Create a sense of security*
Others. LANGUAGE *Stimulate exchange* *Explore alternative perspectives* *Ask provocative questions* *Develop stories*	Others. SOCIAL NORMS *Get users to respect constrains* *Add proportional rewards* *Assign roles* *Include penalties*
Others. SOCIAL PRACTICES *Create team challenges* *Develop competitive mechanics* *Include cooperative mechanics* *Include social interaction* *Build relationships* Others. SOCIAL UNPREDICTABILITY *Leave space to social casualty* *Promote positive group influences* *Include unexpected social behaviours*	Others. SOCIAL FAMILIARITY *Facilitate community building* *Develop connections* *Strength or develop sense of belonging* *Provide welcoming atmosphere* Others. TIME *Develop social relations increasingly*
World. VERIFICATION *Create realistic digital scenarios* *Make assets reliable* *Base reconstruction on true/scientific information* *Let expert validate assets*	World. PHYSICAL CONTEXT *Choose one or more hints:* *Include the three dimensions* *Apply physical laws* *Assign visual properties to objects (material, colours, etc.)*

(*continued*)

(*continued*)

Self. PERSONAL DISPOSITION *Limit Distractions* *Include curiosity triggers* *Encourage extroverted behaviour* *Develop identification* *Support Sense of Direction*	Self. PERSONAL CONTEXT *Create challenges* *Offer Choices* *Address expectations* *Engage users meaningfully* *Adopt valuable concepts*
World. ENVIRONMENTAL UNPREDICTABILITY *Focus on original assets* *Highlight peculiar aspects* *Develop unexpected events* *Underline uniqueness*	World. ENVIRONMENTAL EMBODIMENT *Provide users with appropriate feedback* *Include high level of immersivity (spatial audio, stereoscopy, narrative)* *Let users interact with environment (hybrid and XR -smart objects)*
World. ENVIRONMENTAL FAMILIARITY *Create a comfortable environment* *Include spatial markers* *Develop a safe space* *Strengthen sense of place*	World. TIME *Include natural conditions: weather, season* *Make users perceive time flow*

5 World and Verification Component: Authenticity in Reproductions

In the reproduction of artworks, authenticity is applied by staying true to the historical context and the original artist's intentions. When recreating historical artifacts or artworks, the aim is not to produce exact replicas but to authentically represent the original artwork's essence and reveal the process of reproduction. This allows viewers to observe both the current state of the artwork and its condition at the time of its creation. Three exemplary cases that illustrate the importance of authenticity in restoration are the practice of Kintsugi [Fig. 4], *The Fate of the Animals* by Franz Marc [Fig. 5], and Marcel Duchamp's *The Large Glass* [Fig. 6].

Kintsugi.

Kintsugi, or "golden joinery," is the Japanese art of repairing broken pottery using lacquer mixed with powdered gold, silver, or platinum. Rather than concealing the cracks, this technique highlights them, celebrating the object's imperfections as part of its history. As a philosophy, kintsugi reflects the Japanese aesthetic of wabi-sabi, embracing imperfection and transience. It emphasizes the beauty of wear and tear, viewing cracks and repairs as integral parts of an object's life rather than flaws to be hidden. Overall, kintsugi embodies a profound philosophy that finds beauty and meaning in imperfection, offering a unique perspective on the nature of existence and the passage of time[1]. One of the most famous examples of Kintsugi practice is the Golden Tea Bowl, also known as the "Kintsugi Chawan." This tea bowl dating back to the 15th century, is housed in the Tokyo National Museum in Japan [Fig. 4]. It gained widespread recognition for its

[1] https://en.wikipedia.org/wiki/kintsugi.

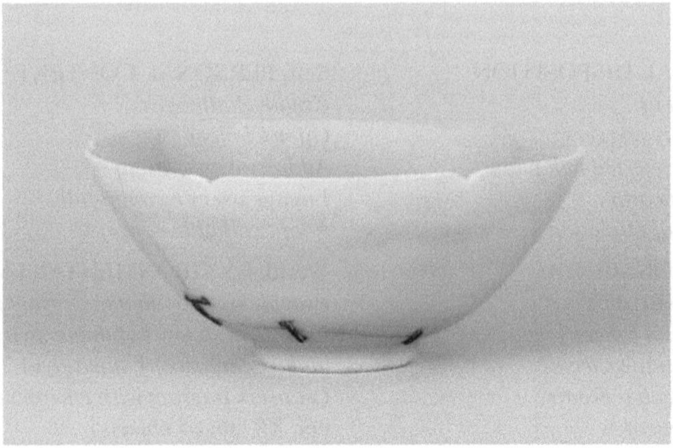

Fig. 4. Celadon Tea Bowl (Bako-Kan) Ryusen, Tokyo National Museum

exquisite repair work, which includes intricate golden seams running along its cracks. The Golden Tea Bowl exemplifies the artistry and philosophy of Kintsugi, showcasing how imperfections can be transformed into sources of beauty and admiration. The process of mending pottery through Kintsugi can be seen as a form of restoration, but with a distinctive philosophical twist. In traditional restoration, the goal is often to restore an object to its original state, minimizing or concealing any evidence of damage. However, Kintsugi embraces the object's history, including its imperfections, by highlighting the cracks with gold or other precious materials. This approach reflects a restoration that not only preserves the functional and aesthetic value of the pottery but also transforms it into something new, celebrating its "flaws" as part of its narrative. Therefore, in a broader sense, Kintsugi could be viewed as a form of "authentic" restoration, as it retains a connection between the original and restored elements, preserving the integrity of the object by respecting its history and the passage of time.

Fate of the Animals.

The artwork "*Fate of the Animals*" is a renowned painting by Franz Marc, created in 1913 during the Der Blaue Reiter movement [Fig. 5]. The artwork, oil on canvas, depicts a chaotic scene of animals amidst destruction, contrasting Marc's usual peaceful animal portrayals. Following damage to the painting's in a 1916 warehouse fire, Paul Klee, a close friend of Marc, undertook its restoration[2]. Klee utilized old photographs and added a brownish tint to the paint, creating a notable variation. Although speculative, some scholars propose that Klee's use of the brown tint was intended to distinguish between the original and restored sections of the painting. Whatever the reasons Paul Klee's restoration techniques on "Fate of the Animals" were, the viewer can discern the distinction between Franz Marc's original work and Klee's restoration efforts. This discernment not only preserves a high degree of authenticity but also enriches the painting's narrative. By referencing the restoration process in the painting's aesthetics, the viewer gains insight into the history and evolution of the artwork, further enhancing its significance.

[2] https://en.wikipedia.org/wiki/fate_of_the_animals.

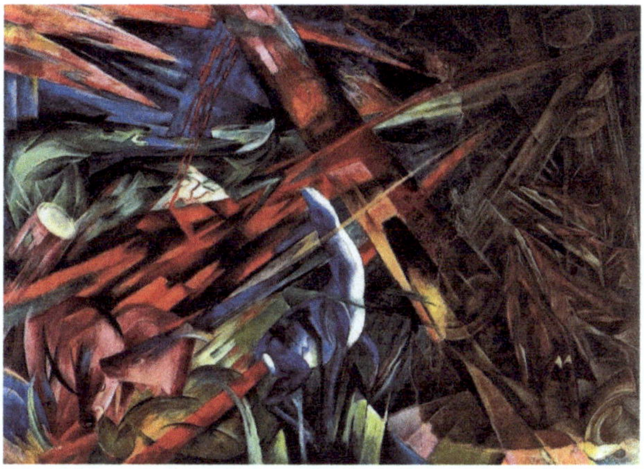

Fig. 5. Fate of the Animals" by Franz Marc, 1913. Oil on canvas, 195 cm × 268 cm (77 in × 106 in). Currently housed at the Kunstmuseum Basel, Basel

In the realm of art, authenticity holds profound significance, shaping our perception and appreciation of objects. In *The Fate of the Animals* by Franz Marc, authenticity is preserved through Paul Klee's act of repainting the damaged areas, while simultaneously lowering the tones of the painting to match how they would have appeared originally. Similarly, the art of Kintsugi embodies authenticity in a different form, emphasizing the beauty of imperfection and the inherent value of repair. Instead of discarding or attempting to hide the damage, Kintsugi artisans embrace the cracks and fractures as integral parts of the object's story while restoring its functionality. In both the Fate of the Animals and Kintsugi, authenticity emerges from a deep appreciation of the object's journey, as each imperfection, whether it's a brushstroke on a painting or a crack in a ceramic vessel, becomes a testament to the object's history and the hands that have shaped it. In essence, the Fate of the Animals and Kintsugi offer profound insights into the nature of restoration in art, in that the goal of a restoration must always involve authenticity, which in the case of the above examples can be understood as keeping the artwork's original essence or features during restoration processes. Such approaches ensure that any restoration allows a clear connection between the original elements and the restored aspects, so that viewers can discern the difference between what was original and what was restored. The emphasis is on maintaining the cultural and historical value of the artwork while allowing it to be appreciated in its restored form.

The Large Glass and its Reproductions.

In the article by Bryony Bery on the reconstruction of Marcel Duchamp's "The Bride Stripped Bare by Her Bachelors, Even" by Richard Hamilton[3], the author traces the journey of the artwork known today as the Large Glass, from its creation to its multiple exhibitions, breakages, repairs, and reconstructions, shedding light on the challenges posed by its physical state and, more importantly, on debates surrounding authenticity of replications and restorations in art history. The timeline of Marcel Duchamp's Large

[3] Through The Large Glass: Richard Hamilton's Reframing of Marcel Duchamp.https://www.tate.org.uk/research/tate-papers/26/through-the-large-glass.

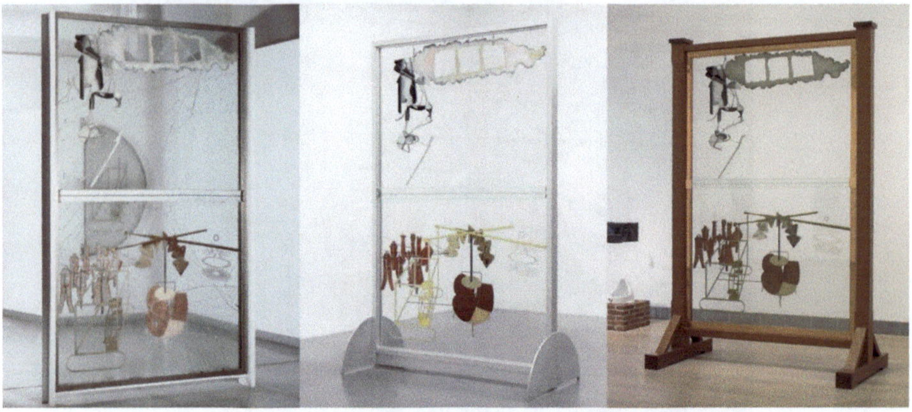

Fig. 6. Marcel Duchamp's 'The Large Glass,' also known as 'The Bride Stripped Bare by Her Bachelors, Even,' [34] was produced from 1915 to 1923. Currently housed in the Philadelphia Museum of Art, Philadelphia (left); two reproductions from Hamilton (centre) and Linde (left)

Glass provides a detailed account of the artwork's journey, encompassing its creation, exhibitions, breakages, repairs, and reconstructions. As a chronicle, the time lines offers insight into the complexities of authorship, authenticity, and conservation associated with an iconic artwork, revealing the evolving narrative surrounding it. Along the same lines, the concept of "original" comes also into question, as the various reconstructions of the Duchamp's Large Glass by artists Ulf Linde and Richard Hamilton provoke discussions on authorship in art, as the points of departure for fabricating a copy of an artwork differ greatly, as there are clear divisions between fabricating a replication by following the visual information provided by the original work or realizing a reproduction following the conceptual basis of the work as the artists had documented it [34]. Therefore, both artists, through their different approaches, illuminate how replication can diverge based on whether one is adhering strictly to the physical and visual details of the original or whether one is engaging more deeply with the conceptual framework laid out by the original artist (Table 1).

Most importantly, Hamilton and Linde's work highlights a key issue in contemporary art: a replica can either strive to mimic the exact visual appearance of the original, or it can aim to capture the original's spirit and conceptual underpinnings. This distinction emphasizes the dual aspects of physical execution and conceptual interpretation in the reproduction of art. These considerations broaden the dialogue on how we perceive and value original works versus their replicas, challenging traditional views on creativity and the transmission of artistic ideas across different contexts and through different media. This is particularly relevant in the context of Duchamp, whose work fundamentally questioned the art world's conventions and the role of the artist. Marcel Duchamp's role in redefining authorship within the art world is underscored his introduction of the Readymade [34] and his provocative artistic strategies challenged the traditional boundaries between the artist's intention and the audience's reception. His works compel us to question not only who makes art, but how the identity of the maker affects our understanding and valuation of art itself. In the case of the replicas by Richard

Hamilton and others, we see Duchamp's influence echoing through their efforts. These replicas further the discourse on authorship by blurring the lines between creation and reproduction, between original and derivative. Ultimately, Duchamp's legacy is that he expanded the framework of artistic authorship to include the ideas of selection, context,

Table 1. This chronological table traces the significant events in the life of Marcel Duchamp's work, "The Bride Stripped Bare by Her Bachelors, Even," known as the Large Glass. From its initial creation to the various phases of damage, restoration, and replication, the timeline highlights the ongoing impact and continued re-evaluation of Duchamp's masterpiece.

Year	Event	Description
1915–1923	Creation	Marcel Duchamp creates "The Bride Stripped Bare by Her Bachelors, Even," commonly known as the Large Glass, using various materials including glass, lead wire, and dust.
1927	Damage	The fragile piece is partially shattered in transit.
1936	Restoration	Duchamp restores the sculpture, declaring the asymmetrical pattern of cracks an improvement.
1954	Replica	Duchamp authorizes a replica of the Large Glass by Ulf Linde, a Swedish artist and collector. Linde's replica is constructed with Duchamp's guidance and approval.
1964	Reconstruction	Richard Hamilton, a British artist, begins work on a reconstruction of Duchamp's Large Glass for a retrospective at the Tate Gallery in London. The project aims to faithfully replicate Duchamp's methods and processes.
1966	Exhibition	Hamilton's reconstruction of the Large Glass is completed and exhibited at the Tate Gallery. It is praised for its meticulous attention to detail and fidelity to Duchamp's original intentions.
1968	Death	Duchamp dies at the age of 86 in New York City. Authorisations of reproductions ore replicas of the Large Glass are no longer possible.
1984	Damage	Hamilton's Large Glass replica shatters while on display at the Tate Gallery in London. The damage necessitates a decision regarding conservation and reconstruction.
1986	Reconstruction	Tate Gallery conservators, led by Christopher Holden and Roy Perry, undertake a complete reconstruction of the damaged lower panel of Hamilton's Large Glass. The reconstruction aims to faithfully replicate Hamilton's original work.
1994	Clarification	Ronald Alley clarifies the attribution of the Large Glass at the Tate Gallery, emphasizing Hamilton's contribution. The piece is described as 3/4 Duchamp and ¼ Hamilton or even 4/5 Duchamp and 1/5 Hamilton.
2011	Death	Richard Hamilton dies on September 13, 2011, in Northend, Oxfordshire, England. Authorisations of reproductions ore replicas made by him of the Large Glass are no longer possible

(*continued*)

Table 1. *(continued)*

Year	Event	Description
2014	Exhibition	The Tate Gallery includes Hamilton's reconstruction of the Large Glass in his retrospective exhibition, acknowledging his significant role in its creation. However, questions of authorship persist, with the work primarily listed under Duchamp's name on the Tate's website.
Ongoing	Debate	The Large Glass continues to be a subject of scholarly debate and discussion, with its replicas, restorations, and interpretations contributing to broader conversations about authenticity, authorship, and the nature of art.

and presentation as artistic processes in their own right. This has paved the way for contemporary art to embrace a broader array of creative expressions, where the concept and context of a work can be as significant as the hand that crafted it. Thus, Duchamp did not just redefine what could be considered art; he redefined the very notion of being an artist, challenging subsequent generations to rethink the creative act [34].

Restoration Across Mediums.

In *The Fate of the Animals*, authenticity was preserved when Paul Klee carefully restored the painting by repainting damaged areas while lowering the tones to match the original. This ensured the emotional and aesthetic integrity of the piece remained intact, allowing viewers to connect with both its past and present forms. Similarly, Kintsugi, the Japanese art of mending broken pottery with gold, celebrates the object's imperfections, highlighting the cracks rather than concealing them, thus preserving the object's historical narrative. However, Marcel Duchamp's *The Large Glass* (also known as *The Bride Stripped Bare by Her Bachelors, Even*) further exemplifies the complexities of authenticity in restoration. After the work was accidentally shattered during transport, Duchamp chose to leave the cracks as part of the artwork's final form, viewing them as an integral part of its evolution. This decision reflected his emphasis on chance and transformation in art, adding another layer to the concept of authenticity by incorporating the artwork's unintended changes into its completed state.

Further complicating the notion of authenticity, several reproductions of *The Large Glass* have been made. Notably, Richard Hamilton, a British artist, created a meticulous reproduction in 1965 after extensive consultation with Duchamp. Hamilton's version captures both the original design and the cracks, thus maintaining the integrity of Duchamp's decision to include the accidental damage. Additionally, in 1985, Swedish artist Ulf Linde also created a faithful reproduction of the work. These reproductions are considered authentic in the sense that they stay true to Duchamp's original intentions and the work's unique evolution, but they also raise questions about the boundaries of originality and authenticity in art restoration and reproduction. All three examples—Franz Marc's *The Fate of the Animals*, the practice of Kintsugi, and Duchamp's *The Large Glass* and its reproductions—demonstrate that authenticity in restoration and reproduction involves more than physical duplication. It's about preserving the cultural and historical significance, the artist's intent, and the artwork's evolving narrative. These approaches allow contemporary audiences to engage with both the current condition of the artwork and its original essence, enriching their understanding of its journey and transformation over time.

In the context of the PERCEIVE project, authenticity of artworks is defined as the faithful reproduction and communication of the original visual and sensory characteristics, particularly focusing on their colour aspects. This includes accurately recreating how artworks originally appeared and might have been experienced in their original contexts, taking into account factors such as environmental lighting and the material conditions of the artifacts. The goal is to preserve not just the physical integrity of the artifacts but also the original sensory experiences they provided, ensuring that the representation in digital or restored formats remains true to the creator's original intent and historical authenticity. PERCEIVE addresses several challenges in colour preservation across various mediums- It enhances methods for predicting and mitigating colour changes in paintings and works on paper due to aging and environmental factors. The project also develops strategies to maintain and accurately replicate original colours and designs in textiles like dresses and embroidery that are prone to fading. Additionally, it focuses on preserving the authenticity of colours in photographic collections, which often degrade over time, ensuring these images remain vibrant and true to their origins. Finally, the project delves into innovative preservation techniques for born-digital AR artworks, tackling the unique challenges of colour use and transparency in these contemporary formats.

Autochromes

Autochromes, an early form of colour photography, can experience several types of deterioration through factors including historic exposure to light and moisture or mishandling. One issue includes greening, shown through the altering the colour balance, often because of exposure to moisture. Tanning refers to a yellowing or browning of the image caused by exposure to heat. This commonly occurred to autochromes plates which were historically projected. Spotting, appearing as small blemishes or discolorations often green or orange in appearance, occurs via exposure to moisture or chemical reactions. Loss of emulsion is another issue, leading to areas where the image is no longer visible. Finally, fading may happen, resulting in a loss of vibrancy and clarity in the colours originally captured. These effects, while not desirable, can significantly impact the image quality, including the loss of the photograph's visual details.

In a discussion within the project team responsible for historical photography, differing opinions arose regarding the appropriateness of using certain restoration methods on historical photographic materials for public presentations, particularly when working with images that have missing emulsion. One concern raised was the challenges of obtaining copyright clearance for such images. While plates may have already received copyright clearance or were no longer in copyright, the more pressing issue centred on how the restoration—particularly of missing emulsion in portraits—might raise ethical and privacy concerns. Additionally, it was suggested that showcasing visible alterations, like fading or spotting, could offer more value by helping audiences understand the effects of time and external factors, like light, on these materials, thus providing an educational opportunity (Fig. 7).

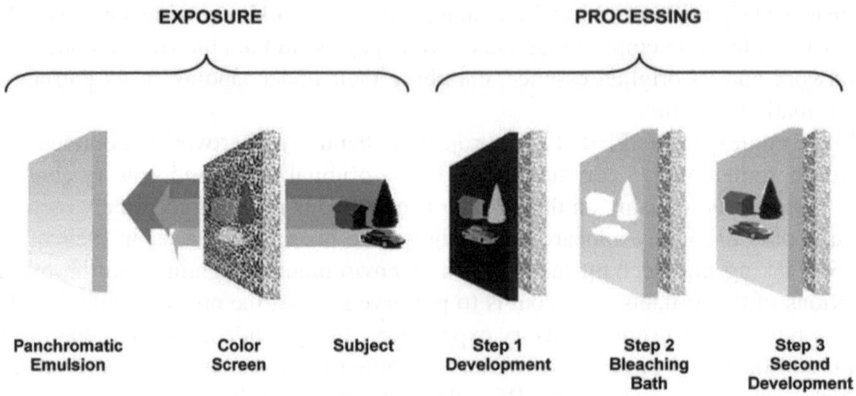

Fig. 7. The image shows the parts of an autochrome plate along with the complete process of exposing and development an Autochrome (adapted from [35]).

Conversely, another perspective advocated for the use of different techniques to present both generated and original digital images, ensuring that viewers can distinguish between them. This approach would emphasize the research aspect, where the exhibition of these images explores restoration processes and invites public engagement. The argument was made that rather than expecting a flawless AI-generated reproduction, the goal should be to create a credible restoration while providing an exhibition method that allows viewers to distinguish between the original and its AI-generated missing parts. Like Kintsugi, this approach marks imperfections to reflect the artwork's journey, supporting the idea of prioritizing authenticity over perfection in public demonstrations. Furthermore, such an approach would adhere to institutional guidelines, which were agreed to be crucial for gaining overall approval. This ensures that both the technical and ethical considerations are aligned with broader preservation standards.

AI and Cultural Heritage

The use of AI in restoring digital images of damaged autochromes presents intriguing possibilities, though this avenue remains speculative and is yet to be fully explored. AI has the potential to address a variety of damages, from missing emulsion and glass cracks to scratches. By recognizing patterns and textures, AI could identify and replace missing emulsion, digitally repair cracks, and remove scratches using advanced image segmentation techniques. For issues such as greening, tanning, and colour fading, AI could analyse the colour distribution in the damaged autochromes, detect the areas affected by these changes, and adjust the balance and saturation to match the original colour patterns. Through sophisticated algorithms, AI may restore the vibrancy and accuracy of the colours, compensating for the effects of aging. This process could offer unique opportunities for institutions to preserve and present historical works that have otherwise deteriorated or are difficult to exhibit due to their fragile state. Furthermore, such an approach would adhere to institutional guidelines, which are, of course, crucial for gaining overall approval and permissions to exhibit. This ensures that both technical and ethical considerations are aligned with broader preservation standards, thus allowing the use of AI to be considered an authentic approach in the digital restoration of autochromes.

By integrating AI into this preservation process, institutions can address both practical restoration challenges and engage audiences with the historical and cultural significance of these early colour photographs in ways that were previously not possible. By combining physical damage repair with colour correction, AI has the potential to enhance the restoration process while preserving the authenticity and historical value of early colour photographs. However, as this approach is still in its early stages, further exploration is required to fully understand its capabilities and implications. This ongoing investigation will not only refine the technical aspects of AI-driven restoration but also contribute to the evolving discussion on the role of technology in preserving cultural heritage, offering a balance between innovation and preservation ethics [36].

Autochrome Research

NTNU university is at the forefront of research into autochrome photograph restoration, using multispectral scanning techniques to virtually separate the photographic emulsion from the dyed starch grain mosaic [35]. By capturing images across various wavelengths and applying advanced algorithms, they can precisely distinguish and analyse each layer of the photograph. This innovative approach enables more targeted restoration efforts, addressing both physical damage and colour degradation. In this process, digital restoration techniques are employed to revive the original colours and materials. Utilizing spectral imaging and advanced image processing, faded colours are rejuvenated to closely match their original vibrancy. This approach preserves the authentic historical context of the photograph and upholds the artist's original vision, offering a more accurate and faithful restoration. To create a physical facsimile of an autochrome, the digital restoration process begins by scanning the original to produce two layers: one for the colour mosaic and one for the existing emulsion. If parts of the emulsion are missing, AI is used to digitally reconstruct the missing sections, which are added as a third layer, completing the restoration.

- A glass plate is created with the potato starch mosaic, made up of the three color components that have been digitally colour-corrected. To produce this facsimile, the original autochrome undergoes a complex scanning process to separate the silver emulsion from the colour mosaic, allowing the mosaic to be corrected before being reassembled.
- A glass plate is created with the black-and-white silver emulsion image, which will be placed over the colour mosaic plate in the final facsimile. The last plate in the process contains any digitally restored areas of missing emulsion, and when this plate is layered over the first two, it completes the restoration, replicating the original autochrome's appearance.
- The last plate which contains the missing areas is created using AI-driven algorithms that analyse the spectral data and by combining that with historical knowledge, it recreates the missing parts (Fig. 8).

Once printed on glass, each of the glass layers is carefully mounted in a display frame for viewing, allowing the viewer to see the results of combining of the layers.

The Autochrome Demonstrator

To demonstrate the research described above, a science demonstrator was developed

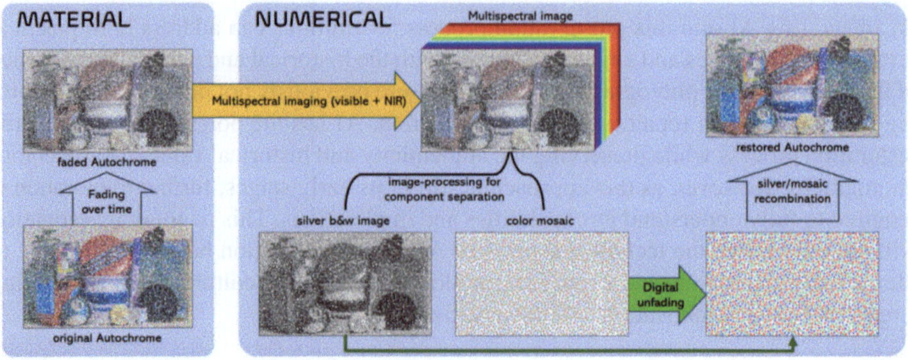

Fig. 8. Illustration of the imaging techniques used to obtain the layers for the proposed presentation of the digital restoration of faded Autochromes

for viewing Autochromes that have been restored using the separated layers. Termed the 'Autochrome Demonstrator', the device is designed to showcase both the fabrication process of autochrome plates and the innovative restoration techniques used in the research. It aims to highlight these technologies while maintaining authenticity in the presentation of the restored images[4]. Integrating the Autochrome Demonstrator into the museum display offers the public a unique opportunity to juxtapose digitally restored images with their authentic counterparts. Viewers can examine the intricate details of both the original Autochrome plates and the digitally restored elements, gaining a deeper understanding of the a new type of restoration process and its impact on preserving cultural heritage. Further, with the Autochrome Demonstrator, the problem of effectively showcasing digitally restored pieces within a museum environment is addresses, as it adheres to standards of curational ethics, which dictate that it is essential to devise presentation strategies that facilitate a comprehensive comparison between the restored and original objects. The Autochrome Demonstrator incorporates several key components to optimize the viewing experience. Firstly, the Lightbox uniformly illuminates autochrome plates, providing diffused light for optimal viewing, with its open design allowing viewers to observe its internal components. Next, the Plate Holder securely holds and aligns plates within the Demonstrator, ensuring accurate positioning for viewing. An Ocular Occluder controls viewing conditions by regulating light entry through a system of pinholes, enhancing the overall viewing experience[5].

The Plates.

The Autochrome Demonstrator intricately showcases the essence of autochrome photography by using a series of optically connecting plates, each playing a pivotal role in unravelling the complexities of this historic photographic technique and the restoration of it using Artificial Intelligence. At its core lies the Colour Mosaic plate, an embodiment of the original colour mosaic captured by the autochrome process. This

[4] Toward Creating an Autochrome Demonstrator for Museums, Retrieved October 2024, from.https://perceive-horizon.eu/index.php/2024/05/30/toward-creating-an-autochrome-demonstrator-for-museums/.

[5] Pinhole Occluder. Retrieved April 2024, from https://en.wikipedia.org/wiki/Pinhole_occluder.

plate serves as the primary visual source, captivating viewers with its vibrant hues and intricate patterns, providing a window into the vivid world captured through early colour photography. Accompanying the Colour Mosaic plate is the AI-Generated Data plate, a testament to contemporary technological prowess. Powered by artificial intelligence, this plate painstakingly reconstructs any missing data that may have faded over time, ensuring the integrity of the image restoration process. By utilizing three separate plates for the colour mosaic, AI-generated data, and emulsion, the Autochrome Demonstrator facilitates a comprehensive viewing experience. It enables viewers to distinguish between the original colour mosaic, any artificially generated data, and the underlying emulsion layer, thus ensuring transparency and authenticity in the presentation of autochrome images. To address any colour fading in the autochrome on view, it's essential to replace the Colour Mosaic with a corrected version where the colour values of the mosaic pattern have been adjusted using AI (Artificial Intelligence) technology. This corrected version ensures that the colours accurately reflect the original intent of the autochrome image, compensating for any degradation or fading that may have occurred over time. By employing AI to enhance the colour values, the Autochrome Demonstrator can present images with improved fidelity and authenticity, providing viewers with a more accurate representation of the original photographic content (Fig. 9).

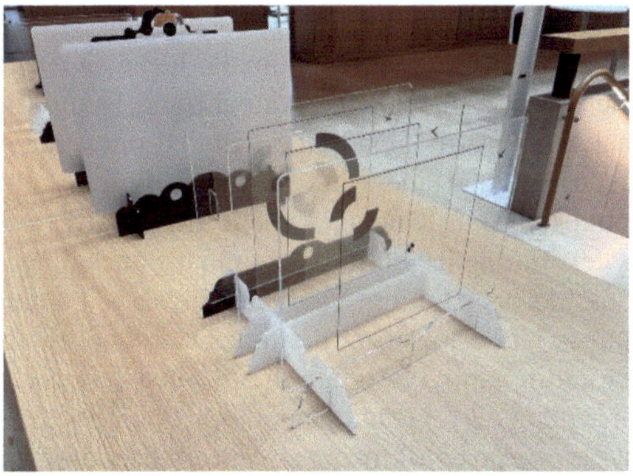

Fig. 9. The image shows Autochrome Demonstrator and how it uses separate layers to stack the components parts of an Autochrome to create a digital restoration

The Ocular Occulders.

The Ocular Occluders are an essential component of the Autochrome Demonstrator. They comes in two distinct types: the Single Occluder and the Double Occluder. The Single Occluder is outfitted with a small 1mm pinhole in its viewer, used for checking the alignment between the plates. Using it, ensures accurate positioning of the plates on the one hand and on the other it enables viewers to observe the alignment with utmost clarity and precision. The Double Occluder however offers a multifaceted approach to viewing autochrome plates. Comprising of two occluders mounted on a stand with a 10cm spacing, it features a 4mm pinhole on one occluder and a circular pattern of 2mm pinholes on the other. The Double Occluder facilitates alignment viewing, similar to its Single counterpart. Additionally, but it also contributes to a seamless re-blending of the emulsion and colour mosaic. This dual functionality, achieved through varying pinhole sizes and a circular pattern, enhances the visual experience, ensuring optimal integration of the emulsion and colour mosaic for viewers of the Autochrome Demonstrator[6] (Fig. 10).

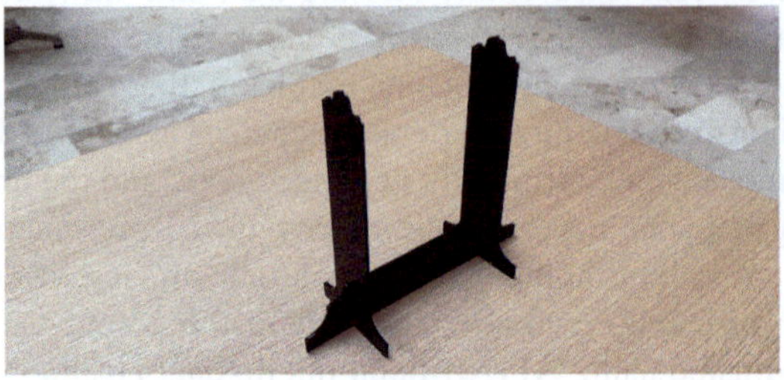

Fig. 10. The Ocular Occluder This is the component used on the Demonstrator to control the viewing conditions. The uses a system of pinholes to regulate the amount of light entering the eye, which in turn enhances the viewing experience. Here it is combined with a second occluder which aides in blending the separated colour mosaic closer to the original

The approach taken by the Autochrome Demonstrator mirrors the principles of authenticity seen in other restoration practices, such as the Kintsugi method and the restoration of Franz Marc's The Fate of the Animals. Just as Kintsugi highlights imperfections to honour the object's history, the Autochrome Demonstrator preserves and showcases the original elements alongside the restored ones, maintaining a clear distinction between what is original and what has been digitally enhanced. This transparency echoes the philosophy in Kintsugi, where the repaired cracks become part of the object's narrative, celebrating its journey rather than concealing its past. Similarly, in the restoration of Franz Marc's painting, Paul Klee preserved authenticity by carefully repainting damaged areas while lowering the tones to reflect the original appearance. In both cases, the restoration aimed to retain the original essence, allowing viewers to appreciate both the artwork's current state and its historical integrity. The Autochrome Demonstrator employs the same ethos, using advanced technology not simply to restore the Autochromes but to enhance the viewer's understanding of their original cultural and aesthetic significance. The ethical dimension of the Autochrome Demonstrator plays an essential role, particularly through its transparency. By allowing viewers to differentiate between original and generated elements, it aligns with ethical restoration practices that prioritize truthfulness. This transparency allows the viewer to see what has been restored versus what remains from the original, promoting an informed engagement with the artwork. The ability to clearly distinguish between original and digitally generated parts prevents any deception regarding the authenticity of the restoration, maintaining integrity. This approach mirrors ethical considerations in the use of AI technologies, ensuring that digital interventions do not obscure the history or value of the original work. Moreover, this method respects the artistic intent of the original creator by highlighting where interpretive decisions were made during restoration, similar to how Paul Klee preserved the essence of The Fate of the Animals and Kintsugi honours an object's

[6] Toward Creating an Autochrome Demonstrator.

imperfections. By presenting both the restored and original elements, the Autochrome Demonstrator fosters a deeper understanding of the artwork's historical and cultural narrative, ensuring that audiences can appreciate the original work while understanding the modern technologies applied to preserve it. In all these examples—whether in ceramics, paintings, or Autochromes—the restoration goes beyond mere replication. It seeks to preserve the authenticity and emotional resonance of the original work, ensuring that the audience can engage with both the restored elements and the object's historical narrative in a meaningful way. This transparent, ethically driven approach allows for a balance between restoration and the preservation of historical truth, honouring the journey of the artwork while engaging with the opportunities provided by modern technology.

6 World and Physical Context: The Authenticity of Original Colour Appearance

How can we enable users to perceive the original appearance of artworks, thus enhancing the authenticity of the experience but also of the digital reproduction itself? Considering the "physical context" of objects requires to deal with components that have an impact or are impacted by it, such as "verification", "embodiment" and "time".

In PERCEIVE, we aim to develop colour restoration approaches as an attempt to visualise the original appearance of artworks that have partially or extensively lost their original colour. Colour change can manifest in several ways (colour loss, fading, alteration, etc.) and the different cases depend on the object typology as well as on various deteriorating factors. Among these factors, light exerts the most significant impact on the colouring materials, but often its impact is worsened due to the synergetic influence of other weathering parameters or conditions. In any case, although each of the case studies dictates a different approach towards the reconstruction of the original appearance, PERCEIVE adopts some basic concepts that are in common and refer to our aim to preserve Authenticity. Based on a set of assumptions, the restoration approaches should be guided by the artist's technique and original intention in creating the artwork, although these are seldom fully known. By nature, such restoration attempts may bring a pluralistic solution to support multiple hypotheses, but they do not match the unique appearance of the original and are thus not authentic by themselves. Nonetheless, to preserve as much as possible the authenticity of the original, which is inextricably linked to the material substance of the object, the restored renderings should build upon residual traces of the original materials, i.e., pigments, colourants (preserved or faded) identifying them as a carrier, a persistent element of authenticity. In other words, we have to collect scientific data about the chemical composition and ageing properties in relation to various stress-inducing factors (light, humidity, temperature, etc.) for the residual materials in the artworks, where they exist.

Depending on the type of artworks, there is a different richness of residual information present. For example, in the case of textiles, the reverse side contains unfaded fibres that facilitate a restoration very close to the original [37, 38]. Paintings have materials that are protected from degradation factors on the edge, under the rebate of the frame [39, 40]. Although the materials under the frame might not be representative of the full pigment palette or composition of the painting, they still represent an additional piece of evidence for data collection or validation of the proposed restoration approaches. In the case of historical photos or film, such as autochromes and lenticular films, the relationship with authenticity is influenced not only by the fading of the dyes, but also by the medium of reproduction, such as the projector, as shown in Fig. 11. Thus, here we are dealing with a possible disconnection between the materiality of the artworks and its fruition means. As far as polychrome statues are concerned, the remaining traces might be very poor, and not even obviously localised. For this reason, it is recommended to investigate polychrome statues with spatial imaging techniques to identify possible areas where residual material is present [41].

The component of "physical context" is strictly connected with that of "verification". In cases, in fact, where residual material is not present or is faint to provide reliable information for the original appearance (e.g., complete detachments in paintings, frescoes or polychrome statues), other contextual objects or other documentary or iconographic sources should be used as supporting evidence: archival records, recurrent art motifs depicted repeatedly, art history books with description of common materials, recipes, and practises typical of a certain art period. Fig. 12. shows the inverse relationship between the amount of remaining original material and the level of assumptions adopted in the design and development of the colour restoration tools.

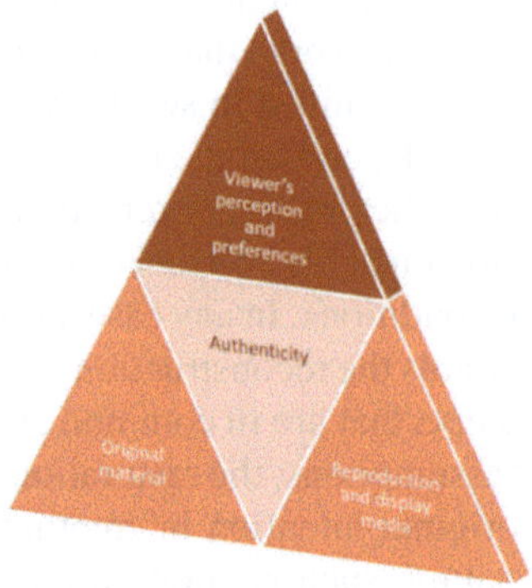

Fig. 11. Representation of the physical authenticity, as it is influenced by the preservation of the original materials in the object, the reproduction media and the viewer's perception.

If even documentary sources are missing, then colour restoration methods relying on generative methods could be employed, as long as full reliability and interpretation process and the level of speculation are disclosed together with the results. Examples of such generative methods are "inpainting", "colourization" and "outpainting".

Inpainting is an image restoration method, where partially missing content is reconstructed based on transfer of information from the regions neighbouring the missing area, or by learning visual representations typical to a specific type of dataset [42]. The missing content lacks information about the structure and colour, and the bigger the missing area the more challenging it is to accurately reconstruct it [43].

Colourization is the method of adding colour to monochrome images that have integral spatial content depicted in grayscale or sepia tones. From a chromatic point of

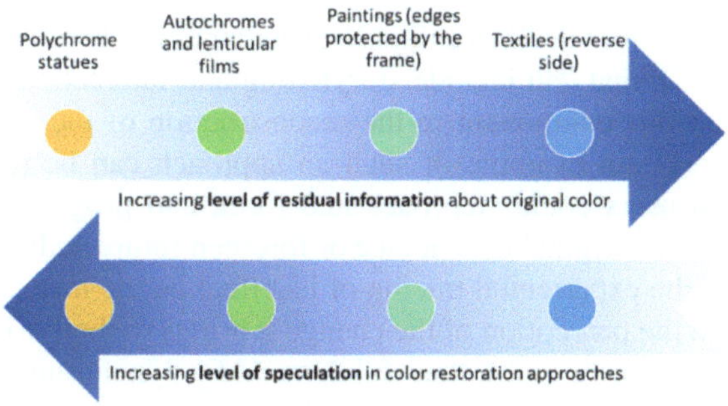

Fig. 12. Representation of the inverse relationship between the existing level of residual material (physical context) and the level of assumptions (verification) in colour restoration approaches.

view, the level of speculation is higher for colourization than for inpainting, as there are no intact coloured areas in the original image to give useful clues.

Outpainting, also known as "image extrapolation", is a method that extends the limited visual field of a given image, and adds the non-captured context, by fully extrapolating the chromatic and structural information [44]. Hence, the creative freedom i.e. the level of speculation (which is inverse to the level of authenticity) is the highest in the case of outpainting, when referring to the two-dimensional image domain. Nonetheless, in the object domain, when extrapolation is desired, the composition of a material and its chromatic information can be revealed from the intact content. From this point of view, outpainting is more constrained in colour terms than it is in structural terms. Fig. 13. Shows a graphical representation of these three image restoration tasks and the level of speculation required for achieving colour and spatial reconstruction.

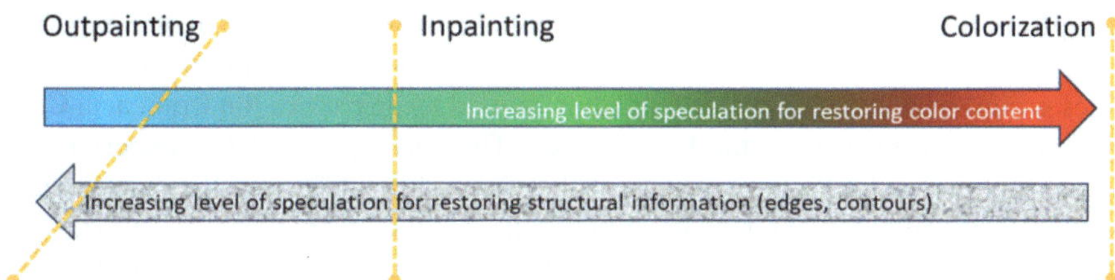

Fig. 13. Graphical representation of inpainting, colourization and outpainting image restoration tasks and their inverse required level of speculation (verification).

The solutions of the colour restoration approaches should be iterative, and not closed-form, as they should allow for continuous improvements and refinements of the data-driven and learning-based algorithms, because of updated data collection experiments and feedback from quality assessments from the stakeholders.

Another important parameter to integrate in these restoration approaches is "time", as it has also an impact on authenticity. Material aspects of the objects in fact vary across time. The signs of use and weathering are testimonies of the age and history, and they are proof of quality and resistance to damage [45]. The fragility of colour and its susceptibility to change, is in itself an evidence of authenticity. Through design strategies (element: "time flow"), that can include storytelling and interaction mechanisms, it is possible to highlight this component in the reconstruction of the original appearance of a coloured artwork. An example of such an approach can be the introduction of a timeline on which users could step back and forth, and progressively regain more complete versions of the original appearance or foreseen future additional damage. The concept of time and the experiential tracing of its effect on the materiality of artworks, step by step, through the perception of the change it brings to their original appearance, can be redemptive in contrast to the stressful demand for synchronisation of perception with the rapid technological evolution of everyday reality.

The aim of this approach remains to communicate knowledge to appreciate authenticity and share/enhance the sense of care and also manage (negotiate with) the change (back and forth in time) that is occurring in any case, (we cannot fully prevent it, maybe slow it down).

7 Examples of Authentic Digital Experiences

How can we design digital experiences that make users perceive them as authentic? To better understand how the authenticity conceptual framework (3.3) can be transformed into a proper application, we analyse here examples of digital or hybrid applications that can be taken as examples of authenticity. In some cases, they are just focused on a few components, in other cases they offer an experience that is more in line with our definition of "authenticity". Following [3], we have tried to identify and represent how the components are distributed, using radar charts.

Title. Brancacci Point of View (POV) (2023)
Description. Brancacci POV is an interactive collaborative and guided hybrid experience developed by the DHIlab of CNR ISPC, together with Florence Municipality [46, 47]. It enables groups of visitors to explore together, guided by an expert, the Brancacci Chapel in Florence, a masterpiece of the Renaissance. The participants of this experience can use a smartphone or tablet during most of the visit, while at the end they are invited to play the role of a restorer, entering the monument as if they were there, by wearing a VR headset and using diagnostic tools such as an Ultraviolet lamp. The expert, during the entire experience, guides the group by explaining step by step the story of the monument and by requesting different tasks to be carried out together. The project presents a direct mapping with the conceptual framework.

Dimensions and Components.

1) SELF: starting task requires to identify with a character (sense of belonging), curiosity-led storytelling; 2) OTHERS: the experience is based on a continuous exchange between the guide and the group, who are in physical co-presence (hybridity) and can exchange thoughts; 3) WORLD: the validation, realistic aspect and embodiment are fully developed (Fig. 14), (Fig. 15).

Web Reference: http://brancaccipov.cnr.it/.

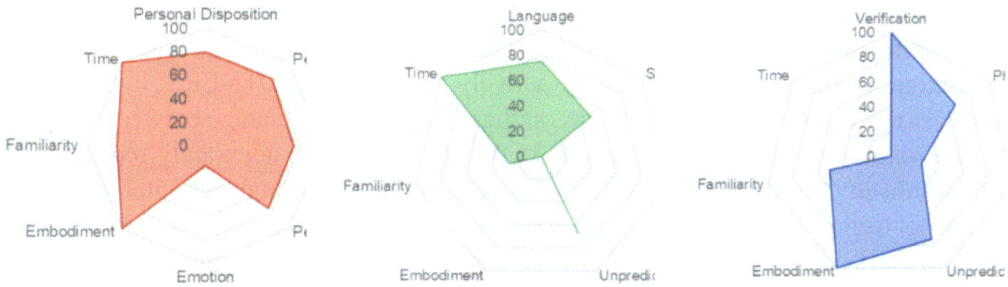

Fig. 14. A representation with a radar chart of the mapping of authenticity components in BrancacciPOV experience along the Self (left), Others (centre) and World (right) dimensions.

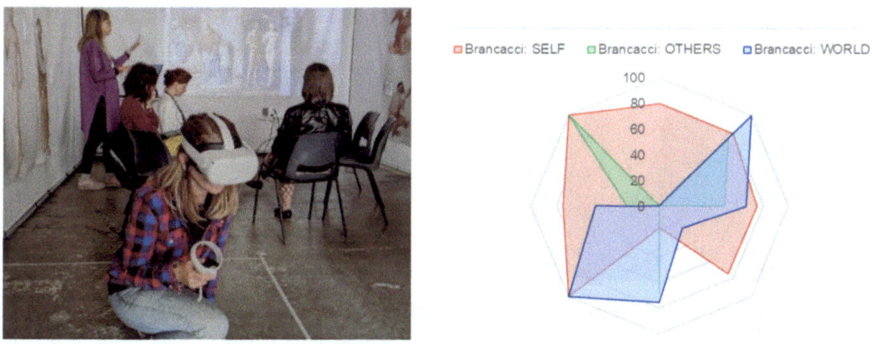

Fig. 15. Brancacci POV Hybrid experience (Photo CNR ISPC) and the general representation of the three dimensions (red: self, green: others, blue: world)

Title. The Corvey monastery XR experience (2023).
Description. The project developed by Fraunhofer IGD, involved the creation of an interactive AR guided tour companion app for the Corvey monastery, situated in a UNESCO World Heritage Site with strict conservation constraints. The environment, specifically the St. John's choir, lacked descriptive historical information and showcased only faint remnants of its original grandeur. The AR application titled "Guarded by Angels in the Celestial City" aimed to reconstruct the rich decorations and significance of the site through intensive scientific research and comparative analysis. It focused on scenarios such as uncovering sinopia, exploring the first Carolingian basilica, and recreating faded wall drawings. The UI prioritized user-friendliness, with features like tracking initialization prompts and a "tablet down" gesture for comfortable viewing. The app utilized AR overlays to present realistic room models and VR perspectives of vanished wall paintings, stucco statues, and scanned objects. Interaction between guide and visitors was facilitated through a shared network, enabling the guide to control content visibility on visitors' devices and tailor the experience based on group interests. The guide accessed a specialized UI to select topics and trigger content, including 3D models, VR objects, animations, videos, and images, enhancing the immersive tour experience [48].

Dimensions and Components. 1) SELF: Personal Disposition, Personal Context, Cognition, Perception and Familiarity are main components (with a prominence of the knowledge acquired by users who are actively involved); 2) OTHERS: is the dimensions less represented, with just a focus on the dialogue and exchange between the guide and the visitors during the experience; 3) WORLD: data are accurately reconstructed and visualised (Verification) and Embodiment is realised through the interaction with the 3d environment and the tablet that is used as a sort of "magic window" through which visitors can see accurate and realistic reconstructions and augmented information (Fig. 16).

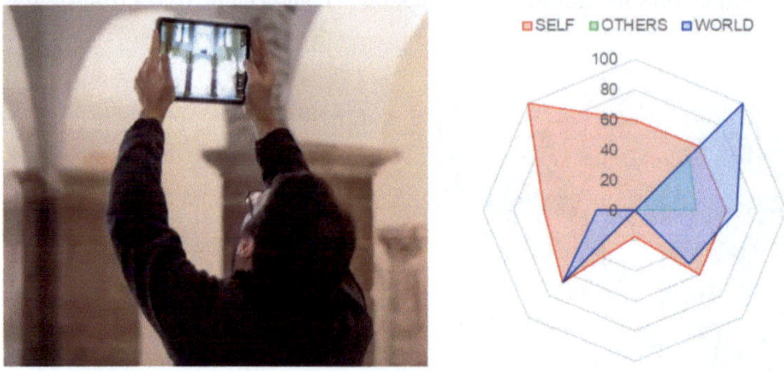

Fig. 16. A user uses the tablet to explore the church, guided by the expert, visualising in overlap the reconstructed paintings and decoration; and the Radar Chart with overlapping dimensions mapping the authenticity of the application.

Title. The Colours of Ara Pacis (2014) / L'ara com'era (2017)

Description. The two projects have offered to the visitors of the monument of Ara Pacis in Rome the possibility to view the hypothetical reconstruction of its colours. In its first version (The colours of Ara Pacis) the reconstruction was projected directedly on the monument; in the second project (L'Ara com'era, *The Altar as it was*) an VR head mounted display, with an external camera mounted on it that transformed it into a Mixed Reality device, was given to the visitors who could physically move in the space around the monument and see not only the reconstructed colours but also animations and storytelling about the origins of Rome and of the family of Augustus, the emperor.

Dimensions and Components. We report here the components that better represent an example of authenticity. 1) SELF: Visitors could perceive deeply the history not only of the monument but also of the Romans, (with a prominence even in this case of the knowledge acquired by users who were passively involved); 2) OTHERS: although the application was single user, the visit was done collectively with the support of audio; 3) WORLD: data have been accurately reconstructed and visualised (Verification) and Embodiment is realised through the immersivity of the MR device although no interaction was possible.

Reference: https://www.arapacis.it/it/mostra-evento/i-colori-dell%E2%80%99ara; Video: https://www.youtube.com/watch?v=Akd5-r1gZKc&t=5s.

Title. Revealing Flashlight (2014)
Description. Application of AR and natural Interaction (Spatial Augmented Reality), that allows visitors to see details and possible reconstructions by pointing their fingers at an object. The application allowed a new interaction and visualization technique in spatial AR, helping to reveal details of cultural artifacts including their reconstructed colours. It locally and interactively augmented a physical artifact by projecting an expressive 3D visualization that highlights its features, based on an analysis of its previously acquired geometry at multiple scales [38].

Dimensions and Components. 1) SELF: personal context and cognition are the most representative components, together with the focus on familiarity; 2) OTHERS: the application is single user, and no social dynamics are supported nor audio feedback; 3) WORLD: data have been accurately reconstructed and visualised (Verification) and Embodiment is realised through the direct and interactive mapping of the reconstruction on top of the real object (Fig. 17).

Reference: https://www.labri.fr/perso/preuter/?page_id=1027.

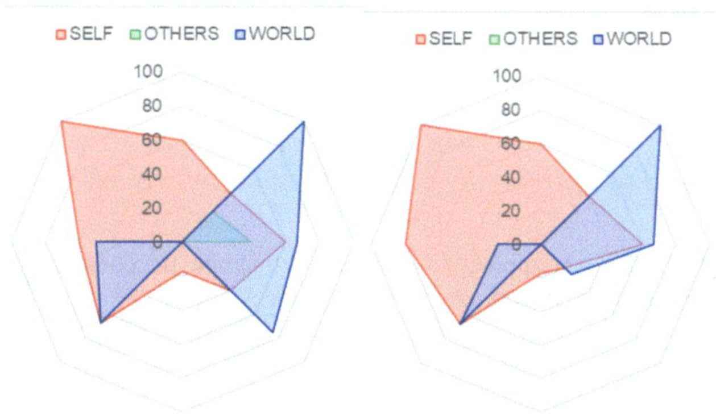

Fig. 17. Radar Chart with overlapping dimensions and mapping the components of authenticity of the application L'Ara Com'Era (left) and of the Revealing Flashlight (right).

Title. Ozan1982
Description. Mobile serious game aimed at engaging tourists spending times during the summer only on the beaches, inviting them to visit the local archaeological museum of Ugento (Salento, Lecce - Italy). The game starts from tangible coloured objects on the beaches that acts as triggers of the game. Through a QRCode the tourists become part of a mistery game that need to be solved only by visiting the museum.

Dimensions and Components. 1) SELF: familiarity, perception and cognition; 2) OTHERS: the application is single user for smartphone, although no social dynamics are supported the game is meant to involve the group of friends or family in the visit to the museum; 3) WORLD: the game blend physical elements and digital ones, moreover it requires to move physically in the space and time.

Title. COSMOTE Chronos

Description. Free mobile AR application released by Greek telecom operator COSMOTE which allows visitors of Greek historical sites to see the buildings and their original colours via AR on their mobile phones.

Dimensions and Components. 1) SELF: familiarity is the most exemplary component, together with the focus on perception and cognition; 2) OTHERS: the application is single user for smartphone, and no social dynamics are supported. A simple AI Avatar is implemented; 3) WORLD: it's the most representative dimension: data have been accurately reconstructed and visualised (Verification). Physical context is realised through Augmented Reality, so the user can interact with 3D objects.

Reference:https://www.cosmote.gr/cs/cosmote/en/cosmote-chronos-akropoli-virtual-tour.html (Fig. 18).

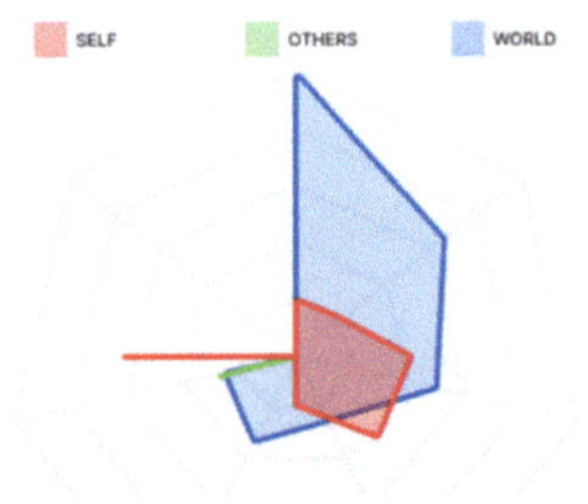

Fig. 18. Radar Chart with overlapping dimensions and mapping the components of authenticity of the application COSMOTE Chronos.

Title. Play a Kandinsky

Description. Digital project funded by Google Arts and Culture whose prime focus is on the study on synaesthesia by Kandinsky. In the selected case study, "Yellow-Red-Blue" (1925), colours are associated to different sounds and emotions. The application allows visitor to engage with the canvas, following a participatory paradigm.

Dimensions and Components. The experience is very engaging from an experiential point of view. 1) SELF: the components of perception and embodiment are highlighted. The experience is single user, so the others dimension was not significantly considered. 3) WORLD: assets are verified (the painter and his works), an impactful atmosphere has been created, and, in addition, unpredictability is significant due to the use of artificial intelligence. Embodiment components were also well implemented using touch interactions.

Reference: https://experiments.withgoogle.com/play-a-kandinsky.

Art installation Title. Immersion Room (Cooper Hewitt, NY)
Description. This is a digital interactive immersive space (Margaret Carnegie's bedroom) that offers a unique experience to view Cooper Hewitt's collection of wallcoverings It also allows visitors to create their own decorative pattern and to project it on the wall of the immersive room.

Dimensions and Components. Immersion Room (Cooper Hewitt, NY) is an immersive room where the dimensions of authenticity that emerge are the Self and the World. 1) SELF: the components (no one component overrides the others) are well distributed among Personal Disposition, Cognition, Perception and Familiarity. 3) WORLD: Verification, Atmosphere, Embodiment and Familiarity (components of strength for an immersive room) are well defined and stand out from the others (Fig. 19).

Reference: https://www.cooperhewitt.org/exhibitions/immersion-room/.

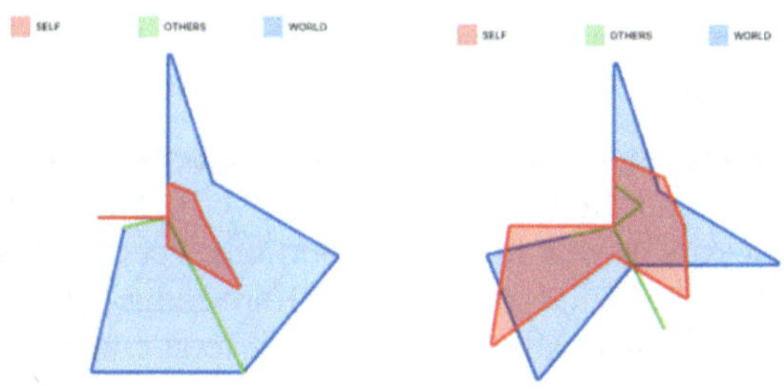

Fig. 19. Radar Charts with overlapping dimensions and mapping the components of authenticity of the Play a Kandinsky experience (left) and Immersion Room (right).

Art installation Title. ArtLens (Cleveland)
Description. Multifaceted AI-based exhibition consisting of several digital products, enabling visitors to engage in different way with the collection: selection of favourite artworks, interactive exploration of the catalogue of the collection, massive support of AI (gaze tracker, motion and pose recognition).

Dimensions and Components. The installation can be dcfined authentic on all dimensions. Indeed, in the design, both the Self, the Other, and the World have been implemented with homogeneity. 1) Self: the components that emerge most are "Familiarity" and the "Embodiment," but also Personal Context, Cognition a Perception. 2) Other: ArtLens stimulates relationships and group. The components of Familiarity and, in addition, "Unpredictability (social)" are also well represented here. 3) World: Verification and Embodiment (environmental) are the most present components. Shortcomings are highlighted on the side of "Time" and "Unpredictability (environmental)".

Reference: https://www.clevelandart.org/artlens-gallery (Fig. 20).

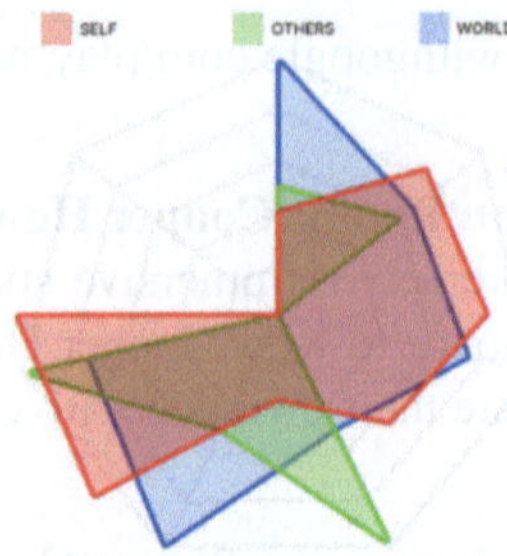

Fig. 20. Radar Charts with overlapping dimensions and mapping the components of authenticity of ArtLens

Project Title. "Virtual Fading" of Art Objects: Simulating the Future Fading of Artifacts by Visualizing Micro-Fading Test Results.

Description. A research project that measures mock-ups of pigments with microfadeometry and then creates a lookup table of colour differences that are matched with certain light doses. This lookup table is then applied to paintings with a block-based composition, where big areas share the same colour. The inference of knowledge from mock-ups to paintings is achieved purely on a colour similarity basis.

Dimensions and Components. As for Light damage calculator this project is focused on objective authenticity and the radar is the same. Verification component is implemented (World) and they are present in small proportions: Personal Disposition (Curiosity), Personal Context (Expectations), Cognition (Knowledge) and Perception (Sight). The dimension of the Other is totally missing, and several components have not been considered.

Reference: [49].

Project Title. Spatio-Temporal Simulation of The Painting: A Japanese Lantern by Oda Krohg

Description: A research project that combines hyperspectral imaging with microfading measurements to restore the appearance of a detail of the painting in the past, as well as to predict its damage in the future, by considering exposure to light as the only aging factor. The simulations are valid for the full spectral space, enabling both colour and false-colour (for specific combinations of wavelengths) visualizations. One of the outcomes of the project is an animation showing the evolution of the painting from past to future.

Dimensions and Components. As for Light damage calculator and Virtual Fading this research project is focused on objective authenticity and the radar is the same. Verification component is implemented (WORLD) and they are present in small proportions: Personal Disposition (Curiosity), Personal Context (Expectations), Cognition (Knowledge) and Perception (Sight). Other is totally missing, and several components have not been considered (Fig. 21)

Reference: [50].

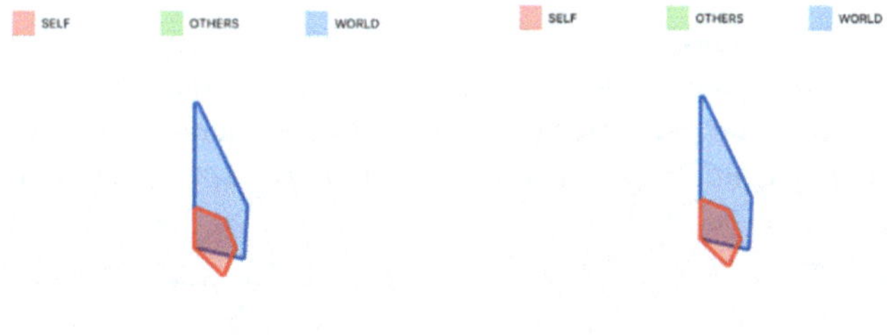

Fig. 21. Radar Charts of the two projects Light damage calculator and Virtual Fading (right).

In the following section we report projects and applications beyond Cultural Heritage and Art, AS videogames and interactive experience with a high level of authenticity thanks to the emotions they arouse and the engaging storytelling.

Game Title. The Vale

Description. The Vale is a story driven, an **audio-**based action-adventure game which sets out to breathe new life into mediaeval combat and fantasy narrative where the player takes the role of a blind princess, Alex. To completely put the player in Alex's shoes, the game utilizes the full potential of 3D audio and haptic controller feedback to deliver gameplay that goes beyond the barrier between player and character, without the use of visuals. This innovative gameplay opens to accessibility and lets visually impaired users participate in an authentic experience.

Main Dimensions and Components. 1) SELF: Personal Disposition, Personal Context and Embodiment (personal) are the more representative components. The multiplayer is not supported yet, so the 2) OTHER dimension is not so relevant. 3) WORLD: the game is audio driven; the environment is not displayed so the Verification is null. However, Atmosphere, Embodiment (environmental) and Unpredictability (environmental) are well designed and implemented.

Reference: https://www.fallingsquirrel.com/the-vale.

Game Title. We Were Here Forever
Description. This game is an experience that conveys not only a sense of authenticity but even cohesion and a sense of care towards the other player is *We Were Here Forever*. It is an online first-person cooperative adventure set in a fictional mediaeval castle for two players only. The game is all about **cooperative** discovery, immersion and teamwork through communication. The player and her/his partner must solve puzzles through smart communication and observation. Trapped in different parts of an ancient castle, the players can only communicate through their walkie-talkies to Fig. Out how to escape.

Dimensions and Components. We Were Here Forever is a highly authentic game. All dimensions have been considered and well implemented. Multiplayer is supported and well-integrated to the game design. SELF: Embodiment (personal) is the highest components but also Time, Personal Context, Familiarity (personal) and Cognition are considerable (Fig. 22).

Reference: https://totalmayhemgames.com/we-were-here-forever-info/.

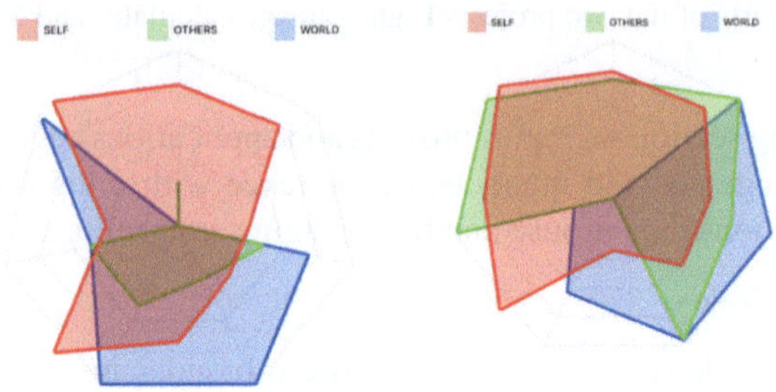

Fig. 22. Radar Charts with overlapping dimensions and mapping the components of authenticity of "The Vale" (left) and "We were here together" (right)

7.1 Towards an Authenticity Index and Evaluation Strategy

To conclude this work, we can define Authenticity in Virtual Experiences as:

a multi-dimensional concept, made of three dimensions (self, others, world), each characterised by specific components and elements, that works by touching the deeper self, through performative actions that transform the unfamiliar into familiar, and by creating social connections and embodiment in the environment.

How can we evaluate to what extent a visitor has perceived the authenticity of a digital experience? How can we measure the authenticity of a digital reconstruction in the World dimension? In this chapter we summarise and propose a meant to analyse the User Experience. It includes: 1) SELF: 35 questions; 2) OTHERS: 23 questions and 3) WORLD: 24 questions. Questions can be answered with Yes/No or Likert scale.

SELF	OTHERS	WORLD
1. *I was concentrated/not distracted* 2. *I was curious* 3. *I acted extroverted* 4. *I felt to be identified in the experience* 5. *I felt not to be lost* 6. *I felt it was challenging* 7. *I felt to have choices* 8. *My expectations have been satisfied* 9. *It has been meaningful* 10. *It was valuable* 11. *I understood* 12. *I felt to have acquired new knowledge* 13. *I can recall the experience* 14. *It made me reflect* 15. *I was aware about myself* 16. *I could listen to audio/sound* 17. *I felt I was tasting at something* 18. *I used the touch* 19. *I used my view* 20. *I felt the smell of something* 21. *I felt Anger* 22. *I felt Disdain* 23. *I felt fear* 24. *I felt Joy* 25. *I felt sadness* 26. *I felt surprised* 27. *I could interact* 28. *I felt immersed* 29. *I felt to be there* 30. *I could easily access* 31. *I felt comfort / I felt in my comfort zone* 32. *I understood the experience* 33. *Elements of the experience were familiar* 34. *I felt to be secure / not harmed* 35. *I perceived an internal evolution during the experience*	*1.I could discuss with others* *2.I could explore alternative perspectives* *3.I was provoked by questions* *4.I was told stories* *5.I was engaged in challenges with others* *6.I was involved in competitive tasks* *7.I was involved in cooperative tasks* *8.I could interact with the others* *9.I could create a relation with the others* *10.I know I should respect social constraints* *11. I obtained rewards proportional to my skills* *12.Participants were assigned a role* *13.I was aware we could get penalties if we were behaving not correctly* *14.The behaviours of the others were unpredictable* *15.I felt the influence of others* *16.The behaviours of the others were unexpected* *17.I have communicated also with gestures with the others* *18.I could hear familiar tone of voices* *19.There was a sense of community* *20.I could develop connections with others* *21.I felt to belong to the group* *22.I felt to be welcomed* *The relations I have created with the group improved during the experience*	1. *Scenarios seem real* 2. *Scene(object seem reliable* 3. *Scene/objects were reproducing the original* *4.scene/objects were scientifically validated* *5.digital environment was 3d* 6. *Physical laws were working there (i.e. gravity, etc.)* 7. *Scene/objects have appropriate properties (as colours, material)* 8. *Lighting created an atmosphere* 9. *Sounds/music created an atmosphere* *10.a mix of light, sound, effects, created atmosphere* *11.visual style contributed to create an atmosphere* *12.all around me was original/novel* *13.I noticed peculiar aspects* *14.Unexpected events happened* *15.There were unique elements* *16.I got appropriate feedback* *17.I felt immersed in the environment* *18.I could interact with the scene / objects* *19.scene (setting) was comfortable* *20.there were elements I could use for orientation* *21.I felt to be in a safe space* *22.I felt a sense of belonging with the place* *23.I noticed natural conditions (weather, seasons)* 24. *I felt time was flowing*

Acknowledgments. This study was funded by the PERCEIVE project, that received funding from the European Union's Horizon research and innovation programme under grant agreement No 101061157; BrancacciPOV was funded by Friends of Florence and Jay Pritzker Foundation, Ugento project was funded by PNRR M1-C3 - Misura 1 - Investimento 1–2.

Authors Contributions:. Giuseppe Città contributed Sect. 1; Samuele Spotti contributed to Sect. 1, 2; Sofia Pescarin contributed Sect. 3, 4 and 8; Arthur Clay contributed to Sect. 5; Irina Ciortan, Sophia Sotiropoulou, Giorgio Trumpy contributed to Sect. 6; Ivana Cerato contributed to Sect. 7.

Disclosure of Interests. The authors have no competing interests to declare that are relevant to the content of this article.

References

1. Pescarin, S., Città, G., Gentile, M., Spotti, S.: Authenticity in VR and XR experiences: a conceptual framework for digital heritage. In: EUROGRAPHICS Workshop on Graphics and Cultural Heritage (2023)
2. Varga, S., Guignon, C.: Authenticity. Stanford Encyclopedia of Philosophy (2020). https://plato.stanford.edu/entries/authenticity/. Accessed 13 Nov 2022
3. Pescarin, S, Città, G, Spotti, S.: Authenticity in interactive experiences. Heritage **7**(11), 6213–6242 (2024). https://doi.org/10.3390/heritage7110292
4. Fleeson, W., Wilt, J.: The relevance of big five trait content in behavior to subjective authenticity: do high levels of within-person behavioral variability undermine or enable authenticity achievement? J. Pers. **78**(4), 1353–82 (2010). https://doi.org/10.1111/j.1467-6494.2010.00653.x
5. Whelan, D.C.: Extraversion and counter-dispositional behaviour: Exploring consequences and the impact of situation-behaviour congruence (Doctoral dissertation, Carleton University) (2014). https://doi.org/10.31234/osf.io/vksc2
6. Wilt, J.A., Grubbs, J.B., Exline, J.J., Pargament, KI.: Authenticity, presence of meaning, and struggle with ultimate meaning: nuanced between-and within-person associations. J. Res. Pers. **93**, 104104 (2021). https://doi.org/10.1016/j.jrp.2021.104104
7. Grandey, A.A., Fisk, G.M., Mattila, A.S., Jansen, K.J., Sideman, L.A.: Is 'service with a smile' enough? authenticity of positive displays during service encounters. Organ. Behav. Hum. Decision Process. **96**(1), 38–55 (2005). https://doi.org/10.1016/j.obhdp.2004.08.002
8. Pescarin, S., Martinez Pandiani, D.S.: The impact of story structure, meaningfulness, and concentration in serious games. Information **13**(12), 567 (2022). https://doi.org/10.3390/info13120567
9. Tiberghien, G., Bremner, H., Milne, S.: Performance and visitors' perception of authenticity in eco-cultural tourism. In: Theming Asia. Routledge, pp. 145–158 (2018). https://doi.org/10.1080/14616688.2017.1285958
10. Mikeska, J.N., Howell, H.: Authenticity perceptions in virtual environments. Inf. Learn. Sci. **122**(7/8), 480–502 (2021). https://doi.org/10.1108/ILS-10-2020-0234
11. Chen, H., Rahman, I.: Cultural tourism: An analysis of engagement, cultural contact, memorable tourism experience and destination loyalty. Tourism Manage. Perspect. **26**, 153–163 (2018). https://doi.org/10.1016/j.tmp.2017.10.006
12. Beverland, M.B., Farrelly, F.J.: The quest for authenticity in consumption: consumers' purposive choice of authentic cues to shape experienced outcomes. J. Consum. Res. **36**(5), 838–856 (2010). https://doi.org/10.1086/615047
13. Rivera, G.N., Andrew, G.C., Kim, J., Vess, M., Hicks, J.A., Schlegel, R.J.: Understanding the relationship between perceived authenticity and well-being. Rev. Gen. Psychol. **23**(1), 113–26 (2019). https://doi.org/10.1037/gpr0000161
14. Goldman, S.: The psychology of worldviews: jaspers/heidegger. EPIS J. **1** (2012). https://doi.org/10.1037/1089-2680.8.1.3

15. Vannini, P., Franzese, A.: The authenticity of self: conceptualization, personal experience, and practice. Sociol. Compass **2**(5), 1621–1637 (2008). https://doi.org/10.1111/j.1751-9020.2008.00151.x
16. Coupland, N.: The authentic speaker and the speech community. In: Llamas, C., Watt, D., (eds.) Language and Identities, pp. 99–112. Edinburgh University Press (2010). https://doi.org/10.1515/9780748635788-013
17. McKinney, S., Perry, S., Katifori, A., Kourtis, V.: Developing digital archaeology for young people: a model for fostering empathy and dialogue in formal and informal learning environments. In: Communicating the Past in the Digital Age: Proceedings of the International Conference on Digital Methods in Teaching and Learning in Archaeology, 12th-13th October 2018, ed. Hageneuer, S., pp. 179–95. Ubiquity Press (2020). https://doi.org/10.5334/bch.n
18. Katifori, A., Lougiakis, C., Roussou, M.: The role of high-fiving for sustaining engagement in social VR experiences. In: CHI '21, May 08–13, 2021, Yokohama (2021) https://osf.io/7b3m4/download/?format=pdf
19. Rickly-Boyd, J.M.: Authenticity & aura. Ann. Tour. Res. **39**(1), 269–289 (2012). https://doi.org/10.1016/j.annals.2011.05.003(2012)
20. Vico, L.: Authenticity and realism: virtual vs physical restoration. In: Di Giuseppantonio Di Franco, P., Galeazzi, F., Vassallo, V. (eds.): Authenticity and cultural heritage in the age of 3D digital reproductions, McDONALD Institute Conversations (2018). ISBN: 978-1-902937-85-4
21. Kronqvist, A., Jokinen, J., Rousi, R.: Evaluating the authenticity of virtual environments: comparison of three devices. Adv. Hum. Comput. Interact. **3**, 2937632 (2016). https://doi.org/10.1155/2016/2937632
22. Griffiths, M.D., Nuyens, F: An overview of structural characteristics in problematic video game playing. Curr. Addiction Reports **4**(3), 272–83 (2017). https://doi.org/10.1007/s40429-017-0162-y
23. Bressani, M., Sprecher, A.: Atmospheres. J. Archit. Educ. **73**, 2–4 (2019). https://doi.org/10.1080/10464883.2019.1560792
24. Zimmermann, F.: Historical digital games as experiences: how atmospheres of the past satisfy needs of authenticity. Game| World| Architectonics **19** (2021). https://doi.org/10.17885/HEIUP.752.C10376
25. Souza, L.H., Kastenholz, E., Barbosa, M.D.L.A., Carvalho, M.S.E.S.C.: Tourist experience, perceived authenticity, place attachment and loyalty when staying in a peer-to-peer accommodation. Int. J. Tourism Cities **6**(1), 27–52 (2019). https://doi.org/10.1108/IJTC-03-2019-0042
26. Gaver, B., Dunne, T., Pacenti, E.: Design: cultural probes. Interactions **6**(1), 21–29 (1999). https://doi.org/10.1145/291224.291235
27. Kassan, A., et al.: Becoming new together: making meaning with newcomers through an arts-based ethnographic research design. Qual. Res. Psychol. **17**(2), 294–311 (2020). https://doi.org/10.1080/14780887.2018.1442769
28. Pescarin, S.: Museums and virtual museums in Europe: Reaching expectations. SCIRES-IT-Scientific Res. Inf. Technol. **4**(1), 131–140 (2014). https://doi.org/10.2423/i22394303v4n1p131
29. Panksepp, J.: Affective consciousness: core emotional feelings in animals and humans. Consciousness cognition **14**, 30–80 (2005). https://doi.org/10.1016/j.concog.2004.10.004
30. Kilteni, K., Groten, R., and Slater, M.: The sense of embodiment in virtual reality. Presence Teleoperators Virtual Environ. **21**(4), 373–387 (2012). https://doi.org/10.1162/PRES_a_00124
31. Ludwig, K.: Contemporary Philosophy in Focus. UK; Cambridge University Press, Cambridge (2003)

32. Pescarin, S., Bonanno, V., Marasco, A.: Social cohesion in interactive digital heritage experiences. Multimodal Technol. Interaction **7**(6), 61 (2023)
33. Darzentas, D., Bodiaj, E., Benford, S., Spence, J., Preston, W., Koleva, B., Adams, M.: GIFT Project Deliverable D6. 4 Package and Release of the Final Toolbox (2020)
34. Duchamp, M.: The Bride Stripped Bare by Her Bachelors, Even (The Green Box). Translated by Richard Hamilton, Percy Lund, Humphries & Co. (1960)
35. Casella, L., Masahiko, T.: Effects of low-oxygen environments in the light fading of six dyes present in the autochrome color screen. J. Am. Inst. Conserv. **51**(2), 159–174 (2012)
36. Amaro, A.C., et al., editors. Heritage in the Digital Era: AI and Its Applications. MDPI (2023)
37. Vannucci, G., Joram, S., Beselin, A., Röhrs, S.: Micro fading test for textile single yarns: a new methodology applied to the reformation tapestry to assess its sensitivity to light. Heritage Sci. **11**(1), 47 (2023). https://doi.org/10.1186/s40494-023-00884-w
38. Watteeuw, L., Iterbeke, H.: Enclosed Gardens of Mechelen: Late medieval Paradise gardens revealed. Hannibal (2018)
39. Kirchner, E., van der Lans, I., Ligterink, F., Geldof, M., Megens, L., Meeden-dorp, T., et al.: Digitally reconstructing Van Gogh's field with irises near arles part 3: determining the original colors. Colour Res. Appl. **43**(3), 311–327 (2018). https://doi.org/10.1002/col.22197
40. Palomero, C.M.T., Soriano, M.: Digital cleaning and "dirt" layer visualization of an oil painting. Optics Express **19**(21) (2011)
41. Ridel, B., Reuter, P., Laviole, J., Mellado, N., Couture, N., Granier, X.: The revealing flash-light: Interactive spatial augmented reality for detail exploration of cultural heritage artifacts. J. Comput. Cult. Heritage (JOCCH) **7**(2), 1–18 (2014)
42. Van Noord, N.: Learning visual representations of style. Tilburg University Research Portal (2018). https://research.tilburguniversity.edu/en/publications/19844a3c-e0ae-4ee5-97a6-523f6b5282d0
43. Ciortan, I.M., George, S., Hardeberg, J.Y.: Colour-balanced edge-guided digital inpainting: applications on artworks. Sensors **21**(6), 6 (2021). https://doi.org/10.3390/s21062091
44. Wang, M., Lai, Y.K., Liang, Y., Martin, R.R., Hu, S.M.: Bigger picture: Data-driven image extrapolation using graph matching. ACM Trans. Graph. **33**(6) (2014). https://doi.org/10.1145/2661229.2661278
45. Strlič, M., Thickett, D., Taylor, J., Cassar, M.: Damage functions in heritage science. Stud. Conserv. **58**(2), 80–87 (2013)
46. Pescarin, S., et al.: Hybrid XR collaborative and guided experiences in cultural heritage: Brancacci POV prototype. In: EUROGRAPHICS Work-shop on Graphics and Cultural Heritage, pp. 157−161 (2023). https://doi.org/10.2312/gch.20231171
47. Ferdani, D., et al.: Brancacci chapel in florence: surveying and real-time 3d simulation for conservation and communication purposes. Int. Arch. Photogramm. Remote. Sens. Spat. Inf. Sci. **48**, 535–540 (2023)
48. Olbrich, M., Zapf, A., Stiegemann, C., Pröbe, A.: Large room scale augmented reality in an unaltered world heritage site. In: EUROGRAPHICS Workshop on Graphics and Cultural Heritage (2023)
49. Morris, H.R., Whitmore, P.M.: Virtual fading" of art objects: simulating the future fading of artifacts by visualizing micro-fading test results. J. Am. Inst. Conserv. **46**(3), 215–228 (2007). https://doi.org/10.1179/019713607806112279
50. Ciortan, I.M., Poulsson, T.G., George, S., Hardeberg, J.Y.: Tensor decom-position for painting analysis. part 2: spatio-temporal simulation. Heritage Sci. **11**(1), 84 (2023). https://doi.org/10.1186/s40494-023-00913-8

Sense of Care in the Design of Digital Heritage Applications

Manuele Veggi[1,2], Vanessa Bonanno[3], Ivana Cerato[1], Irina-Mihaela Ciortan[4], Arthur Clay[5](✉), Gabriele Fiorenza[3], Delfina Sol Martinez Pandiani[3,6], and Sofia Pescarin[1]

[1] CNR ISPC, Florence, Italy
{manueleveggi,ivana.cerato,sofia.pescarin}@cnr.it
[2] University of Roma La Sapienza, Rome, Italy
manuele.veggi@uniroma1.it
[3] University of Bologna, Bologna, Italy
{vanessa.bonanno,gabriele.fiorenza}@studio.unibo.it
[4] NTNU ColourLab, Gjøvik, Norway
irina-mihaela.ciortan@ntnu.no
[5] HSLU, Lucerne, Switzerland
arthur.clay@hslu.ch
[6] University of Amsterdam, Amsterdam, The Netherlands
d.s.martinezpandiani@uva.nl

Abstract. Cultural heritage care emphasizes preservation, ethical duty, empathy, and relational bonds, with museums fostering deeper connections and historical understanding. This paper develops a design approach to interactive media applied to coloured collection, by focusing on a "care theory" for cultural heritage. The work also analyses art from diverse periods and media, exploring how various engagement types—visual, emotional, interactive—evoke care in viewers. Multimedia art similarly fosters empathy and reflection. The study utilized a Cultural Probe Kit to analyse caring behaviours, revealing traits and triggers that prompt a caring attitude. A proposed framework outlines emotional and empathetic aspects, emphasizing meaningful connections between caregivers and cultural heritage, enhanced through storytelling, engagement, and interactive technologies. Museums, as social spaces, benefit from user participation, fostering care and meaning making. Two prototypes like *ColorColab* and *MyTISSE* are described, focusing on their capacity to engage visitors with hands-on, multi-sensory technology, enhancing their personal connections to artifacts and social interactions. The chapter closes testing the framework and caring trigger on a number of different interactive projects.

Keywords: Participatory Design · Care Theory · Interactive Technology

1 Literature Review on "Care Theory"

It has been outlined that "*the discourse on cultural heritage is mainly centred on the concepts of ownership and conservation, of national identity, (…) of preservation and exploitation*" [1]. This sensitivity has a long history and can be traced back at least to the

S. Pescarin et al. (Eds.): *Chromatic Visions*, LNCS 15750, pp. 128–161, 2026.
https://doi.org/10.1007/978-3-032-07792-9_4

famous letter by Raphael to Pope Leo X [2]. In this document, which undoubtedly laid the foundations for modern sensitivity to cultural heritage, the artist invites the Bishop of Rome to take *care* of the remains of classical monuments and artworks.

This caring approach stems mainly from the need and the duty to preserve what Raphael defines as "little remains" from risks and degradation factors and it is often driven by an intergenerational ethical imperative. However, despite the relevance of the "care theory" towards cultural heritage, a formalization of this concept is still absent. Major research has indeed been carried out only in the field of health care or pedagogy. The "care theory" in the cultural domain is thus often reconstructed through philosophical, literary, and art-historical sources, as well as psychological and neuroscientific studies.

Indeed, one of the most relevant frameworks for analysing and understanding the concept of care derives from ethico-philosophical studies. For instance, in one of his recent essays, *Philosophy of Care*, Borys Groys reconstructs the origin of "care" from ancient Greece, particularly Plato, to modern times [3]. If we limit ourselves to the history of 20th century philosophy, Martin Heidegger offers an interesting interpretation analysis of this concept from a metaphysical and existential perspective. In his speculation, the human being is defined as a Dasein, literally meaning "being (sein) - there (da)", which is continuously threatened by death and hence lives in a "basic state-of-mind of anxiety" [4]. His existence is permanently projected into the future, which presupposes that he continuously cares for his life. Consequently, "Care" (in German: *Sorge*) is the Being and the source of the Dasein.

Although this reading is particularly evocative, the modern interpretation of care from an ethical, moral, as well as political perspective is greatly influenced by the pivotal contribution of Carol Gilligan. She seems to go back to the tragic conflict between Antigone and Creon, as the concept of care is in dichotomy with the concept of justice. This analysis has been summarized as follows [5]: according to Gilligan, some people base ethical decisions on principles of justice, equality, impartiality, and rights. This is the justice perspective. But others base their decisions on a care perspective, in which the need to preserve relationships and minimize hurt takes precedence over considerations of justice and rights. The care perspective places special significance on attachment and compassion, Gilligan writes, "the moral injunction not to act unfairly toward others, and not to turn away from someone in need, capture these different concerns".

Gilligan's reflection is indeed based on a "**care theory**", which Soto [6] describes as follows: "in its purest form, [it] assumes that the actors [caregiver and care-receiver] are willing to develop relationships with each other and that care will emerge from those relationships". The goal of this intimate bond is to "strive for competence because we want to do our best for those we care for". An essential component of this framework is therefore the necessity for a **mutually fulfilling relationship**; when this equilibrium is disrupted, caring mechanisms falter [7]. In human-to-human relationships, exemplified by the attachment between a parent and a child, the robustness of the bond is evident, as observable actions trigger reciprocal responses (e.g., a crying child prompts parental attention). Conversely, in human-to-object relationships, this responsive reciprocity is lacking due to the inherently weaker connections between individuals and inanimate entities.

For the understanding of the "care theory" within cultural heritage, a crucial observation is provided by **Joan Tronto** and her essay *Moral Boundaries* [8]. Here, the author provides a definition of care that extends the horizons of Gilligan's theory also to include objects and introduces an intergenerational ethical dimension that is vital to the preservation of cultural heritage. This updated definition highlights "the ways in which care can function socially and politically in a culture" (103): *"On the most general level, we suggest that caring can be viewed as a species of activity that includes everything that we do to maintain, continue, and repair our 'world' so that we can live in it as well as possible. That world includes our bodies, ourselves, and our environment, all of which we seek to interweave in a complex, life-sustaining web"*.

In this perspective, caring is no longer seen as a disposition or a "perspective", but as a **set of practices**. The author also describes a possible relevant framework for describing the relationship between caregiver and care recipient. It is based on the key concept of **responsibility** and foresees a phase of "**care-receiving**", where "*the object of care (responds) to the care it receives. For example, the tuned piano sounds good again, the patient feels better (…). This phase provides the only way to know that caring needs have actually been met*" [8].

This claim paves the way to an essential question for the understanding of caring for cultural artefacts as the identification of this possible "reward" may not be immediate. Groys suggests a persuasive solution to this *impasse* based on a double interpretation of each artwork [2]. In the first one, cultural artefacts are reduced to a "*mere* object", i.e. only their pure material aspects (e.g. its support and the materials used in its production) are considered. The cultural value is instead the core of the second perspective, which recognises the power of each work of art to unveil the "world" and the context in which the artist lived. While the former can be easily recognised, the latter one requires a deeper understanding, which may derive from different sources. For the purpose of the PERCEIVE project, the most relevant one is *museum mediation.*

Groys devoted a previous essay on curatorial practices to this topic [9]. In his interpretation, museums can be compared to hospitals as institutions devoted to care. This claim does not hold exclusively to conservation professionals, rather it engages civil society as a whole (bold added): "*Artworks seem to be genuinely sick and helpless - **the spectator has to be led to the artwork, as hospital workers might take a visitor to see a bedridden patient**. It is in fact no coincidence that the word "curator" is etymologically related to "cure." Curating is curing. The process of curating cures the image's powerlessness, its incapacity to present itself. The artwork needs external help, it needs an exhibition and curator to become visible." **The medicine that makes the sick image appear healthy - makes the image literally appear and do so in the best light - is the exhibition**.*

This approach, which focuses more on the message of the artworks rather than on its mere material aspect, is also at the basis of Montanari's *sentimental approach* to cultural heritage, according to which an artwork is the "witness of the presence of another Time within our Time" [1]. His approach seems the most effective to sum up these considerations on the ethical and political practices of care. The function of Cultural Heritage in a mass democracy is precisely that of allowing not only those who do intellectual work, but potentially all citizens, to 'detach themselves from the present', to

come into contact with the 'force of the human in man': the human that is in ourselves, and in the other humans we are given to meet. This is the condition for walking 'at a human pace' again, that is, for building a more human world. Consequently, resolving the challenge of art collections' safeguarding suggests the potential of strengthening the relational bond between them and individuals, thereby **diminishing the perceived distance between the viewer and the piece**. This can be encouraged by fostering a deeper understanding of the artwork, facilitating a transformation of the encounter into a personalised experience, and ultimately empowering citizens to actively engage with and contribute to our cultural heritage through collaborative and co-creative endeavours.

Numerous factors contribute to the cultivation of caring attitudes and can be embraced: attachment, personification, uniqueness and idiosyncrasies, sense of wonder, compassion, perception of fragility, and sense of justice. Noddings, for example, elucidates the intrinsic drive towards moral conduct, rooted in the ethics of care, particularly empathy [7]. Importantly, research on **empathy** underscores its malleability throughout life, influenced by a range of strategies [10, 11].

Scholarship within the Cultural Heritage domain has delineated a distinct form of empathy known as "**historical empathy**" characterised as a cognitive and affective process involving engagement with historical figures [12]. This form of empathy can be nurtured through a "facilitated dialogue" approach, which encourages individuals to share their perspectives, personal beliefs, and experiences while also challenging them to critically examine presented information and consider alternative viewpoints. This methodological approach was implemented in the *BrancacciPOV* project, utilising an "interpretation-reflection loop" framework within a hybrid virtual reality collaborative and guided experience [13].

In this context, our focus is on understanding how to elicit and assess a caring attitude specifically towards cultural heritage artefacts. We define the term ***caring attitude*** as "**a deliberate and action-oriented event that occurs between two actors**, where one needs help and protection and the other provides whatever they need" [14]. Given the absence of specific studies on how to evoke and measure this caring attitude in the context of (coloured) cultural heritage, in the next section, we detail our dedicated study and exploration. These efforts were based on the analysis of alternatives perspectives from art and multimedia and on the use of a Cultural Probe Kit, facilitated by a Master's thesis at the University of Bologna.

2 Alternative Perspectives on Care from Art and Multimedia Application

Art. The research enquiry of this section is to explore a selected number of artworks stemming from various periods and genres that address the subject of care. The study aims to analyse how these artworks trigger a sense of care in viewers and how different forms of engagement in the artworks contribute to this experience. The intention of the artist and the reception of the viewer of the artist's intentions differ by various degrees. The intention of the artist is described as the element of care present in the artwork, but the viewer's reception of it is defined as what triggers a sense of care in the viewer. The research is based on critical analysis and interpretation of the artworks, aiming to shed light on how artists connect with their audiences, evoke emotions, and foster a sense of care through their creations. The methodology used in this study involves a selection of artworks from contrasting periods, genres, and diverse media that explore the theme of care, with the goal of presenting a comprehensive review and analysis of these artworks, focusing on their forms of engagement with viewers. The core material used in this study is based on a selection of artworks from contrasting periods, genres, and which made use of diverse media. The artworks' level of engagement is categorised into five key aspects, or forms of engagement: visual, emotional, interactive, intellectual, and conceptual, which are described in the following paragraphs.

The Factor Engagement. All acts of care are inherently tied to engagement. Whether it is showing kindness, compassion, or concern, caring requires **active involvement and interaction with others**. To care about someone or something, one must be willing to invest time, attention, and genuine interest in their well-being. Care is not a passive emotion; it necessitates acting and providing support. This engagement can be driven by **various factors**, such as biological needs, a sense of justice, or a profound connection with someone or something. Ultimately, care is a dynamic and interactive process that fosters empathy, understanding, and a sense of responsibility towards others and their welfare. A sense of care can be generated by engagement with artwork. Art has a unique ability to evoke emotions, prompt introspection, and foster connections with the subject matter or themes portrayed. When viewers actively engage with art, whether through contemplation, interpretation, or participation in interactive installations, they may experience a range of emotions, including empathy, compassion, and concern. Artists are able to infuse their works with **elements that resonate with human experience**, addressing issues related to society, nature, personal struggles, or emotional states. By exploring these themes, art triggers a sense of care by encouraging viewers to reflect on their own experiences and bring about empathy for others, and perhaps even contemplate challenges and solutions. Therefore, the act of engaging with artwork allows individuals to establish a connection with the art's message or the artist's intentions, creating an emotional response that may translate into a genuine sense of care and appreciation for the artwork's subject matter or the broader human experience it represents. A comprehensive review of artworks listed in the table is presented here, focusing on their forms of engagement with viewers. Each artwork's level of engagement is categorized into five key aspects: visual engagement, emotional engagement, interactive engagement, intellectual engagement, and conceptual engagement. Through this examination, we aim to shed light on the diverse ways in which artists connect with their audiences, eliciting emotions, contemplation, and participation. The artworks' varying degrees of engagement offer valuable insights into the profound impact art can have on evoking a sense of care and fostering connections between the artwork, the artist, and the viewer.

- **Visual Engagement**: Approximately 40% - Visual observation is the most common form of engagement, as viewers interact with the artworks primarily through visual perception, interpreting and contemplating the imagery presented.

- **Emotional Engagement**: Approximately 30% - Many artworks evoke emotional responses from viewers, fostering a sense of empathy, compassion, and care. Emotional engagement is prevalent in artworks that explore themes of vulnerability, loss, love, and human experiences.
- **Interactive Engagement**: Approximately 15% - Several artworks incorporate interactive elements, inviting viewers to actively participate in the artwork. These interactions may involve touch, movement, or engagement with digital interfaces, creating a more immersive and participatory experience.
- **Intellectual Engagement**: Approximately 10% - Some artworks prompt intellectual engagement as viewers reflect on the symbolism, themes, and deeper meanings conveyed by the artists. Intellectual engagement often arises in artworks that address social issues, cultural heritage, and the human condition.
- **Conceptual Engagement**: Approximately 5% - A few artworks involve conceptual engagement, where viewers contemplate the underlying concepts and ideas explored in the artwork. These may involve reflections on care, self-awareness, and the relationship between art and technology

The Factor Medium. Based on the ranking of forms of engagement in the artworks mentioned, it does not necessarily suggest that any particular media is inherently better for engagement. Different forms of media offer unique opportunities for artists to engage with their viewers. For example, interactive installations and augmented reality art provide direct and immersive engagement, while traditional media like painting and sculpture can evoke emotional and intellectual responses through their visual representations. Ultimately, the effectiveness of engagement depends on how well the artist leverages the chosen medium to convey their intended message and connect with the audience. A skilled artist can create engaging and impactful artworks across a wide range of media by understanding the strengths and limitations of each medium and using them to complement the overall artistic expression. Rather than focusing on which media is better for engagement, artists prioritise understanding their artistic vision and based on that that, artists select the medium that aligns best with their artistic intent, and which will maximise the potential for engagement with goal of creating meaningful and memorable experiences for the viewer. To provide clarity and streamline the analysis of the artworks, three pieces were deliberately chosen from contrasting periods and media. By delving into "The Gleaners" by Jean-Francois Millet, "Tent" by Tracey Emin, and "Ecce Homo" by Arthur Clay and Ingo Lie, readers will gain profound insights into the themes of care and empathy beautifully depicted in these iconic masterpieces. As viewers engage with these images, they can experience the care factor inherent in the artworks and connect emotionally with their content. Each piece evokes a profound sense of care and appreciation for the diverse facets of human existence, sparking contemplation and fostering a deeper understanding of the artists' intentions. Depicted in Fig. 1 is the iconic artwork "The Gleaners" by Jean-Francois Millet which captures a profound moment in the lives of three peasant women. Depicted gathering leftover crops from a field after the harvest, the painting's significance lies in its portrayal of caring for the less fortunate and the essence of community support. The artwork's aspect of care is evident through the women's act of gleaning. With unwavering determination, they toil in the fields, labouring to provide sustenance for themselves and their families.

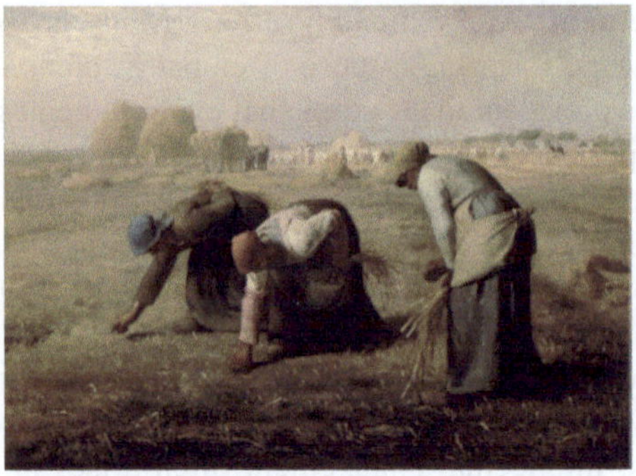

Fig. 1. "The Gleaners" by Jean-Francois Millet, 1857. Oil on canvas. Located at Musée d'Orsay, Paris, France.

Their **dedication and effort symbolise the essence of care**, reflecting the laborious journey of those who face challenges in meeting their basic needs. As viewers encounter "The Gleaners," a profound trigger of care and empathy is invoked brought on by how the painting conveys the dignity and perseverance of the working class, inspiring feelings of compassion and recognition for those facing hardships.

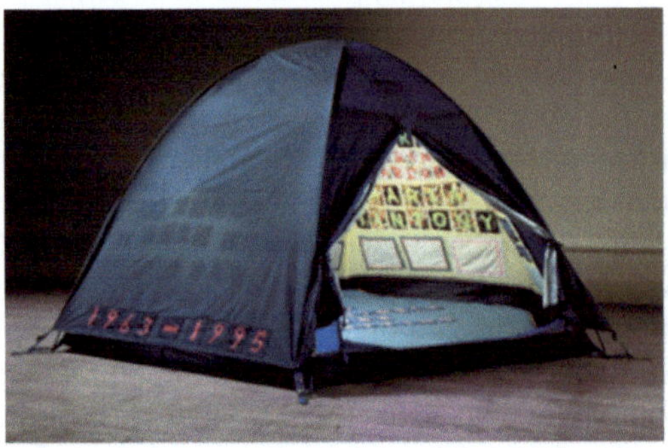

Fig. 2. "Tent" by Tracey Emin, 1999. Tent with appliquéd text and light. Destroyed.

The artwork "Tent" by Tracy Emin that is depicted in Fig. 2 is a conceptual artwork exploring themes of intimacy and emotional care. Through the text appliquéd on the tent, Tracey Emin reveals the names of everyone she has ever slept with, along with intimate reflections. The artwork invites viewers to encounter the **personal and confessional nature of the text, evoking feelings of care and empathy**. The tent serves as a private space where the artist reflects on her past relationships, revealing the vulnerability and emotional connection involved in intimate experiences. Viewers may gain a deeper understanding and compassion for the complexities of human relationships and the emotional care involved in such encounters.

The above two artworks contrast the use of media in artworks created more than a century apart. In modern art, however, an even larger contrast of approach can be seen when the use of electronic media is introduced in the creation of an artwork. The

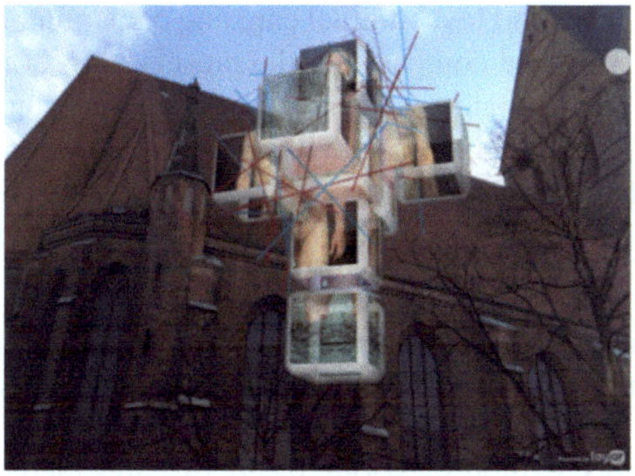

Fig. 3. "Ecce Homo" by Arthur Clay and Ingo Lie, 2017. Augmented Reality (AR). Present Location: Am Markte, Hannover, Germany.

artwork in Fig. 3, "Ecce Homo" is an Augmented Reality (AR) artwork that delves into the impact of the digital age on human self-perception. Created by artists Arthur Clay and Ingo Lie in collaboration with Virtuale Switzerland, the piece presents the artist's self-portrait as an apparition above church ruins. The self-portrait is intricately divided into eight cubes forming a three-dimensional cross, intersected by red and blue lines. This innovative use of colour-theoretical and cosmological concepts highlights the interconnectedness of humanity with the surrounding world while symbolizing the loss of holistic self-awareness in the modern digital era.

The artwork contemplates the aspect of care by depicting a concern for the profound changes in human perception caused by the digital age. It prompts viewers to consider how technology and digitalization influence their self-awareness and connection to the world. "Ecce Homo" triggers this in the viewers as they explore the work on their own smart devices and witness the impact of modern technology on self-perception as become conscious of the fragmented self-portrait and the symbolic nature of the artwork, encouraging contemplation about one's own interactions with technology, imagination, and self-awareness in the rapidly evolving contemporary world. The following summaries provide an overview of the diverse artworks listed in the table, which all explore the theme of care. The referenced traditional artworks capture emotions of vulnerability, highlight maternal love, and depict acts of compassion. In contrast, modern art explores recognition, empathy for refugees, and the significance of human connection. Additionally, electronic artworks delve into emotional care, the complexities of relationships, and the power of empathy. Each of the works cited in the table evokes deep emotions and fosters a sense of care and appreciation for the diverse facets of human existence. Through their powerful imagery and symbolic representations, these artworks invite viewers to contemplate the significance of compassion and support for one another in shaping meaningful human connections.

Care Factor Summary for the Traditional Artworks referenced. The presented artworks, created using a diverse range of media such as painting, sculpture, and photography, share a common aspect of triggering care in the viewer. Each artwork explores various facets of care, including emotional care and understanding for those experiencing turmoil and vulnerability, maternal care and love, mutual care and camaraderie in a collective effort, ultimate acts of caregiving and compassion, the preservation of innocence, the significance of medical care and empathy, and the importance of caring for the less fortunate and providing community support. Through their **powerful imagery and evocative representations of human experiences**, these artworks evoke **deep emotions, empathy, and contemplation, fostering a sense of care and appreciation for the diverse facets of human existence** and the significance of compassion and support for one another.

Care Factor Summary of Modern Artworks Referenced. The presented artworks, spanning a range of media from mixed media installations and street art to documentary photographs and performance art, share a common aspect of triggering care in the viewer. Each artwork explores various facets of care, including **recognition and appreciation** for women's achievements, **compassion** for refugees and the vulnerable, **emotional care in the face of loss and adversity**, the **significance of presence and human connection**, nurturing and sustaining loving relationships, and the empathy and understanding required in intimate encounters. Through their powerful imagery, symbolism, and interactive elements, these artworks evoke deep emotions, empathy, and contemplation, fostering a sense of care and appreciation for human experiences, connections, and vulnerabilities.

Multimedia Applications. Focusing on a peculiar aspect of multimedia application, the following summaries provide an overview of the diverse games listed, which all explore the theme of care. The referenced games capture emotions of maternal protection, empathy for vulnerable characters, and the significance of making moral choices to safeguard a group. In contrast, other games focus on emotional care, nurturing virtual relationships, and the power of cooperation and support. Each of the games cited evokes **deep emotions and fosters a sense of care and responsibility for the well-being of characters and virtual worlds**. Through their immersive gameplay and storytelling, these games invite players to contemplate the significance of compassion and empathy in the context of survival, adventure, and human connections. "Shelter," developed by Might and Delight, is a compelling survival game that immerses players in the role of a mother lynx (Fig. 4). In this heart-warming experience, players must navigate a challenging and hazardous wilderness to protect and guide their vulnerable cubs. The game's central theme revolves around **maternal care and the instinct to safeguard** the young lynx cubs from predators and environmental threats. As players search for food and shelter, they experience a profound sense of responsibility for their offspring, evoking strong feelings of empathy and protectiveness. The emotional narrative and depiction of the mother-offspring bond in "Shelter" create a touching experience that resonates with players, forging a deep connection between the player and the virtual world of the lynx family."

The game "This War of Mine," developed by 11 Bit Studios, is a gripping survival experience set in a war-ravaged city, where players take on the role of a group of civilians striving to endure amidst the brutal conflict (see Fig. 5). The game confronts players with the stark realities of war and the weighty moral decisions they must confront to safeguard their group. In the game's exploration of care, players are immersed in the challenging context of survival, where **every decision holds profound consequences for the well-being of their group**. The fundamental tasks of providing sustenance,

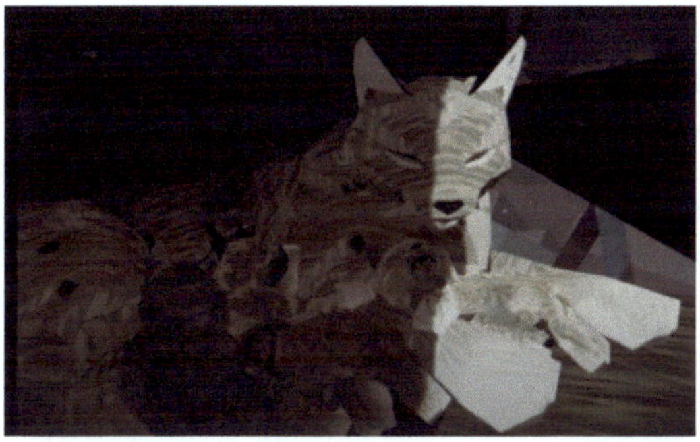

Fig. 4. Screen shot of trailer for the video game 'Shelter' by Might and Delight.

medicine, and shelter become crucial in the face of **morally complex dilemmas**. As players navigate this emotionally charged landscape, they may develop a strong sense of care and responsibility for the survival and safety of their group members. The game's power lies in its **poignant storytelling**, which presents players with moral quandaries that evoke empathy and concern for the welfare of the characters they control. As they grapple with these moral challenges, players may experience a deeper appreciation for the intricacies of care in the context of survival and the profound impact of their decisions on the lives of those they lead through the harrowing landscape of war.

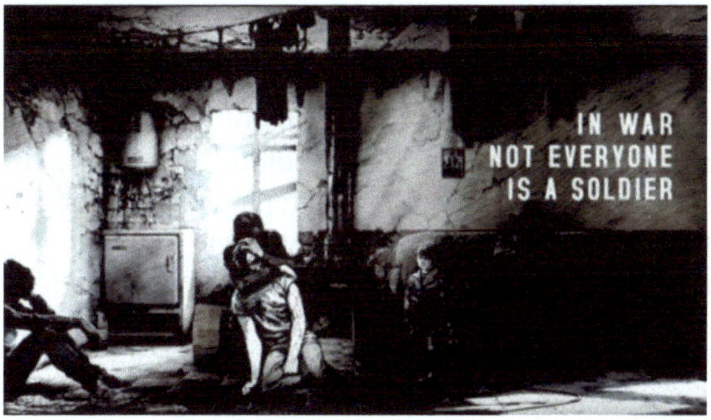

Fig. 5. Screen shot of trailer for the video game 'This War of Mine' by 11 bit studio

Care Factor Summary in Multimedia. All the featured games found in the list share qualities with those cited above although they encompass a variety of game genres and platforms, share a common theme of triggering care in players. Each game explores different aspects of care, from nurturing virtual farms, guiding young cubs through perilous environments, cooperating with strangers on a mystical journey, to forming bonds with mythical creatures. The emotional narratives, gameplay choices, and virtual connections evoke deep feelings of empathy, attachment, and responsibility. Through their immersive experiences, these games foster a sense of care and appreciation for the diverse facets of human existence and the virtual worlds they offer. These features can be traced also in other classes of multimedia products: creators can craft impactful art and digital applications that foster empathy and care by following various approaches. Cultivating empathy, addressing social issues, fostering inclusivity, and utilizing interactive elements are some of the guidelines that can elicit a sense of care in the audience: besides the videogame "This war of mine" by 11bit studio, "The Key" by Celine Tricart is a meaningful example. It consists of a VR immersive experience where players are led by the narrative to experience the **sense of loss** to reach understanding and a caring attitude toward refugees (Fig. 6).

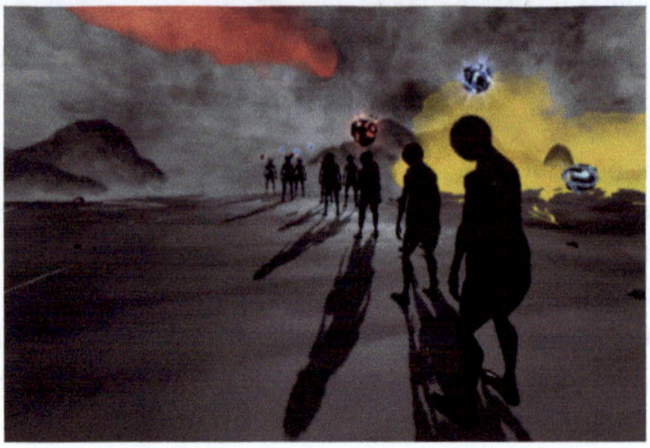

Fig. 6. Screenshot of the VR experience "The Key" (https://thekey-vr.com/)

A sense of care might be triggered also through colour reconstruction tasks. In this sense, a good example is that of **learning-based colourization methods** where, typically, the reference datasets consisting of pairs of colour and monochrome images are obtained by digitally converting available colour images to **neutral tones**. However, if the aim is to recolour historical photographs, then we ought to be aware that several pioneering photographic techniques were prone to a loss of information and a misrepresentation of reality. For instance, the wet-plate photography collected by early European settlers in New Zeeland failed to capture the tattoos (known as "as tā moko") of the Māori indigenous community. As a matter of fact, photographer Michael Bradley recently raised awareness about this important cultural symbol lost in the meanders of a colonization process, with his project entitled *Puaki*, where he took portraits of Māori people, with modern colour photography, as well as the old wet-plate technique. These **pairs of portraits** would represent a more caring and empathetic option for a dual colour-monochrome image dataset, towards use in colourization methods (Fig. 7).

3 Qualitative Analysis on Behaviours and Perception of Care

With the aim of improving the understanding of caring, an experimental analysis was conducted in 2022. A specific Probe Kit survey was designed to gather inspirational information to define caring attitudes more accurately and to comprehend their peculiar characteristics and triggers [15]. According to Burrows, Mitchell, and Nicolle [16], a **Cultural Probe Kit** is a tool for generating ideas used in the design process. The

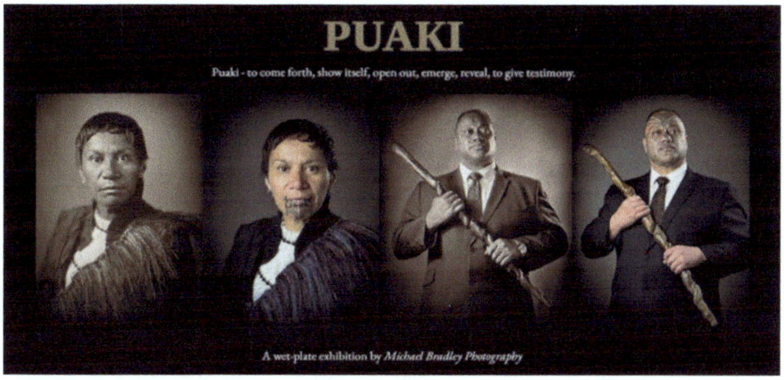

Fig. 7. Screenshot of the main website of the "Puaki" exhibition by Michael Bradley, 2020 (https://www.puaki.com/)

probes are boxes given to participants that contain evocative tasks and any kind of artefact (such as a map, postcards, cameras, diaries, activity notebooks), allowing them to document specific sensations, feelings, or encounters. Various qualitative techniques, including surveys, the compilation of a diary, and the reading and writing of stories, were employed in this particular use case. Twenty participants from the first year of the Master's program "Digital Humanities and Digital Knowledge" at the University of Bologna took part in this exploration. They were questioned, through a diary (Fig. 8) and an activity notebook, on a range of topics, including their **daily routines, attitudes toward caring** for cultural heritage, and **their thoughts on stereotypes** that may have an impact on such attitudes. The information gained has improved our knowledge of the **traits and triggers of the sense of caring**. The findings include: 1) **peculiar traits**, which are the qualitative traits that characterise the caring attitude, and 2) **triggers**, which are behaviours, actions, and circumstances that cause it to develop. With regard to the peculiar traits, we understood how the caring attitude is greater when it is directed **towards something that one owns, even metaphorically**. In fact, the diary entries and responses to the essay prompt, "Write a life episode in which you felt such an urgency to care for something, that you were encouraged to act, to do something," suggested that some of the objects the participants cared for were **metaphorically theirs** because they were **emotionally attached to them**.

Furthermore, the resulting data clarify that caring attitudes are solely **associated with the emotional sphere** and specifically related with **empathy** and the positive and negative feelings experienced during the caring activity. **Emotional storytelling** is another conceivably important trigger; indeed, it has the power to manipulate the activation of the sense of care through the use of a skilfully prepared story. Additionally, from the results we have understood how developing a caring attitude towards something or someone is more likely when you **already have a personal relationship** with them (in fact, 70% of the participants claimed that they already knew the person or item they cared about when compiling the diary).

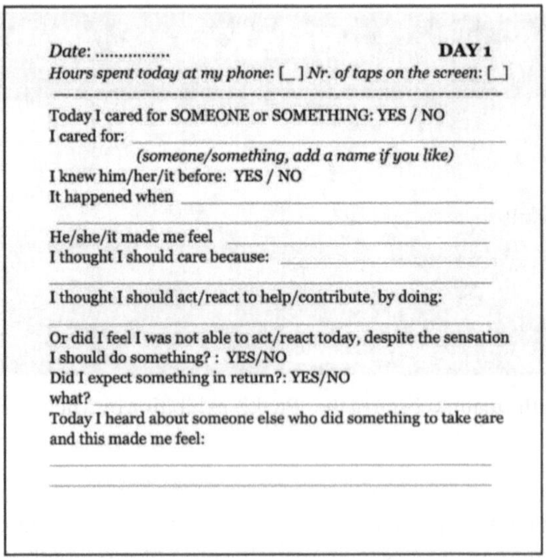
Date: .... **DAY 1**
Hours spent today at my phone: [_] *Nr. of taps on the screen:* [_]

Today I cared for SOMEONE or SOMETHING: YES / NO
I cared for: __________
(someone/something, add a name if you like)
I knew him/her/it before: YES / NO
It happened when __________

He/she/it made me feel __________
I thought I should care because: __________

I thought I should act/react to help/contribute, by doing:

Or did I feel I was not able to act/react today, despite the sensation I should do something? : YES/NO
Did I expect something in return?: YES/NO
what? __________
Today I heard about someone else who did something to take care and this made me feel:

Fig. 8. Layout of the diary with questions related to caring attitude.

4 A Preliminary Conceptual Framework of Care

The threefold analysis previously described in this chapter has led to the identification of recurrent components that can support the definition of a first framework describing the breadth of "sense of care" in the cultural sector. These **components of "care"** are yet not defined as a single characteristic: conversely, in analogy with the methodology used in the analysis of "Authenticity" (see Chap. 3), we adopted the "family resemblance" principle, defined by Wittgenstein in his *Philosophical Investigations*. In this perspective, each of them can be conceived as a set of semantically close traits.

4.1 Single Components and Triggers

Literature review, art and multimedia contributions, and experimental surveying demonstrated how caring for CH is multifaceted and articulated in several components. Scholarly research on "care theory" has emphasised the importance of the relationship between the caregiver and the care-receiver. As outlined by Gilligan, this bond can be established on the basis of different needs. On the one hand, the "justice perspective" plays a significant role as laws, social norms or expectations may ultimately lead citizens to show interest in CH. On the other hand, as in interpersonal relationships (e.g. parent and child), an emotional and personal involvement brings about deeper connections due to the "special significance of attachment and compassion". The more intimate sphere of the caregiver, i.e. identity, emotion, and cognition, are here at stake and the resulting "sense of care" is stronger. Studies also outlined the relevance of the care-receiver, i.e. the cultural artefact or site. Best exemplified in "The Keys", caring is often associated with a dimension of loss, weakness, and, in general, frailty. This emerges clearly also in the definition provided by Tronto's *Moral Boundaries*, which identifies the main goal of care as "maintaining, continuing, and repairing our 'world'". At the same time, the reading of the "The Gleaners" (see Fig. 1) puts forward an additional component: effort. Particularly in the case of inanimate care-receivers, care is expected to emerge when the uniqueness of the artefact and the time and effort required to create it are explained. On the basis of these factors, caregivers feel responsible for what they care about and are willing to take action to satisfy its needs. These activities are heterogeneous, and literature traditionally refers to them with the term of "caring attitude". They are usually associated with the time and devotion spent to properly "take care" of the artwork and the most intuitive example is the conservation and restoration activities daily made by cultural heritage professionals. Nonetheless, at a closer look, it appears clear how caring attitudes also entail behaviours and actions within civil society's reach: for instance, simply worrying for the potential loss of a cultural artifact, visiting cultural institution, individually researching CH-related topics and donate time (e.g. volunteering, petitions, subscription to associations as ICOM) and money for this cause. The analysis of works of art presented in thi chapter described the involvement of the visitor and the artwork in terms of engagement (taxonomised in five distinct categories: visual, emotional, interactive, intellectual, and conceptual). This concept plays a crucial role in the museological perspective on care embraced by the PERCEIVE Project, since it is often associated with the relationship between audience's participation in cultural institutions [17]. As a matter of fact, the recent Routledge *Dictionary of Museology* [18] defines "engagement" almost as a synonym of "participation" and mentions some possible "caring attitudes" for visitors of cultural institutions:

ENGAGEMENT (AUDIENCE) (fr. Engagement (des publics), Participation (des publics)), n. – Encouraging someone to participate, take an interest and become involved in an organisation's project or work. Museum audience engagement or participation occurs in several areas: firstly – social networks, with the creation of communities of users who are taking an increasingly active interest in the museum, and visiting and interacting with it regularly; and secondly – direct participation, which can be seen at community museums and Eco museums (in this case, we usually refer to actors and volunteers rather than visitors).

The theoretical enquiry on care also stressed how these attitudes should be reciprocated through feedback. Tronto underlines how a response should be provided by the care-receiver to communicate that the initial needs have been actually met [8]. Following the reflection by Groys and Montanari [1, 3], this phase coincides here with the unveiling of the message of a work of art, its artistic, historical, aesthetical value together with its power to make visitors "come into contact with the 'force of the human in man': the human that is in ourselves, and in the other humans we are given to meet" [1]. A final remark pertains to potential triggers for sense of care. Albeit literature review provided meaningful food for thoughts to this regard, Sects. 1 and 2 identified several elements which may support designers and CH professionals to enhance civil society attachment to cultural heritage. As exemplified above, they can act on different care components, yet they often focus on caregivers (e.g. the visitors of a cultural institution or venue) and aim at bringing about an emotional, behavioural, or cognitive transformation. A first selection of potential triggers is provided in the following – not exhaustive – list:

- *Cultivate Empathy*, understanding emotions, experiences, and struggles of others, and let art communicate and connect with viewers on a deeper emotional level.
- *Focus on Human Stories* connected to the artwork, its subject, message, and material history. The exploration of themes of vulnerability, compassion, love, and support should resonate with viewers' emotions and life experiences.
- *Address Social and Real-World Issues*: the mediation process should address social issues that require care and attention. Raising awareness about topics may foster a sense of collective responsibility and care. "Darfur is Dying" provides an example of a game that explores care in the context of the hardships faced by refugees.
- *Foster Inclusivity and Embracing Diversity*, which can promote a culture of care and understanding among diverse audiences.
- *Use Symbolism and Metaphor*, which can be powerful tools for conveying care in artworks. Employing visual metaphors and symbolic representations that communicate the themes of care, nurturing, and emotional support can elicit contemplation and emotional responses.
- *Disseminate Cultural Heritage Conservation Methods* illustrating to the visitors the daily effort CH professionals make to preserve artworks and cultural sites.
- *Evoke Personal and Collective Memories*, which can evoke a strong sense of care and nostalgia, fostering an emotional connection that transcends time and place.
- *Inspire Action* in the audience. Engagement should motivate positive change, encouraging individuals to be more caring and compassionate in their everyday lives and communities.
- *Emphasise Nurturing and Building Relationships*, e.g. incorporating elements requiring care for virtual entities, such as animals, crops, or characters, to develop meaningful relationships with them. "Stardew Valley" and "Animal Crossing" are excellent examples of games that emphasise nurturing and fostering connections with virtual entities.
- *Introduce Emotional Storytelling*, crafting compelling narratives that tug at visitors' heartstrings and present moral dilemmas. "Undertale" and "This War of Mine" showcase the power of emotional storytelling, where players' decisions impact the well-being of characters and evoke empathy and attachment.
- *Encourage Cooperation and Teamwork*: promoting cooperation and mutual support allows visitors to form meaningful connections without traditional communication. "Journey" beautifully illustrates how cooperation with strangers can evoke feelings of care and camaraderie.
- *Showcase the Impact of Choices*: implement gameplay mechanics that highlight the consequences of players' decisions, encouraging them to think about their responsibility and the implications of their actions. The game "Papers, Please" serves as an example of a game that explores care in the context of moral dilemmas and the responsibility of making life-changing decisions for others.
- *Engage with Humour and Playfulness*, as the incorporation of light-hearted and playful elements into the game to evoke feelings of amusement and affection. "Untitled Goose Game" effectively uses humour to explore care in a playful manner.

These triggers can also be easily set in a digital ecosystem: interaction can significantly enhance the sense of care by inviting viewers to actively engage with the artwork.

Incorporating elements that encourage participation, self-reflection, and introspection can indeed deepen the emotional connection between the artwork and the audience.

4.2 PERCEIVE Care Framework

On the basis of the caring components and triggers identified in the previous section, it is possible to draft a preliminary framework adapting the multifaceted nature of the "care theory" to the purposes and goals of the PERCEIVE Project. The resulting reference definition of care expands the concept beyond emotional consideration, describing it as an: active process of protection initiated by various factors like biological needs, ethical values, and perceived urgency and which may be catalysed by specific triggers. It consists of a set of attitudes which are made by caregivers to satisfy a situation of urgency of the things or the person they care about. In response to each attitude, care-receivers provide an implicit or explicit feedback to communicate that the initial needs have been met.

Figure 9 articulates this proposal in the conclusions drawn from the state of the art. Firstly, caring for CH is conceived as a process of four main stages (labelled in this study as care components). As seen, the essential prerequisite is the presence of a relationship between caregiver and care-receiver. This bond can derive from both external factors (law or society: the "justice perspective") or internal motivations. In the previous paragraph, it has been outlined how this latter category may produce a stronger sense of care as it taps the deepest chords of the caregiver. Emotion, identity, and cognition are indeed at stake as seen in the CPKs, care often stems from emotional involvement and from potential threats to personal possessions. Moreover, a cognitive appreciation of an artwork inevitably leads to the creation of similar connections.

The essence of the care theory is then the perceived urgency to act. The frailty and the potential loss of the artwork has an impact on the caregiver who feels responsible for the care-receiver. This eventually can lead to a variety of responses, defined in literature as caring attitudes and which may considerably differ in terms of purpose, effort, or involvement. In a museological perspective this coincides with audience engagement. This concept entails multiple dimensions, describing how a visitor relates with a work of art: this spectrum ranges from simply worrying for the preservation of an object to more pro-active behaviours, such as the intellectual, cognitive, or conceptual appreciation, as well as visitor's emotional involvement. Museum studies shed light on how engagements (and hence similar caring attitudes) can be actually associated with participation. This concept, which will be better described in Sect. 5, is broad and identifies multiple contributions to cultural institutions and their mediation activities. According to Maieresse, *visitors' behaviour* plays an essential role [18]: *Encouraging visitor participation is a proactive approach distinct from simply offering cultural content for consumption. This methodology invites visitors, or the individuals comprising what is usually called 'the public', to consciously bring something of themselves to the museum in exchange for what they receive. Conversely, the museum must recognise the value that its visitors bring to its various missions.*

Lastly, the process culminates in the reward, which coincides with the phase of "care-receiving" mentioned by Tronto [8]. As shown, in the context of the cultural heritage, this can be associated with multiple outcomes: following the analysis of Groys [2, 3] and the *sentimental approach* by Montanari [1], it may correspond to the decoding of the true meaning of an artwork and its "force of the human in man" (see *supra*). This phase is implicit in the caring process *per se*, even though the caregiver may not seek an explicit reward for their attitude (as seen in the CPKs). As this ultimately contributes to strengthen the relationship between the visitor and the artwork, caring can be seen as a cyclical process:

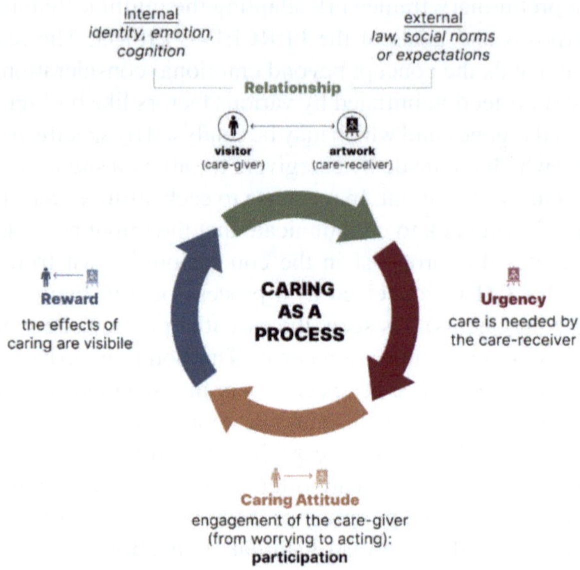

Fig. 9. Caring Components in the preliminary PERCEIVE Care Framework.

Each phase of the caring process can be catalysed by specific **triggers**. The heterogeneity of these solutions is shown in the first list presented in Sect. 4.1, while a possible mapping between triggers and components is proposed in Fig. 10. Below. In this Sankey diagram, triggers have been grouped into five core classes: in this preliminary framework each of them can operate on or more component of the framework:

- *Collaboration and Relationship*, entailing all potential actions which can foster a deeper cooperation among the visitors. Eventually, these triggers aim at bringing about deeper transformation in the audience, with potential impacts on civil society as a whole, enhancing social inclusion and cohesion (for the analysis of social cohesion in interactive experiences on Digital Cultural Heritage [19, 20]. Possible examples for this class are fostering inclusivity and embrace diversity, emphasizing nurturing and building relationships, encouraging cooperation and teamwork, having the right mindset, and redrawing in/out group boundaries.
- *Embodied Engagement*. In the Blackwell companion to Digital Humanities, Kenderdine recognised immersive and embodied experiences as the most relevant characteristics of contemporary approaches to digital cultural heritage [20]. Relying on the thought of the French phenomenologist Merleau-Ponty, who stressed the intimate relationship between perception and cognition [21], *embodiment* recognises

the impact of engaged bodily interaction with artefacts for the creation and manipulation of meanings. More specifically, "in terms of the trichotomy of the system-user-spectators, [it] implies a reciprocal relationship with the context, encompassing users, interactive systems, spectators, co-users, physical surroundings, and the meanings ascribed to these entities" [20]. For a more detailed analysis on the impact of embodiment in Digital Cultural Heritage, also refer to [23]. Inspiring action, engaging with playfulness, facilitating a creative engagement through art practices, create simulations, and simulating effort can be effective strategies to create an embodied experience. In addition, research in the domain of interaction design and human computer interaction managed to develop more organic design programs or theories on this type of engagement, e.g. [13, 24].

- *Empathy*, which, as seen in Sect. 4.1. is an essential component for the care theory as it can strengthen the relationship between caregiver and care-receiver. Studies have shown that this personal trait may be changed and enhanced to specific strategies [11, 25], also when dealing with inanimate objects or with distant people in time and space. They include perspective taking, perspective giving, active listening, focus on human stories, fostering historical empathy. In addition, in a study on visual arts, Iosifyan demonstrated the correlation between the use of "theory of mind" stories, i.e. a narration able to depict the mental state of the artist during the creation of the work of art, and an increasing aesthetic appreciation [18].
- *Meaningfulness*: as it will be later discussed in Sect. 5, visitors are seeking *meaning* in cultural experiences. Moreover, the care theory assumes that the care-receiver is important (i.e., meaningful) for the care-receiver, which is at the core of the *caring relationship*. Any trigger stressing the relevance of the artwork visitors are looking at may eventually result in stressing their responsibility to act and, thus, in a general improvement of their sense of care. As analysed throughout this chapter, these factors may concern either more personal / psychological elements (labelled in this study as *factors of internal meaningfulness*) or be related to a "justice perspective" (*factors of external meaningfulness*). Some possible examples are addressing social and real-world issue, evoking personal and collective memories, showcasing the impact of choices, focusing ethics / social norms, or getting visitors reflect on themselves (identity, emotion, and cognition) starting from the observation of the artwork.
- *Storytelling and Mediation.* An essential element in cultural heritage experiences is the narrative framework, which can catalyse profound emotional or cognitive processes in the visitor. From a museological perspective, this coincides with the concept of *mediation*, "to increase the sharing of artworks and knowledge, provide concrete solutions to socioeconomic and sociocultural inequalities in accessing culture and, by extension, encourage public participation". Traditionally, it is indeed associated with "an act of language, whether verbal or non-verbal or the means by which a heritage, museum or contemporary art professional connects an artefact (work of art, object, historical account, natural space, etc.) with an audience, in an institutionalised setting" [12]. Several techniques may be adopted to foster care, such as using symbolism and metaphors, disseminating cultural heritage, introducing emotional storytelling, filling visitors with wonder and enchantment, choosing appropriate media to convey information, creating relaxed mood, facilitating dialogue [17], highlighting the

risk of the loss or of an irremediable damage of the artefact, and improving visitors' knowledge on the work of art.

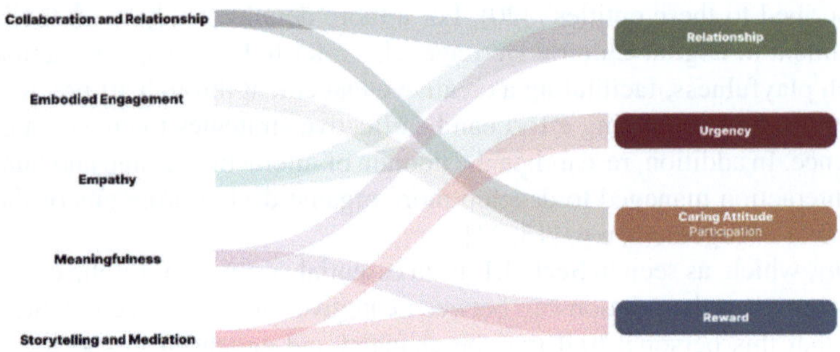

Fig. 10. Triggers of Caring

Lastly, conceiving care as a circular process provides useful food for thought in the designing and in the preparation of visitors' experiences pursuing this goal. Indeed, any component can be identified as a possible starting point. As a result, **caring relationships may be both the premise and result of a *caring prototype*.** For instance, engagement can work as an elicitor of visitors' attention towards the needs of a cultural artefact (*caring attitude*): if its meaning and relevance are effectively communicated during the experience, visitors may result in feeling a deeper connection with the artwork. In this perspective, interactive technologies offer a unique opportunity to realise a similar behavioural change.

4.3 Tool for the Design of Experiences Strengthening Visitors' Sense of Care

In analogy with the methodology followed for the study of "Authenticity" [26], an extension of the *Visitor Toolbox* has been put forward, including a new possible deck of cards for the "Audience Goals" stage and the "Ideation" phase. We include here the description of the different elements that we have included.

Stage 3. Audience Goals

1. Make Visitors Reflect on Irreversibility
2. Use Universal Concepts to Trigger Ethical Behaviours
3. Stimulate Visitors' Individual Memories
4. Encourage Introspective Contemplation for Personal Growth
5. Assign Tasks Involving Caring Actions
6. Promote Non-Verbal Interactions
7. Increase Empathy
8. Extend Knowledge
9. Boost Enchantment
10. Promote Citizens' Protection Behaviours
11. Make Citizens Learn about the Importance of Coloured Collections
12. Make Citizens Learn about Heritage Protection

13. Promote Citizens' Respectful Behaviours Towards Local Heritage and Coloured Collection
14. Make Visitors Willing to Pay for Sustainable Services and Products
15. Involve Visitors in Taking Care Activities at the Museum

Stage 5: Ideation

1. *Simulation.* Sense of awareness is stimulated in visitors through visual simulation of irreversible changes to something important.
2. *Ethic.* Visitors are involved in metaphors and stories fostering connections to ethical themes felt as priorities for them (human rights, climate, war, gentrification, etc.).
3. *Personal.* Visitors are engaged in experiences that evoke individual memories about life aspects such as family, known places, and relevant objects.
4. *Self-reflection.* Visitors are involved in experiences that prompt introspective thoughts and inner exploration encouraging self-awareness.
5. *Action.* Experience encourages the visitor to use symbolic tools or resolve tasks performing caring actions to something or someone.
6. *Embodiment.* In the experience, visitors' body movements can trigger actions and events, enhancing cognitive processes through non-verbal interactions.
7. *Historical Characters.* A narrative that mixes historical facts and evocative plot empowers a deep emotional connection between visitors and cultural heritage.
8. *Empathy.* The visitors spend time in contexts unusual to them or outsiders, doing theme-specific activities (volunteering in museum guides, laboratories, workshops, etc.).
9. *Perspective Taking.* The visitor adopts another person's viewpoint improving understanding and empathy from a diverse perspective (different expertise, social group, religion etc.).
10. *Active Listening.* A specific character's story is presented to the visitors which are actively invited to empathise with the protagonist's sentiments and thoughts.
11. *Theory of Mind Stories.* Visitors are told a story with characters acting with specific behaviours and are invited to imagine motivations, feelings and sentiments behind these characters' actions.
12. *Perspective Giving.* Visitors are invited to give their perspective on certain arguments.
13. *Having the Right Mindset.* People's Empathy isn't immutable: it's a fluid trait that could be increased. It can be reinforced by proposing to visitors, stories with deep-changing characters and self-improvement processes.
14. *Redrawing In/Out Group Boundaries.* Visitors imagine themself acting inside (increasing their sense of belonging) or outside a group (reflecting and reacting to a sense of rejection).
15. *Social Norms.* Visitors are informed about the emphatic actions and reactions of others strangers, producing a positive mirror-effect on their behaviour.
16. *Appropriate Media to Convey Information.* Information that is far from visitors' understanding (too complex data and articulated reasoning) are humanized by adopting specific media and correct communication strategies.
17. *Art Practices.* Visitors are involved in artistic activities crafting self-made artworks or and collaborative performance.

18. *Relaxed Mood.* Provide a relaxed and positive environment to the visitors adopting strategies to avoid anxiety and reduce preoccupations that otherwise may reduce empathy.
19. *Enchantment.* Develop a sense of wonder in the visitors.
20. *Facilitated Dialogue.* Include active dialogues with experts/guides and among participants; provocative questions and alternative perspectives are proposed to visitors.

5 Participation: A Care Component in Cultural Interactive Experiences

A crucial contribution for the understanding of *participation* in the museum and cultural context has been put forward by Nina Simon, with the publication of the *Participatory Museum* [17]. Her definition relies on an audience-centred institution, where visitors can "construct their own meaning from cultural experiences" (II). User's voice should hence be listened to and guide exhibition and interactive experiences design in the cultural sector. In this perspective, traditional approaches are overcome as (2):

In traditional exhibits and programs, the institution provides content for visitors to consume. Designers focus on making the content consistent and high quality. (...) In contrast, in participatory projects, the institution supports multi-directional content experiences. The institution serves as a "platform" that connects different users who act as content creators, distributors, consumers, critics, and collaborators.

Therefore, thanks to this participatory approach, museums have become a system of relations among the institution, its personnel, the collection and the audience. Therefore, not only are museums the places where our cultural heritage is preserved, but also catalysers of social experiences. As a result, *musealia* are considered as social objects, i.e. "engines of socially networked experiences, the content around which conversation happens" [27]. In this perspective, the importance of direct involvement of visitors is indeed crucial, as "demonstrations that involve 'guests from the audience' or encourage small groups of visitors to handle objects allow visitors to confidently connect with objects in a personal way" (152). To allow audiences a more significant participation in cultural institutions, direct interaction with the collection is hence essential. In addition, the participatory experience may rely also on other approaches to ignite an active critical process in the visitors. For instance, asking questions is by far one of the most implemented. The author specifies that *perspective-giving* strategies may be effective only if they fulfil two main requirements (140):

- The question is open to a diversity of responses. If there's a "right answer," it's the wrong question.
- Visitors feel confident and capable of answering the question. The question draws on their knowledge, not their comprehension of institutional knowledge.

Recent technological advances have afforded new participatory and interactive approaches to knowledge and increased the tangibility of cultural objects and ideas. Some of these approaches have allowed for the combination of cognitive, sensorial, historical, and artistic aspects into one experience. As such, the advent of new technologies has paved new avenues for tackling education about cultural artefacts, their interpretations, and their implications. This is especially true of coloured cultural artefacts. In this section, we will investigate the ways *digital technologies can be used in an interactive media participatory design perspective.* In particular, we believe that a fruitful approach is through participatory and interaction design.

The attempt to more fully integrate the use of technology with the materiality of a heritage display is part of a wider line of debate on making the museums and their artefacts more engaging for visitors, overcoming the separation between technology and heritage holdings. Increasing interest in the physicality of the experience with cultural objects (hybridity) is also shown by many works in museum studies [28, 29] as well as the rise of public installations focused on multiple levels of engagement [30, 31]. Petrelli and colleagues [31] believe that "the digital and the material can become components for the design of a holistic visitor experience that crosses the digital-material boundary. The challenge is in weaving the digital and the material to create seamless immersive and novel visitors' experiences". There are numerous recent examples of new museological approaches that take a decidedly multi-disciplinary approach with human and ludic factors in computing, such as adding content layers on exhibitions through technology, enhancing audience immersion through digital techniques, developing virtual museums and serious games [32], personalising hybrid museum experiences through digital gifting [33], controlling interactive exhibitions through the use of tangible smart replicas or tangible data souvenirs, and more.

5.1 Alternative Texturizations and Personalization: ColourColab (2020)

In 2020, a prototype for an interactive art installation was designed within a Master's thesis of Digital Humanities and Digital Knowledge (DHDK) at the University of Bologna: *ColourColab* [34]. *ColourColab* was a proposed interactive installation cantered around colour and multiple classical sculptures or their copies. The experience would enable users to select one or several polychromy visualizations of the sculptures, either through an online tool or an augmented reality (AR) interface (Fig. 11). A range of pre-made palettes would be available in the lab for users to explore and experiment with. Some palettes are grounded in scholarly research on ancient pigments. Another category of pre-made palettes is based on iconic paintings, art history masterpieces, or artist-commissioned works. A third type of available palettes is inspired by the top fabric patterns of the year and the most prevalent colours and trends in the modern fashion world. Additionally, a database of user-uploaded palettes is provided for inspiration. Importantly, users would have the option to dynamically create their own palette, including colours from their own skin, body, clothes, and surroundings. *ColourColab*'s educational purpose in the context of museums relates to the notion of awakening, which aims to arouse curiosity, to lead to questioning and developing the capacity to think.

5.2 Fostering Caring Attitudes and Participation Through Art Practices

As seen, sense of care is a broad concept: it entails several attitudes and interacts with multiple possible cognitive goals for interactive applications. Recent studies stressed the essential role of curiosity, as "we are curious for things we care about" [35]. Curiosity (see

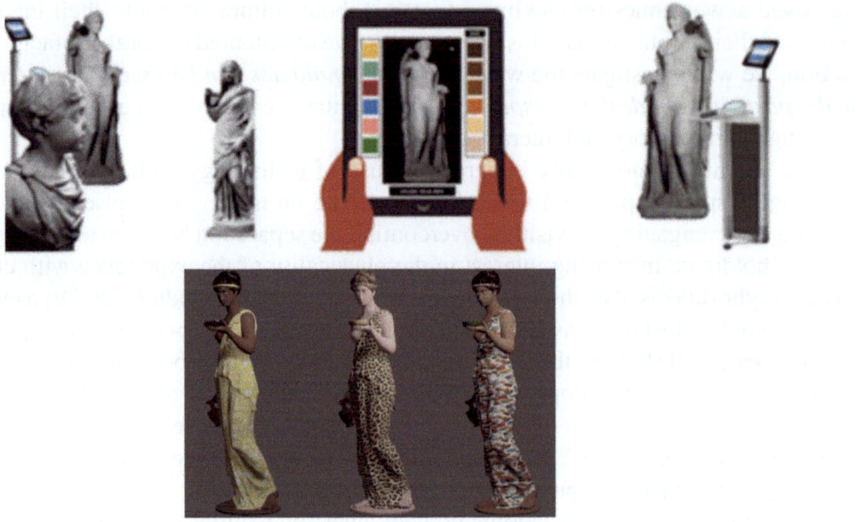

Fig. 11. The design of the ColorColab experience, allowing users to explore colourisation of multiple sculptures with a variety of palettes that the user may choose from or create.

also Chap. 5) has been indeed defined as a quest for information to close a knowledge gap [36]. However, if this gap is enhanced and put into relations with high level "Concepts", which embrace "life themes" of "high levels of intensity" and "self-related information", interest can be elicited, which may succeed in catalysing a deeper behavioural change in the visitor [36].

In turn, triggering interest towards an artwork within an interactive application may succeed in allowing the audience to live a meaningful and authentic experience. Chen and Rahman [37] indeed showed that visitors intrinsically seek for meaning as meaning-making is proved to facilitate the understanding of what has been lived. More specifically, meaning-creation is inherently linked to the establishment of mental connections [38]. This active process is also considered to stem from several needs: among these, major emphasis has been laid on the key terms of Self and the identity.

This pursuit of interest and meaning creation processes as a cognitive goal of an interactive application share with contemporary musicological approaches the emotional and psychological involvement of the visitors, who should be able to "construct their own meaning from cultural experiences" [17]. The proven effectiveness of hands-on activities in this domain paved the way to Bilda and colleagues' study on creative engagement, i.e. the adoption of art practices at the basis of visitors' participation: this solution entails a *transformative dialogue* between the user and the interactive systems and results in "a shift in intentions and expectations", associable also to an actual behavioural change [39].

To verify the possibility of implementing a similar approach to polychrome's collections, a first caring prototype is being developed, which focuses on a peculiar asset of museum collection, i.e. conservation data. They are indeed a precious resource, enabling visitors to explore hidden details lying under the pictorial film. Nonetheless, also when exhibiting these objects, museums often rely on a monodirectional top-down approach: specific details are for instance shown to the visitor, who is now able to read in that specific fragment of data the history of the painting (the most intuitive case is compositional arrangements implemented by the painter). Albeit it might succeed in the acquisition of limited pieces of knowledge, this approach ignores other potentially more participatory strategies to the works of art, which may elicit visitors' sense of care and interest, as well as, eventually, catalyse collaborative processes of meaning creation. In this analysis, we have tried to exploit the current research in this field to guide the design of a first UX to pursue these goals in the cultural sector. The case study has been drawn from the collection of an Associate Partner of the PERCEIVE Project, the Art Institute of Chicago (AIC) and is a masterpiece of Fauvism, an historical avant-garde who gave a prominent role to the semantic of colours: *Bathers by a River* by Matisse (1869 – 1954). The painting, which has been thoroughly described by D'Alessandro [40], was conceived together with *La Musique* and *La Danse II* and is characterised by a difficult gestation (1909 – 1917). Physicochemical analyses (X-Radiography and cross-section photomicrographs) demonstrated in fact the presence of at least six versions: the bright, "aggressive", typically Fauvist colours were turned in darker and paler hues, and the position of the characters and of the natural elements was radically changed. Among the most plausible justifications for this change, two are the most convincing: i.e. the contact with Cubism and African art, and the historical context. In fact, the Great War might have had a serious impact on the painter's psychological status and concerns might be reflected also in the choice for darker hues and in a more rigid and vertical position of the female figures.

The developed prototype, named *MyTISSE*, focused on the X-radiography (Fig. 12) as a starting point for creative engagement, as participants are expected to draw their reinterpretations (which results in a trigger of type "Embodied Engagement", see 4.4) directly on an image of X-Ray scan[1]. A first testing session was hence organized to prove the soundness of the initial research question on a qualitative sample of twenty-four master students, PhD candidates and young researchers of the field of digital humanities at the University of Bologna. Participants, who had no previous knowledge of the painting, were asked to read and sign an Informed Consent form and they were given a tablet and an interactive pen. The device contained the application Ibis Paint X with a preloaded image of the X-Ray scan and two-colour palettes, reflecting the initial (reconstructed) and the final version of the painting (see a more detailed description of the testing protocol and of the retrieved results [41]). The initial drawing and painting phase being concluded, participants were provided a brief explanation on the history of the painting (which, in the PERCEIVE Care Framework, corresponds to the trigger "Storytelling and Mediation" acting on the "Reward" component) and were later asked to answer a brief summary. The data were later saved as *.csv file and processed in Python.

The huge differences in the reinterpretation (e.g. Fig. 13) show how the activity is participatory as "open to a diversity of responses" and the obtained results from the survey eventually show how through creative engagement it is possible to foster audience's care towards the collection and impact on visitor beliefs. The survey indeed asks visitors to summarise the experience and, secondly, provides a selection of relevant elements,

[1] https://archive.artic.edu/matisseinteractive/p0.html.

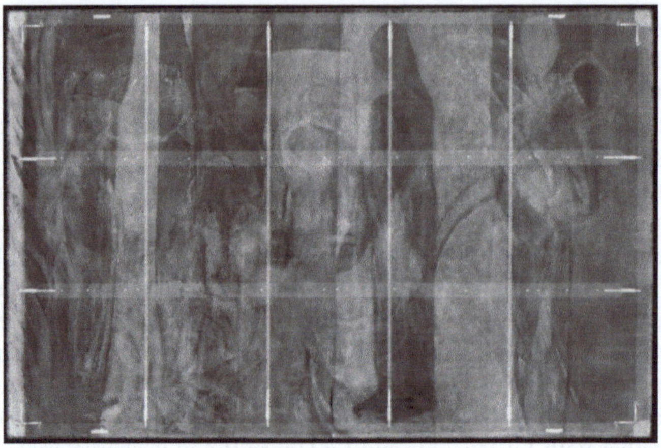

Fig. 12. X-radiograph of Matisse, *Bathers by a River*, after [40] © Art Institute of Chicago

whose salience should be assessed by the participants, re-ordering them from one to five: these tasks are aimed at detecting which aspects of the experience are perceived as more important.

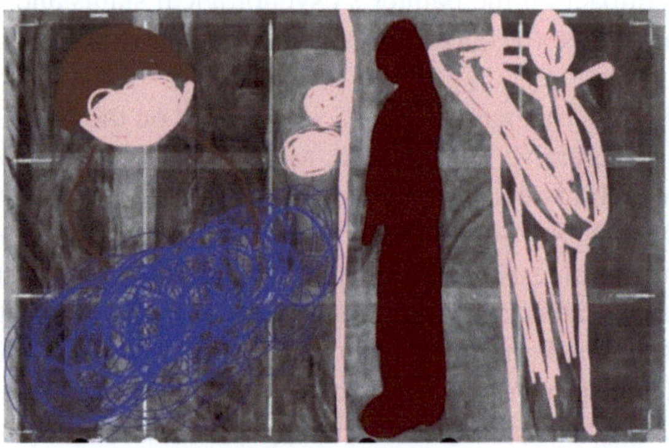

Fig. 13. Reinterpretation of Matisse, *Bathers by a River*, Participant C3

The results are hence consistent with the initial hypothesis: the most frequent words in the summaries are *painting*, *Fig. / shape* (here considered synonyms, given the context), *colour* and *interesting*. Similarly, the answers to the second question show the importance of the change of colour, the possibility to reconstruct the different versions of the painting through conservation data and the impact of the war in the composition of the *Bathers*. On the contrary, a secondary role was played by the contact with other artistic avant-gardes (e.g., Cubism) and the fact that it was conceived together with other two famous paintings. To efficiently assess the development of caring relationships with the artifact, a fictitious scenario was provided in the survey: the artwork would be undergoing an irremediable corrupting process and, in a few years, the damages to the pictorial film may be unrepairable. Participants are asked to state how sorry they would be for this loss on a Likert scale (values from 1 to 5). Their answers demonstrate anew the **creation of an intense sense of care**, since all the values are either high or extremely high. The survey also invited participants to describe the major takeaways from this experience: almost one participant out of four indeed claimed to have received a benefit. In addition, a closer look at visitors' feedback shows how *MyTISSE* succeeded in filling knowledge gaps and questioning visitors' beliefs on art, colour, and cultural experiences, as shown by the following answers:

- *Yes. I think it helped me to approach art in a new, different way, a way I had never thought of before. I'll probably attend exhibitions and visit museums with a different mindset from now on!! (Participant A1).*
- *It helped me to fill the curiosity that this exercise arose (Participant A4).*
- *It was helpful - also through the use of this questionnaire- to put down the feelings that this painting gives me, because I often do not consciously think about that when I look at pieces of art. (...) It is inspiring, thought provoking and admirable the way a painting of a casual, carefree activity such as bathing by a river, brought me a sense of sadness and fear considering the style of the painting but especially the context of the artwork. (Participant A6).*
- *I think it did somehow. I think that it has helped me with questioning my own perception of colour and of different ideas. It certainly has me question my own idea of art and colour (Participant C2).*
- *It helped me in developing a sense of care for art and overcoming stereotypes about styles and colours associated with an author. (Participant C4).*

This enthusiasm towards this kind of activity is also mirrored by other two explicit questions, which reported a unanimous agreement: the first one concerned the possibility for museums to encourage similar activities, while the second one asked to the participants if they would be curious to see the reinterpretations of the other visitors. In particular, this latter result paves the way to possible pursuit of **collective participation**, a direction which is particularly promising to convert the cultural institution into "a social place, full of potentially interesting, challenging, enriching encounters with other people" [17].

6 Examples of Digital Experiences Enhancing Sense of Care

To verify the soundness of the framework described in Sect. 4, in the following paragraph we scrutinise a restricted selections of different projects and multimedia products (part of them have been already mentioned in Sect. 2), exemplifying different strategies through which sense of care towards historical sites and CH – and in particular coloured collection - may be strengthened. In this overview, the Sankey diagram proposed in Fig. 10 Has been adopted as a useful guide to identify the most relevant features of these projects.

Title. Animal Crossing: New Horizons & The Metropolitan Museum of Arts
Description. *Animal Crossing*, by far one of the most beloved videogames, is a social simulation game, where the player can choose an avatar. Thanks to a collaboration with the Metropolitan Museum of Arts in New York, players can now customise their virtual

homes directly browsing from the collection of the museum. The Digital Department of this institution has indeed released in open access more than four hundred thousand images of their collection, alongside with other institutions as the Cincinnati Art Museum or The Getty.

Caring Trigger. ***Meaningfulness*** As other social simulation video games, *Animal Crossing* encourages players to take care of their own rural villages and homes. The main assumption of this joint project is that by using these paintings, even just for a simple decorative purpose, players can establish a potential caring relationship with them: artworks are now considered as players' own possessions. As seen during the analysis of the results of CPKs, this factor can be considered as a trigger of *internal meaningfulness.*

Reference: https://www.metmuseum.org/it/perspectives/animal-crossing-new-horizons-qr-code.

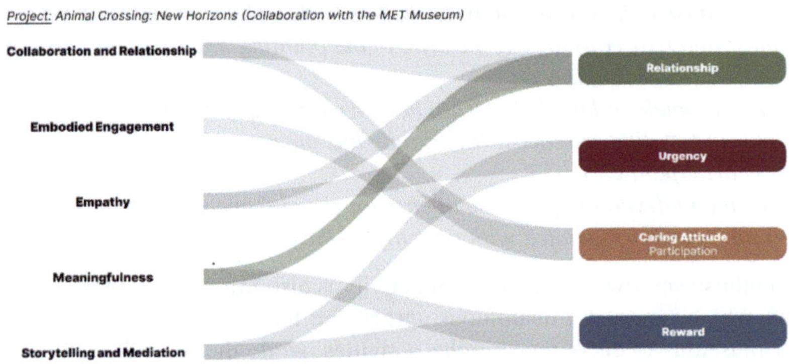

Fig. 14. Visualisation of caring triggers and components in Animal Crossing: New Horizons (joint project with the Metropolitan Museum of Art in New York)

Title. Colour The Temple (2015) – Chroma AR (2022)
Description. The Metropolitan Museum of Art has also developed interactive experiences on coloured reconstructions of ancient art and architecture, developing a tool that used projected light to digitally restore polychromy on The Temple of Dendur. Besides, the projection also presented animated versions of the stories depicted on the Temple. After this first experience, the MET has developed a new AR experience for the visitors of the exhibition the *Chroma*, with reconstructions of ancient sculptures and introducing a new polychromy reconstruction of the MET's Archaic-period Sphinx finial.

Similar Projects. In the past decade, several projects have been proposed to access lost polychromy of archaeological findings and ancient artworks. Possible examples have been already mentioned in Sect. 3.6: *The Colours of the Ara Pacis* (2014), *L'Ara com'era* (2017), and different exhibits of *Keys to Rome* [42]. Concerning this latter exhibition, the two installations by the MET are similar to *Revealing Flashlight* (see again Sect. 3.6) and *AR-tifact*, an AR installation[2] on iPad through which the visitors could see a cultural artifact or a fragment of its, together with its reconstructive details, such as the polychromy (Fig. 15).

Caring Triggers/Component. Embodied Engagement. AR and gesture-based applications fit well with the definition of embodiment (4.4) and are increasingly being adopted by museum to allow visitors to unveil the hidden secrets of their collection. Sarah Kenderdine provided a thorough description of how a similar AR application perfectly exemplifies embodied museography, focusing on the *Pure Land AR* exhibit, which allows visitors to discover the inaccessible Mogao Grottoes (Gansu Province, China) by means of a tablet [20]. The relationship between similar installations and the PERCEIVE caring framework finds support in the study by Groys on curatorial practices (see Sect. 1). Here the author states that "the artwork needs external help, it needs an exhibition and curator to become visible" and that "the medicine that makes the sick image appear healthy - makes the image literally appear and do so in the best light - is the exhibition". In these projects, embodied engagement is a caring practice because it allows visitors and curators to work together to restore and make *visible* a feature of the artwork that would otherwise be lost forever.

Reference: https://www.metmuseum.org/it/perspectives/color-the-temple (Colour the Temple); https://www.metmuseum.org/it/exhibitions/listings/2022/chroma/chroma-ar (Chroma AR).

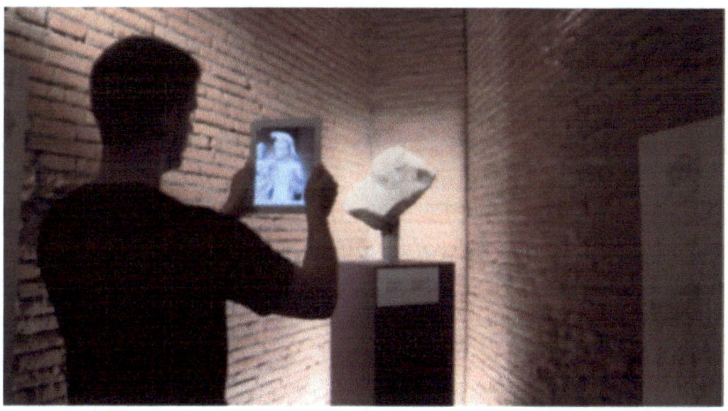

Fig. 15. AR-tifact AR mobile applications

As shown in. Fig. 14 and Fig. 16, these projects seem to implement one single trigger (or class of triggers) to ignite a reaction of the visitor on one of the four components of the PERCEIVE care framework. The following examples, on the other hand, show how this differentiation between classes of triggers is not immediate and can be milder. In particular, video games and narrative experiences in which historical or cultural heritage plays a major role (and not only a "decorative" purpose as in *Animal Farm*) provide food for thought in this perspective. In addition, these two examples will show how these multiple triggers are not isolated: on the contrary, a first trigger may catalyse an action or a behaviour in the visitor which in turn serves as a new trigger for another component of the framework.

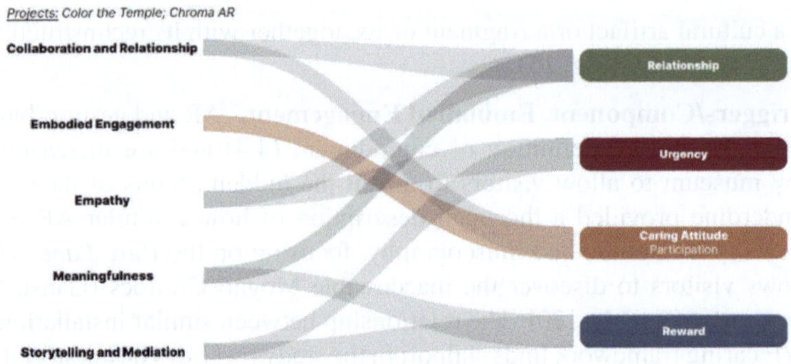

Fig. 16. Visualisation of caring triggers and components in Colour the Temple and Chroma AR

Title. Father and Son
Description. Launched by Archaeological Museum in Neaples (MANN) in collaboration with TuoMuseo and Università della Campania Luigi Vanvitelli in 2017, it is the first worldwide museum game. It tells the story of a MANN archaeologist and his son he has never met, Michael, who is also the main character of this game. This adventure through the museum takes the player through different historical contexts, corresponding to the three main collections of the museum, i.e. Pompeii, Ancient Egypt and the Farnese Collection.

Caring Triggers/Component. *Father and Son* obtained a considerable success in the audience, which is in great part due to the deeply emotional narrative framework. It has been indeed outlined that [43]: "*The video game is characterized by a preeminent narrative dimension, which supports and accompanies the journey of the main character, a journey that is not only physical, but also, and above all, psychological: the pain for the loss of a hitherto distant and little understood parent is in fact diluted through the emotions and feelings of universal value that the protagonist experiences during this journey. (…) This emotional aspect turned out to be of great importance for its success, arousing the involvement of the user and, often, a deep empathy between the player and the protagonist of the story*". As in traditional first-person games, also in *Father and Son* the visitor identifies with the protagonist: the overlapping of the player's identity with the young orphan Michael lays the basis for the strong ***empathic bond*** described by Solima above. This is an essential condition as this encourages the visitor to proceed in the subsequent phases of the game. ***Storytelling*** is thus used either to describe a situation of loss or to set new challenges: Michael (and the player / visitor) feels responsible for completing these tasks and is therefore encouraged to act. The interactive experience as a whole provides in turn a new trigger: the direct involvement is a factor of ***internal meaningfulness***. This ultimately strengthens the relationship between the player and the protagonist of the story and "to a certain extent, by extension, - as claimed by Solima [43] - between the player and the museum".

Reference: https://mann-napoli.it/father-and-son-the-game/

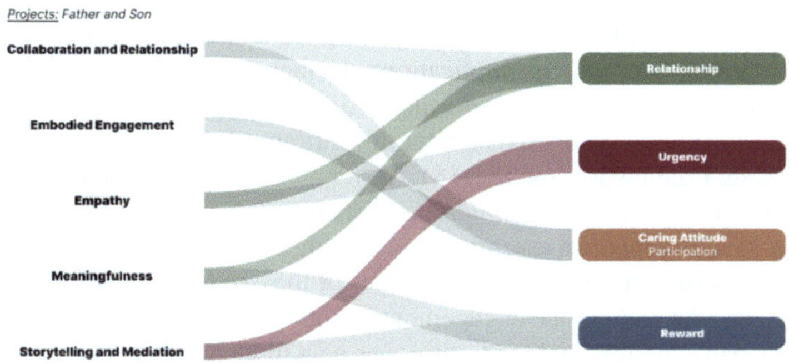

Fig. 17. Visualisation of caring triggers and components in Fathers and Son

Title. "Memoria EMigrazioni"

Description. The Galata Museo del Mare (Sea Museum) in Genoa, Italy recently opened a new section "Memoria EMigrazioni" (Memories and Migrations). This section delves into the history of migration in Italy, starting from the Italian emigrants to North and Latin America in the 19th and 20th centuries to the current immigrants from Africa and Asia. The permanent exhibition combines archival document, objects belonged to the migrants, and interactive installations. In one of these a vertical screen portrays an officer in the Registry Room at Ellis Island in New York Harbour: this was where migrants were inspected and, if they did not meet all the criteria for admission, they could be rejected and forced to return home. The visitor is here asked to answer some specific questions in the shoes of an Italian migrant at the beginning of the last century. The entire museographical project of this exhibition is based on a strong identification between the visitor and the migrants. The experience commences with visitors receiving a replica passport once/which belonged to one of the many migrants whose story and journey across the ocean is today documented in the archive. Immediately adjacent to the aforementioned video installation, which features orange walls (Fig. 17), are drawers labelled with the names of migrants. When these drawers are opened, the migrant stories are revealed, corresponding to the respective passports' name. Finally, the experience concludes with an interactive kiosk that enables visitors to engage with the archives of the CISEI, the International Center for Studies on Italian Emigration (http://www.memoriaemigrazioni.it/), and to find out whether any of their ancestors emigrated to the America in the past centuries.

Similar Projects. The videogame *Paper Please* (https://papersplea.se/) offers a similar experience: the player is an immigration inspector controlling the flow of people entering the Arstotzkan side of Grestin from Kolechia. His job is using only the documents provided by travellers and the Ministry of Admission's primitive inspect, search, and fingerprint systems to decide who can enter Arstotzka and who will be turned away or arrested. The theme of migration is moreover at the centre of the VR experience *The Key*, discussed in Sect. 2

Caring Triggers/Component. The theme of migrations, presented in a complex narrative and interactive visit in "Memoria EMigrazioni" starting from single museum objects and archival documents, is of topical relevance. As a matter of fact, it merges different perspective and food for thought deriving from history, social sciences and has profound social and geopolitical implications. In their participatory nature, museums and cultural institutions become platforms of discussion of these themes and design complex experiences which involve almost the entirety of the triggers described in the PERCEIVE Care Framework (see Fig. 18). As in *Father & Son*, a deep identification and the establishment of a *strong empathic* bond between the visitor and the protagonist of the narration, i.e. migrants, are essential. Storytelling enables to strengthen this intimate relationship and structures the different sections of the exhibition, inviting the audience to directly engage throughout the engagement. Quoting Groys, also in historical heritage museums this participation enables to discover new hidden elements on the life of the migrants: they provide again triggers of *internal meaningfulness*, which strengthen the relationship between the visitor and the museum. Owing to the peculiar nature of this collection, this resulting intimate relationship provides important food for thought for the audience, which in the future is expected to be more empathetic for the situations of migrants and refugees. As anticipated in Sect. 4.2, similar behavioural triggers relying on the establishment of interpersonal relationship are potentially aimed to a stronger social cohesion.

Reference: https://www.metmuseum.org/it/articles/animal-crossing-new-horizons-qr-code.

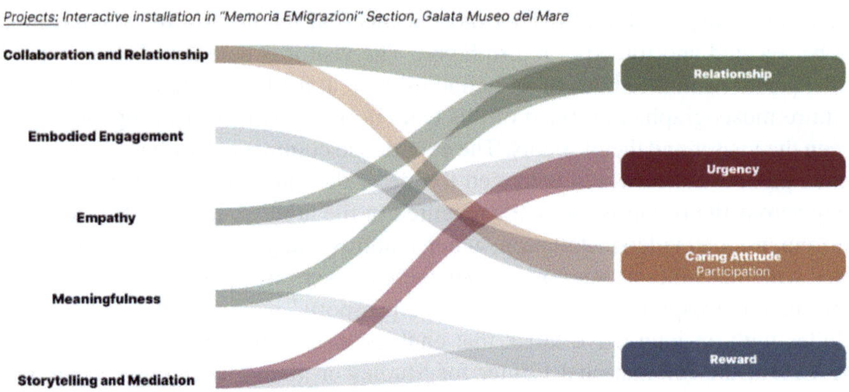

Fig. 18. Visualisation of caring triggers and components in "Memoria & Migrazioni"

Acknowledgments. This study was funded by the PERCEIVE project that received funding from the European Union's Horizon research and innovation programme under grant agreement No 101061157.

Authors Contributions. Gabriele Fiorenza, Delfina Sol Martinez Pandiani, Sofia Pescarin, Manuele Veggi contributed to Sect. 1; Arthur Clay contributed to Sect. 2; Gabriele Fiorenza, Manuele Veggi, Sofia Pescarin contributed to Sect. 3; Arthur Clay, Sofia Pescarin and Manuele

Veggi contributed to Sect. 4; Delfina Sol Martinze Pandiani, Sofia Pescarin, Manuele Veggi contributed to Sect. 5; Vanessa Bonanno, Ivana Cerato, Delfina Sol Martinez Pandiani, Manuele Veggi contributed to Sect. 6.

Disclosure of Interests. The authors have no competing interests to declare that are relevant to the content of this article.

References

1. Montanari, T.: Se amore guarda: Un'educazione sentimentale al patrimonio culturale. Einaudi, Torino (2023)
2. The Letter to Leo X by Raphael and Baldassare Castiglione, c.1519. In: Vaughan H, Hicks P. Palladio's Rome: A Translation of Andrea Palladio's Two Guidebooks to Rome. Yale University Press, New Haven (2006)
3. Groys, B.: Filosofia della cura. Timeo, Palermo (2023)
4. Heidegger, M.: Being and Time. Blackwell, Malden (Mass.), p. 228 (2007)
5. Andre, C., Velasquez, M.: Men & women; justice & compassion. Issues Ethics **3**(1) (1990). https://www.scu.edu/mcae/publications/iie/v3n1/men.html
6. Soto, N.E.: Caring and relationships: developing a pedagogy of caring. Villanova Law Rev. **50**(4), 859 (2005)
7. Noddings, N.: Educating Moral People: A Caring Alternative to Character Education. Teachers College Press, Williston (2002)
8. Tronto, J.C.: Moral Boudaries: A Political Argument for an Ethic of Care. Routledge, New York (1993). https://doi.org/10.4324/9781003070672
9. Groys, B.: On the Curatorship. In: Art Power. MIT Press, Cambridge (2008)
10. Bazalgette, P.: The Empathy Instinct: How to Create a More Civil Society. Hachette UK (2017)
11. Zaki, J.: The War for Kindness: Building Empathy in a Fractured World. Crown, New York (2019)
12. McKinney, S., et al.: Developing digital archaeology for young people: a model for fostering empathy and dialogue in formal and informal learning environments. Commun. Past **12**, 179 (2020). https://doi.org/10.5334/bch.n
13. Pescarin, S., et al.: Hybrid XR collaborative and guided experiences in cultural heritage: brancacci POV prototype. In: Bucciero, A., et al. (eds.) GCH23: Eurographics Workshop on Graphics and Cultural Heritage. The Eurographics Association. https://doi.org/10.2312/gch.20231171 (2023)
14. Chaskin, R.J., Rauner, D.M.: Youth and caring. Phi delta kappan **76**(9), 667–674 (1995)
15. Pescarin, S., et al.: PERCEIVE cultural probe kit on caring attitude in XR experiences: activity book, diary and art practices for designers to explore and solicit care in cultural heritage domain. In: Zenodo (2023). https://zenodo.org/records/8285611
16. Burrows, A., Val, M., Colette, N.: Cultural probes and levels of creativity. In: Proceedings of the 17th International Conference on Human-Computer Interaction with Mobile Devices and Services Adjunct (MobileHCI '15), pp. 920–923. Association for Computing Machinery, New York (2015)
17. Simon, N.: The Participatory Museum. Museum 2.0, Santa Cruz (2010)
18. Mairesse, F.: Dictionary of Museology. Routledge, London (2023). https://doi.org/10.4324/9781003206040
19. Pescarin, S., et al.: social cohesion in interactive digital heritage experiences. Multimodal Technol. Interact. **76**(6), 61 (2023). https://doi.org/10.3390/mti7060061

20. Kenderdine, S.: Embodiment, entanglement, and immersion in digital cultural heritage. In: Schreibman, S., Siemens, R., Unsworth, J., (eds.) A New Companion to Digital Humanities. John Wiley & Sons, Chichester (2015). https://doi.org/10.1002/9781118680605.ch2
21. Merleau-Ponty, M.: Phenomenology of Perception. Routledge, London (2006)
22. Dourish, P.: Where the Action Is: The Foundations of Embodied Interaction. MIT Press, Cambridge (Mass.) (2001). https://doi.org/10.7551/mitpress/7221.001.0001
23. Johnson, M.: The Meaning of the Body: Aesthetics of Human Understanding. University of Chicago Press, Chicago (2010). https://doi.org/10.7208/chicago/9780226026992.001.0001
24. Höök, K.: Designing with the Body: Somaesthetic Interaction Design. MIT Press, Cambridge (Mass.) (2018). https://doi.org/10.7551/mitpress/11481.001.0001
25. Iosifyan, M.: Theory of mind increases aesthetic appreciation in visual arts. Art Percept. **9**(2) (2021). https://doi.org/10.1163/22134913-bja10011
26. Pescarin, S., et al.: Authenticity in the design of digital heritage applications. In: Pescarin, S., Barandoni, C., Clay, A., Papadopoulos, G., Sandu, I.C.A. (eds.) Chromatic Visions. LNCS, vol. 15750, pp. 81–127. Springer, Cham (2026). https://doi.org/10.1007/978-3-032-07792-9_3
27. Dann, R.: Touch in museums: policy and practice in object handling. Senses Soc. **7**(1), 99–102 (2012). https://doi.org/10.4324/9781003135616
28. Dudley, S.H.: Museum materialities: objects, sense and feeling. In Museum materialities. Routledge, London (2013)
29. Brignull, H., Rogers, Y.: Enticing people to interact with large public displays in public spaces. In Interact **3**, 17–24 (2023)
30. Ciolfi, L.: Supporting affective experiences of place through interaction design. CoDesign, **3**(S1), 183–198 (2007). https://doi.org/10.1080/15710880701309308
31. Petrelli, D., et al.: Exploring the aesthetics of tangible interaction: experiments on the perception of hybrid objects. In: Proceedings of the TEI'16: Tenth International Conference on Tangible, Embedded, and Embodied Interaction, pp. 100–108 (2016). https://doi.org/10.1145/2839462.2839478
32. Pescarin, S., Martinez Pandiani, D.S.: The impact of story structure, meaningfulness, and concentration in serious games. Information **13**(12), 567 (2022). https://doi.org/10.3390/info13120567
33. Back, J., et al.: GIFT: Hybrid museum experiences through gifting and play. In: Workshop on Cultural Informatics co-located with the EUROMED International Conference on Digital Heritage 2018 (EUROMED 2018), Nicosia, Cyprus, November 3, 2018, vol. 2235, pp. 31–40 (2018)
34. Martinez, P.D.S.: A Collaborative Colour Laboratory: Using 3D Modelling, Texturization, and AR to Challenge White Supremacist Uses of Ancient Classical Sculptures. In Building Community Online (2020)
35. Phillips, R.: Curiosity: care, virtue and pleasure in uncovering the new. Theory Cult. Soc. **32**, 149–161 (2015). https://doi.org/10.1177/0263276414565718
36. Hidi, S.E., Renninger, K.A.: On educating, curiosity, and interest development. Curr. Opin. Behav. Sci. **35**, 99–103 (2020). https://doi.org/10.1016/j.cobeha.2020.08.002
37. Chen, H., Rahman, I.: Cultural tourism: an analysis of engagement, cultural contact, memorable tourism experience and destination loyalty. Tourism Manage. Perspect. **26**, 153–163 (2018). https://doi.org/10.1016/j.tmp.2017.10.006
38. Baumeister, R.F., Vohs, K.D.: The pursuit of meaningfulness in life. In: Snyder, C.R., Lopez, S.J. (eds.) Handbook of positive psychology, pp. 608–618. Oxford University Press, Oxford (2002)
39. Bilda, Z., Edmonds, E., Candy, L.: Designing for creative engagement. Des. Stud. **29**, 525–540 (2008). https://doi.org/10.1016/j.destud.2008.07.009

40. D'Alessandro, S.: Matisse Paintings, Works on Paper, Sculpture, and Textiles at the Art Institute of Chicago. Art Institute of Chicago, Chicago (2019)
41. Veggi, M., Pescarin, S.: Creative engagement and meaning creation: a first experimental protocol on interactive cultural experiences for conservation data. Digit. Appl. Archaeol. Cult. Heritage **32**,(2024). https://doi.org/10.1016/j.daach.2024.e00321
42. Pescarin, S., (ed.) Keys to Rome. Roman culture, virtual museums. CNR ITABC (2014)
43. Solima, L.: Il gaming per i musei. L'esperienza del Mann. Economia della Cultura **3**, 275–290 (2018). https://doi.org/10.1446/91289

Extending Civic Participation (in Public Spaces)

Arthur Clay[1](✉), Vanessa Bonanno[2], Federica Bonifazi[2], Chiara Gemma Fedon[1], and Sofia Pescarin[2]

[1] HSLU, Lucerne, Switzerland
{arthur.clay,chiara.fedon}@hslu.ch
[2] CNR ISPC, Florence, Italy
{federicabonifazi,sofia.pescarin}@cnr.it

Abstract. In this chapter we analyse how digital applications dedicated to coloured collection could enhance civic participation through the Open Space Museum prototype, curiosity and social cohesion. The first section deals with the Open Space Museum, designed by Arthur Clay and N55, that features a portable, modular exhibition showcasing diverse artworks to foster public engagement and artistic interaction. It includes Case Study and Chroma Elements, AR technology, and engagement applications, creating a versatile, immersive, and pandemic-ready cultural space that extends traditional museum hours. The second part focuses on how curiosity can foster civic participation by prompting exploration and learning. It transforms from a negative trait to a valued tool for motivation, encouraging people to resolve uncertainties and seek knowledge. Berlyne's concept of curiosity defines it as a response to environmental complexity and gaps in knowledge, leading to active inquiry. This positive feedback loop aids engagement in cultural heritage, supported by PERCEIVE's curiosity-driven interactive design. CNR researchers conducted a four-day campaign at MANN in June 2023, digitizing statues with polychromy traces and analyzing public curiosity. Visitors participated in guided tours, viewing diagnostic techniques and 3D acquisition, which sparked high interest in ancient sculpture coloration, diagnostic methods, and digital reconstruction. The last section is dedicated to Social cohesion, essential for stable relationships in society. Encouraging civic participation through cultural activities strengthens democracy and fosters community connections, bridging social boundaries and empowering citizens to feel engaged and cohesive within their communities.

Keywords: Participation · Curiosity · Social Cohesion

We have already introduced the concept of "Participation" in Chapter 4. In this chapter we further develop it, introducing the idea of how to extend civic participation through Open Space Museum and Social Cohesion practices in the digital domain. Furthermore, we analyse the curiosity triggers that could contribute to engage citizens in these practices.

S. Pescarin et al. (Eds.): *Chromatic Visions*, LNCS 15750, pp. 162–183, 2026.
https://doi.org/10.1007/978-3-032-07792-9_5

1 Designing an Open Space Museum

The Open Museum project is a collaborative effort between Arthur Clay and the N55 group. It involves creating a portable and modular exhibition space designed to showcase diverse artworks from the PERCEIVE project. The main objective is to engage the public and provide a unique place for artistic interaction. Additionally, the Open Museum will offer a tranquil retreat for visitors, making it a versatile and immersive cultural experience. Especially in today's world, where pandemics and health concerns have become a significant part of our lives, the design of the Open Museum takes into account the need for safety and adaptability. The structure is thoughtfully planned to ensure that visitors can enjoy the artworks and cultural experiences while adhering to necessary health protocols. The use of modular components allows for easy reconfiguration to accommodate social distancing and crowd management. Moreover, the Open Museum is not limited by traditional museum operating hours. It offers a novel approach by opening its doors when other museums might be closed. This extended accessibility allows for increased public engagement, reaching a broader audience and providing an alternative platform for museums to showcase their holdings. By collaborating with established museums, the Open Museum becomes a collaborative extension, enriching the cultural landscape with diverse artworks and fostering creativity and cross-exhibition opportunities. The urgency to create a pandemic-ready structure, combined with its extended operating hours and collaboration potential with established museums, makes the Open Museum an innovative and timely solution to address the challenges faced by the art and cultural community today. It offers a safe, immersive, and inclusive cultural experience, ensuring that artistic expression and appreciation can continue to thrive even during uncertain times.

1.1 The Elements of the Open Museum

The key components of the Open Space Museum are its so-called Case Study and Chroma Elements. In this section, the integral role that these elements play within the project will be explored, in order to reveal how they were designed to showcase diverse artworks and provide an immersive experience for visitors for each one of them. In brief, the Case Study Elements act as gateways to artistic exploration and scientific understanding, while the Chroma Elements delve into the interplay between natural light and digital artworks. Combined, they let visitors discover the diverse scenarios of the PERCEIVE Project in an engaging and playful manner (Fig. 1).

The Case Study Elements are integral components of the Open Museum project, designed to showcase diverse artworks that serve as case studies for the research conducted within the PERCEIVE project. These elements utilise space frame structures made square aluminium tubing, providing a durable and visually appealing framework for housing the artworks. At present, there are eight unique Case Study Elements, each with made up of cubes with a volume of 1.5 cubic meters. Visitors can explore the beauty of these artworks while gaining an understanding of their ephemeral nature and the scientific aspects that contribute to their preservation. The Case Study Elements can be exhibited individually or combined into the larger Open Museum structure, providing a socio-cultural area for shared experiences.

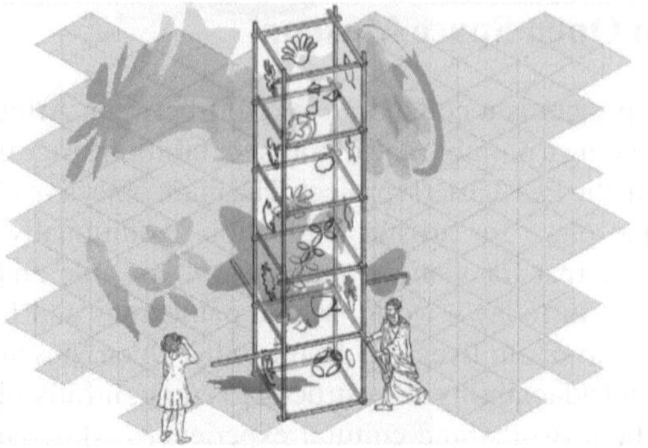

Fig. 1. A design sketch of the Chroma Tower depicting the coloured plates with their design motifs, accompanied by the capstan used to rotate the tower. The rotation alters the mixture of coloured light produced as sunlight passes through the Chroma Tower.

The Chroma Elements serve the distinct purpose within the Open Museum project of showcasing aspects of presenting and preserving Born Digital Artworks, particularly AR Artwork. Resulting from a study involving select AR Artworks, these elements let the visitor explore the interplay between natural light and the virtual assets in the AR artworks. The interplay is made possible by the transparency factors in the coloured plates of the Chroma Tower and the in the textures of the assets used in the AR artworks, which when combined leads to a clearer understanding of how light can have an impact on colour perception.

The AR Portal Elements act as gateways to AR Artworks, enabling viewers to experience digital artworks on their smart devices. The Chroma Tower Element serves as a "colour perception modifier," enhancing the visual experience and interaction with digital artworks featuring transparency. The experiences with the Chroma Tower contribute to further research on presenting and preserving Born Digital Artworks, providing valuable insights into the interconnection of natural light and digital artwork aesthetics. Unlike the Case Study Elements, the Chroma Elements are exhibited independently, creating a unified and coherent Open Museum Space for a cohesive and enriching cultural journey (Fig. 2).

1.2 Case Study Elements Examples

In this section, we explore two case study elements that are integral parts of the Open Space Museum: the GREATWALL CLIMB and the GREATWALL PRINT.

The case study element designed for the Open Museum referred to as the GREATWALL CLIMB is a unique addition to the Open Museum project, presenting a captivating case study on polychromic sculptures from ancient times that were apparently coloured. The element uses a series of space frame elements to form a climbing wall. The wall features hand-foot grips made from casts of various parts of ancient sculptures. Perforated plates on the space frame facilitate the secure attachment of the grips, allowing visitors to climb the wall safely. The artworks used in this Case Study element are three Venus

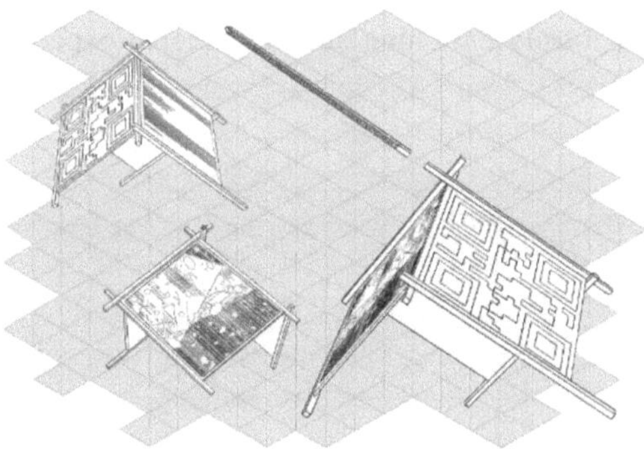

Fig. 2. A design sketch of an AR Portal which acts as gateway to each of the AR artworks on display. The plates of the AR Portals provide access to the AR artworks via QR Codes, information about the artworks, and reflect and refract the coloured light emitted by the Chroma Tower.

sculptures. This tactile learning approach communicates the beauty of ancient sculpture and enriches the methods used to teach about the history of ancient arts, making it particularly effective for engaging youth (Fig. 3).

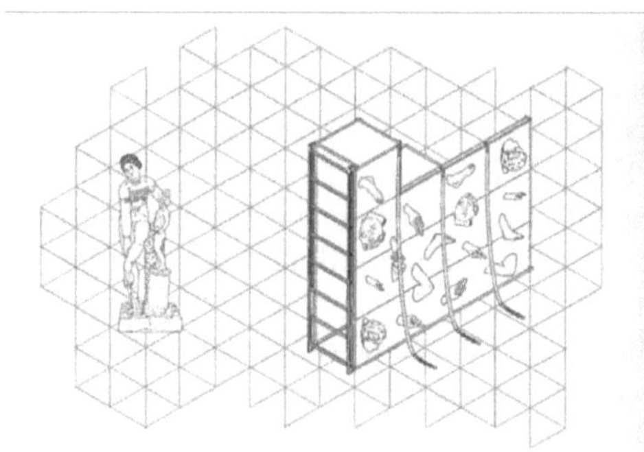

Fig. 3. Sketch of the Case Study Element, the GREATWALL CLIMB, with hand-foot grips segmented from three different Venus sculptures to create climbing routes. By physically interacting with the grips, climbers gain a deeper understanding of the sculptures' intricate details, size, and textures, fostering a tactile appreciation for ancient artistry, creating a truly enriching and educational experience within the museum."

Visitors to the GREATWALL CLIMB element not only enjoy a thrilling climbing experience but also benefit from an educational encounter with the sculpture parts. By physically interacting with the hand-foot grips, climbers gain a more attentive perspective on the sculptures' intricate details, size, and textures. Being able to come in close contact with these artistic elements provides a unique opportunity for tactile learners to better understand the beauty and craftsmanship of ancient sculptures. As climbers navigate the climbing routes, they can examine each grip closely, marvelling at the skill and artistry of the sculptors from ancient Rome and Greece. This hands-on encounter offers a deeper appreciation for the artworks, as visitors actively engage with history, culture, and aesthetics, creating a truly enriching and educational experience within the Open Space Museum (Fig. 4).

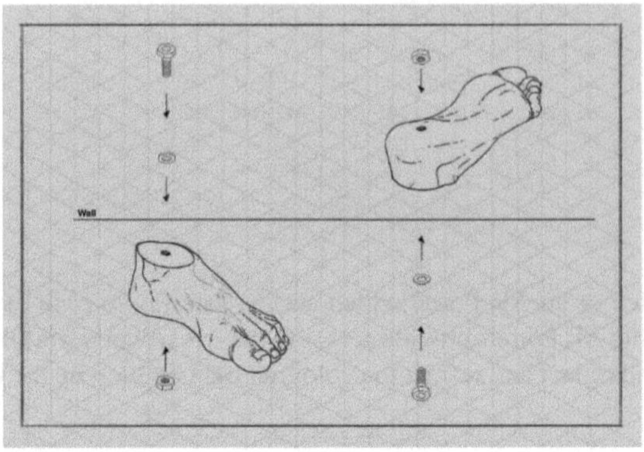

Fig. 4. A climbing hold made from cast plastic with intricate montage details. The hold is won by segmenting digital copies of Venus sculptures and then using a 3D printer to create it.

The case study element designed for the Open Museum referred to as the GREATWALL PRINT is like the GREATWALL CLIMB in that it uses a number of space frame elements to form a wall. Instead of hand-foot grips, a series of panels are mounted to the wall. The GREATWALL PRINT is the element that is dedicated to presenting case study assets from the PERCEIVE scenario on painting. Here, frescos from Pompeii are used which have undergone a loss of colour or which have suffered loss of media due to damage over time. These frescoes are in different areas within Pompeii and are known for their historical and artistic significance. However, due to time and environmental factors, these frescoes have experienced deterioration and loss of data, resulting in areas where restoration is needed to gain a full appreciation of their grandeur.

The Case Study Element presents a novel approach to showcasing frescoes from Pompeii that have been affected by time and damage. The method creates an engagement that places visitors into the role of the heritage scientists on a journey of restoring the frescoes to see how they might have looked at the time of their creation. This is achieved by printing opaque layers on both the front and back surfaces of a two-way mirror. On the front surface of the mirror, a print of the original fresco is overlaid with transparent areas indicating those sections of the frescoes that are missing. On the back surface, the missing parts of the fresco are generated using Artificial Intelligence and printed. When a strong light is directed onto the mirror's front surface (such as the light from a personal smartphone) the mirror is made translucent and the missing sections of the fresco become visible and seamlessly integrate into the fresco, effectively 'restoring' it (Fig. 5).

In this manner, viewers can observe present and restored versions, gaining insights into restoration processes. Authenticity is also present as the visitors can discern between what media is historical and what has been added to it to produce the impression of a restoration. The presentation of the frescoes using the Open Museum offers visitors a sense of authenticity and care by showcasing cutting-edge technology and providing

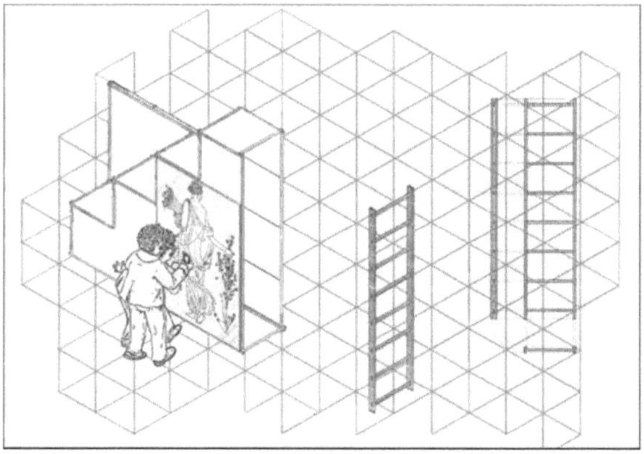

Fig. 5. Design sketch of the GREATWALL PRINT, a Case Study Element in the Open Museum. Using innovative technology, opaque layers on a two-way mirror reveal missing sections of the frescoes when illuminated by light, seamlessly integrating restoration efforts. Visitors witness both present and restored versions, gaining insights into restoration processes while discerning historical authenticity.

a unique educational presentation. It allows visitors to engage with history, restoration processes, and the artistic significance of the frescoes from Pompeii. The element's integration within the Open Space Museum project enhances the immersive cultural experience, making it an enriching and captivating attraction for visitors.

1.3 Engagement Applications

The Open Museum's Engagement Applications represent a groundbreaking approach to enriching the museum experience, fostering a hybrid experience between the real and the virtual that results in profound engagement with cultural heritage for visitors. By scanning the QR code located on the space frame of each Case Study Element, visitors can easily gain access the application. These applications are seamlessly woven into the museum fabric, inspire exploration, learning, and active contribution among visitors.

Like all the engagement applications for the Case Study Elements, the applications for the GREATWALL CLIMB and GREATWALL PRINT feature a user-friendly interface that greatly enhances the usability of the applications, which were designed to ensuring a seamless transition between the physical offerings of the Open Space and the virtual one provided by the application. Overall, by integrating digital applications with the physical engagements offered by the Open Space Museum, the museum's design can be seen as hybrid, which significantly enhances the overall visitor experience (Fig. 6).

The Engagement Application for the GREATWALL CLIMB allows users to identify each hand-foot grip object on the climbing wall, including those inspired by "Venus in a Bikini" and the "Statue of Isis," provides detailed information about these objects' origin and significance, offers instructions for wall climbing, and showcases diverse climbing routes. The interactive application adds a layer of engagement and educational value to the unique GREATWALL CLIMB experience. The application for the GREATWALL Print guides the visitor through the virtual restoration process. By simply using a fingertip

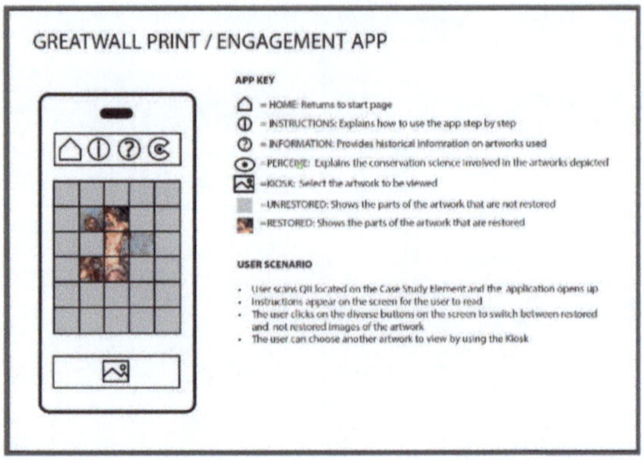

Fig. 6. A mock-up GUI for the GREATWALL Print Engagement Application which is designed to guide the visitor through the virtual restoration process. By simply using a fingertip to glide over the image of the fresco on the screen, visitors gain awareness of how it is possible to provide digital restoration of the frescoes without infringing on the authenticity of the fresco.

to glide over the image of the fresco on the screen, visitors gain awareness of how it is possible to provide digital restoration of the frescoes without infringing on the authenticity of the fresco. Although not all the frescos could be shown in the Open Space Museum, all of them can be seen using the Engagement Application.

1.4 Post Pandemic Museum

In 2020, museums, like many other public spaces and businesses, faced significant challenges due to the COVID-19 pandemic. When museum doors closed in response to the crisis, several notable changes and developments occurred, with one of the most demanding being the crisis of adapting to the absence of physical space and the lack of experience in utilizing alternatives. The Open Space Museum stands out as a model of virtuous practices to adopt in the context of this new pandemic and, more broadly, in the post-COVID-19 era. Its design not only delves into the socio-technical value of museum spaces but also highlights their contemporary, hybrid nature. This approach adeptly navigates the complexity of closed-door policies due to pandemic restrictions, showcasing the museum's ability to blend physical and digital elements to ensure continued engagement and accessibility for visitors.

It is evident that the post-pandemic landscape has witnessed a significant shift toward a comprehensive and inclusive approach. This transition demands that a museum space seamlessly merge digital and physical experiences to cater to a diverse audience, directly responding to the lessons learned during the pandemic, particularly in the realm of digital innovation, which experienced accelerated development during that time. The Open Museum focus on virtual replicas and the potential of digital twins to replicate physical objects and processes, the approach further demonstrates how these technologies can create immersive, engaging, and informative museum experiences, offering innovative ways to present exhibition content and enabling fresh interactions with visitors. Extending civic participation in public spaces through art is a transformative approach to fostering engagement, inclusivity, and pride in communities and the Open Space Museum project serves as a protype example of this in terms of how art can create dynamic and immersive cultural spaces, transcending boundaries, and encouraging active participation.

Through the Case Study and Chroma Elements, the Open Space Museum offers visitors to creatively engage across the five Scenarios of the PERCEIVE Project by integrating a wide range of case study elements dedicated to painting, sculpture, photography, textiles, and digital arts and providing novel ways to interact with them. Two Case Study Elements, along with the Chroma Tower and AR Portals, were discussed in detail as examples. Along with these, it also linked Engagement Applications with these elements, proposing a hybrid entity that extends the physical with the virtual, making digital experiences tangible and accessible. These Case Study Elements, like all the others, not only engage visitors with historical as well as modern artworks, but also enhance their understanding of the significance in the use of technology in cultural heritage. Therefore, all the elements of the Open Space Museum provide immersive encounters with art, providing opportunities for both tactile learning and virtual exploration. The use of modular components allows for easy reconfiguration, ensuring that the Open Space Museum can be implemented in a wide variety of spaces, large and small. With its extended operating hours, it provides a valuable solution for the art and cultural community during challenging times. By offering an alternative platform for museums to showcase their holdings and collaborating with established institutions, the Open Museum becomes a versatile and inclusive cultural space, promoting artistic expression and civic engagement beyond traditional museum hours.

In conclusion, the Open Space Museum's innovative and timely approach, prioritizing safety, and accessibility, stands as a possible model for museums in the future. As a prototype, it showcases the potential of art to create vibrant and engaging public spaces that celebrate creativity, culture, and community, even in the face of the pandemic. It offers a transformative and adaptive experience, inviting visitors to actively participate in shaping the cultural narrative and fostering a sense of ownership in their public spaces (Fig. 7).

2 Curiosity as a Trigger to Extend Civic Participation

2.1 Curiosity and Interest as a Trigger to Extend Civic Participation

Extensive research has unveiled the multifaceted aspects of curiosity and its positive influence in contributing to learning and engagement. Curiosity can be employed as an effective trigger to generate active involvement in individuals, as it can be a powerful motivator of behaviour, initiating actions directed at exploring topics and environments and always aimed at resolving some states of uncertainty and acquiring new knowledge [1]. It was only in the '50s that the meaning behind curiosity shifted from the negative connotation of lacking self-restraints and prying into others' matters, to the positive meaning associated with having inquiring minds, raising questions and seeking answers, doubting and learning [2]. As soon as this shift occurred, researchers and experts from a wide variety of fields worked to give it a valid and complete definition. A definition that has remained as the base ground, even for latest analyses, was provided by Berlyne,

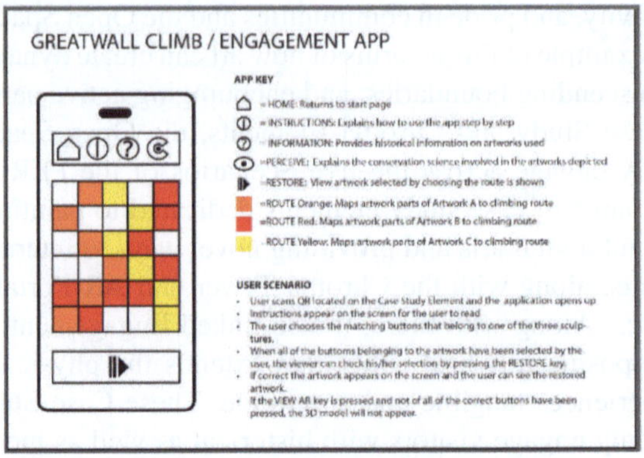

Fig. 7. A mock-up GUI for the GREATWALL CLIMB Engagement Application allows users to identify each hand-foot grip object on the climbing wall from each of the Venus sculptures that were used to create them. The app provides detailed information about these objects' origin and significance, offers instructions for wall climbing, and showcases diverse climbing routes. Selecting all parts of a single sculpture and pressing the "play" button triggers the sculpture to appear as a 3D model on the user's screen.

a British and Canadian psychologist, involved in research surrounding the topics of experimental and exploratory psychology.

"*Curiosity is a state in organisms induced by environmental situations rich in complexity, incongruity, doubt, difficulty, and similar conditions all of which are characterised by the fact that they arouse a state of uncertainty and response conflict. The aroused state can be observed as emotional tension and physiological and neurological arousal [...] that leads to the exploration of the source of the disturbance.*" [2].

Later definitions have been built upon the former, followed by multiple classifications aiming at differentiating among theories and dimensions, with these depending on the object of curiosity, the type of stimuli or the source of the drive. A relevant distinction occurs between *trait curiosity* [3] that is deeply dependent on one's personality, and *state curiosity* [4] that is strongly connected to environmental features and is activated by specific and directed intellectual stimulation caused by unsolved questions and the opportunity to resolve them. Due to its situation-specific form and its dependency on external stimuli, *state curiosity* can arise regardless of ones' innate characteristic and can be directed towards specific topics as long as its triggers manage to produce some sort of disequilibrium or a state of arousal in the individual [2]. This state of arousal is perceived as a keen sense of urgency, a drive to act and to explore the object that raised the curiosity; the state of arousal persists often accompanied by feelings of excitement, agitation, and delight. What Berlyne defined as "state of uncertainty", later developed in the concept of "knowledge gap" (introduced by [4]) that is indeed a perceived "gap" in ones' knowledge base. It refers to the lack of information pertaining to a specific topic or the recognition that there is a disparity between what is known and what one desires to know. The presence of the gap alone doesn't necessarily evolve into curiosity. For curiosity to arise, the subject must also *acknowledge* the gap and recognize the existence of missing information or pieces of information regarding specific aspects of a certain topic.

Indeed, it is this process of recognition [4;5] that ensures the development of a sense of curiosity that leads to start seeking for more information and improving one's knowledge to solve and fulfil the perceived gap. The knowledge gap functions similarly to Berlyne's uncertainty, evolving into a drive to acquire new information through the emergence of exploratory behaviour. Curiosity can be described through three key attributes:

1. The presence of an information gap and a conviction that the gap can be resolved, meaning that there's the perception to have enough capabilities or competencies to fill it [6]
2. Curiosity results in heightened arousal, culminating in a strong sense of urgency to close the gap, which can manifest as excitement, delight, or agitation. Individuals feel compelled to investigate the object of their curiosity until it is either resolved or their attention is diverted.
3. Curiosity is frequently expressed through questioning, doubting or exploratory conduct.

Following these three attributes as references, the development of an individual's curiosity begins with a trigger or multi-triggered scenario [1], which is followed by a reaction, and, finally, concluded in resolution once new information has been gained. Curiosity may be linked to personal, already existing factors, or by contextual and situational aspects [1]. Once curiosity has been triggered, the need to close the gap causes the individual to actively start seeking answers and improvements. This information-seeking behaviour is mainly manifested through question asking and active exploration. Curiosity is resolved only once the acquisition of new knowledge is added to one's knowledge base and closure of the precedent lack of information is then completed. However, as individuals delve deeper into the topic that aroused their curiosity, not only may new questions arise, highlighting new gaps to fill, but also newly discovered personal interest can develop and deepen through the experience [7]. Curiosity is therefore an iterative and active process (Fig. 8), as demonstrated by the positive feedback loop based on reward learning-reinforcement learning model, combined with the features of the knowledge acquisition process [8]. Through this iterative approach that each time leads to resolution of some level of uncertainty, learning occurs, and as new knowledge is gained, visitors become invested. By captivating visitors in this self-rewarding loop [8], triggering curiosity results in a way to also engage the visitors of cultural heritage settings and enhance their participation, hooking their attention and evolving in an active and extended research for new knowledge. They become actively involved in looking for relevant resources, engaging in discussions, and seeking opportunities for learning experiences. Precisely because of this nature, curiosity should be considered as fundamental in the design of interactive media and, specifically taken as reference when designing the PERCEIVE experiences foreseen by the end of the project.

In particular the focus involves what to be considered and used as triggers that could activate curiosity, transforming them into strategies and, specifically into visual stimuli, interactive mechanics and narrative scaffolding. Among the many classifications that curiosity has undergone, some (an overview of the different studies [9]) have indeed

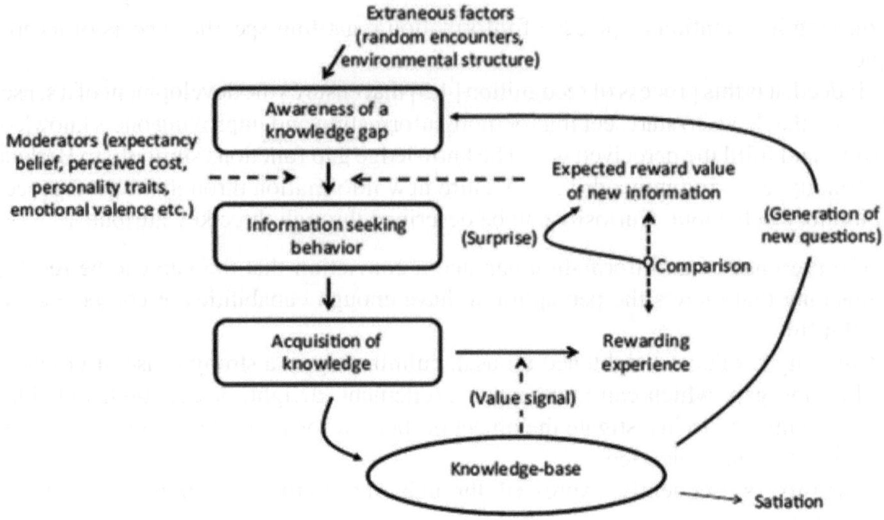

Fig. 8. Summarisation of the positive feedback loop based on reward learning-reinforcement learning model proposed by [8], combined with the features of the knowledge acquisition process.

focused on the reasons and motives that could stimulate curiosity and evolve in an explorative behaviour. The identified triggers include:

1. *Incomplete* and *poorly organised* information, ambiguity, incongruity and complexity: intended as the necessity to acquire more complete and clearer information about a topic.
2. *Unexpectedness* of an information (in the contrast with one's predictions and expectations): this trigger is usually connected to novelty and surprise and leads the individual to search specifically for the reasons why the perceived discrepancy has occurred.

What lastly resulted as an aspect to be further explored regards what can transform curiosity into interest, for the benefit and deeper engagement of PERCEIVE users and in general European Citizens towards coloured collections and their fragility. In particular, curiosity could support and enhance the effects of the caring and authenticity experiences, foreseen in the PERCEIVE prototypes. The relationship between curiosity and interest is of particular importance in educational research. Although, largely used as synonymous or interchangeable concepts, from a theoretical perspective, curiosity and interest are recognised as separate but related constructs. Curiosity typically represents an initial spark, an intrinsic desire to explore and understand, whereas interest represents a more sustained and emotionally charged engagement with a topic. Curiosity is seen as a catalyst for initiating learning. It triggers questions, exploration and active information seeking. As such, curiosity can be a driving force in the early stages of engagement with a subject. Interest, on the other hand, tends to develop as individuals delve deeper into a topic, making emotional connections and finding personal relevance. In order to better understand curiosity triggers and mechanisms that could transform curiosity into interest in the domain of Cultural Heritage and specifically in that of coloured collections, it was designed a Cultural Probe Kit.

2.2 Cultural Probe Kit Exploration on Curiosity and Interest

Cultural probe kits have largely been used as design tools and they typically consist of a set of items and materials that may help the researchers understand their target audiences better, with the overall aim of giving the participants space to document their observations and activities, and to guide and inspire them to reflect on some issues. Probe Kits typically include several key elements, such as diaries (either blank or with guided tasks) for observing and recording information related to behaviours and thoughts; maps for understanding one's relation to surrounding space; disposable cameras for capturing and documenting places or situations; and photos, videos, and postcards to inspire and guide research topics; surveys or questionnaires which can be used to gather specific data [10]. The findings and responses collected through Probe Kits are typically intended for a qualitative analysis rather than quantitative and are mostly used to provide cues and hints for designers to reflect on.

The Curiosity Probe Kit created for PERCEIVE, included a single notebook, proposing written and online tasks, incorporating a variety of media such as images, Instagram posts and short videos to engage participants. It was thought to be divided into three main sections. The first one included an introductory questionnaire and specific tasks for the following days. The second section acted as a diary, guiding participants over eight days to observe and record aspects of their daily lives. The third last part contained advanced tasks related to cultural heritage and the theme of lost polychromy in ancient sculptures. The Curiosity Probe Kit was given to a group of students enrolled in the Interaction Media Design course taught by Professor Sofia Pescarin as part of the Master's Degree program in Digital Humanities and Digital Knowledge of the University of Bologna during the academic year 2022–2023. The main objective of the study using the Curiosity Probe Kit was to collect some *qualitative* data regarding the target group's relationship with curiosity in their daily lives and in cultural heritage related situations. The study focused on three main research questions:

Q.1 - Are there triggers of curiosity that work better than others in cultural heritage contexts?

This question seeks to identify which specific triggers (among the ones found in the literature and designated as novelty, unexpected, incongruity etc. or others that may eventually be added by participants) are more effective in piquing curiosity within cultural heritage contexts, with a particular emphasis on lost polychromy.

Q.2 - Are there aspects of the research on polychromy that specifically act as triggers of curiosity?

This question delves into the specific elements related to the topic of lost polychromy that might stimulate curiosity (history, diagnostics, meaning of colours etc.). Identifying these aspects can help in tailoring exhibits or educational materials to maximise their impact in eliciting curiosity.

Q.3 - Does previous knowledge of the topic influence the levels of triggered curiosity in cultural heritage settings?

This final question aimed to provide insights into the role of prior knowledge in triggering curiosity, It investigates whether individuals with differing levels of prior knowledge about specific topics demonstrate varying degrees of curiosity, aiming t understand the best practices for designing museum experiences to engage audiences with diverse backgrounds. The first section of the Curiosity Probe Kit opens with a general overview on the participants' background and follows with a 20 questions analysis to investigate both their levels of trait curiosity, and the predisposition of their curiosity to be triggered by certain stimuli. Moving on, the second activity of the book was intended as a first moment for the participants to reflect around the concept of curiosity, asking them to recall a time of their life, recent or not, when they felt particularly curious about something. Following, the third activity involved telling a story, but this time it had to be about a time when the participant had attended a cultural event and was particularly curious about something. The participant was then asked to read or retell the story to one of their friends or university colleagues. The fourth presented the picture of the Hercule Farnese held at the National Archaeological Museum of Naples (MANN) and a brief description of what was depicted, the task for the participant was to read two different captions and reflect if their curiosity was aroused and eventually by what. Whereas, the fifth activity was again focused on cultural heritage events, and asked to visit a monument or a museum and to take a picture of something that triggered the participants' curiosity, explaining what it depicts and why it was stimulating. Lastly, the final activity was a role reversal activity in which the participants interviewed a friend and a family member and asked them to describe/narrate an episode or experience that made them feel particularly curious.

The second section of the Cultural Probe Kit was organised as a diary that participants could fill in every day for a week following the structure and the inputs given.

The third and last section of the Probe Kit was the one more centred on the topic of lost polychromy. Here the objective of the investigation was to gather data about the aspects of the research that may attract visitors, giving insight on what triggers to use to raise their curiosity and how to communicate them best. The section was divided into three activities, in the form of questionnaires to be completed online.

The **first activity** is opened by a black and white reproduction of a drawing made by the French author Quatrémere de Quincy in the 19th century, of what he thought the inside of the Greek temple of Olympia could have looked like. After looking at the black and white version, the participants were explained that originally the drawing had colours that have now faded, and they were set before a multiple-choice decision of what the original drawing could look like between five different options of recoloration, of which just one was correct. Once they chose a possible version, the participants were redirected to an image of the original drawing and were asked to look at it and analyse it compared with their choices and finally they were asked to rate their level of curiosity for the topic.

The **second activity** focused specifically on diagnostic analysis for the identification and reconstruction of polychromy on ancient sculptures by a team of researchers at the Museum of Fine Art of Boston and the polychromy of a replica of the Athena Parthenos. A description of the workflow they followed and on the outcomes of their research was given to the participants in the form of a summary of the presentation the researchers made at the 2022 *Polychromy Roundtable* held in Rome. The participants were once again asked to rate their level of curiosity.

The **third activity** in the diary consisted of a last moment of reflection directed on why the participant think the pigments and colours that were applied on classical statues are now completely gone, after a brief overview on the possible motives that caused that, including the meticulous removing of traces of colour during the 18th century due to ideological and cultural reasons; concluding with a last question on the general level of interest. The probe kit in its integrity can be downloaded at [11].

2.3 The Expert Room: Exploring Curiosity on Lost Polychromy at the Museum

In June 2023 CNR researchers launched a four-days mission at the National Archaeological Museum of Naples (MANN) to digitise and carry out an archaeometric campaign of some of the displayed statues that were chosen as case studies for PERCEIVE thanks to their traces of polychromy. Concurrently with the CNR mission, it was decided to organise two experiments to analyse the levels of curiosity raised in the general public of the museum and its triggers towards the topic of ancient polychromy in Roman sculptures.

In particular the CNR mission included the display hosted in room 78 (closed to the public for the occasion), part of the exhibition *"L'altro MANN. I depositi in mostra"*, which contains exhibits many statues examples of the same iconography of the *Anadyomene Venus*, according to various iconographies, as well as two statues of *erotes* and one of a small *satyr*. All artefacts in this display showed some traces of polychromy, particularly visible in the case of the *Lovatelli Venus*, but easily recognizable also in the other sculptures (traces between the hair of other Venuses and on their robes, colour around the eyes and lips of the erotes and the satyr). As for the second set of statues, those found inside the temple of Isis during its excavations, it was only possible to complete the diagnostic work on the *Venus Anadyomene*, while the *Isis with the sistro* and the *Dyonysus with panther* will require a second mission to be finalised.

While CNR researchers were conducting diagnostic analysis and 3D acquisition of the sculptures in exhibition, room 78 was closed, precluding the visitors' access. To repair in some way the disservice, and to retrieve useful information from the visitors, on the last day of the mission, for just a couple of hours, it was decided to allow small groups of visitors to enter the room and get a glimpse of what the experts were doing inside through a free 15-min guided tour (Fig. 9).

To guarantee the highest possible uniformity among the various visits, a specific procedure was defined on how to approach them, and guidelines were laid down on what to say:

1. Each tour began with a brief introduction to the research in progress and the theme of the PERCEIVE project: the main aims and objectives of PERCEIVE were first presented, highlighting the role of digitization and artificial intelligence and giving an overview of the different scenarios.
2. After this opening presentation the visitors were guided towards the first stop, where experts were working on the photogrammetric acquisition of some of the statues on display, while giving an overview on the different 3D acquisition techniques that

had been generally applied to the research project (both photogrammetry and laser scanner). Some preliminary output of the digitalization, point clouds and preview renderings of the 3D models were briefly showed.

3. The second stop, focused on the diagnostic processes: the visitors were guided inside the "darkroom" where experts were performing some multispectral imaging techniques, VIL (Visible Induced Luminescence) and UVF (Ultra-Violet Fluorescence).
4. To conclude, in order to allow the experts to continue working in the "darkroom", the visitors were led outside and taken to the last stop. Here, since the visitors had at this point in the visits a good overview on how diagnostic techniques worked, it was the moment to also show some of the results of the imaging analysis, explaining how they look like, what kind of information they help in collecting and how they are interpreted.

Despite its brevity, the goal of the tour was to give an overview of the whole facets and aspects of the research process of polychromy identification and recognition, explaining how much information can be acquired using different techniques and how difficult it is to imagine possible reconstructions, highlighting the workflow's obstacles and challenges, but yet suggesting the aid that digital tools could provide to speed and help the process. This first event went under the name of *Expert Room - Waiting for the European Days of Archaeology*, six slots of time of twenty minutes each were organised, and booking was mandatory to access the event, a maximum of five people were allowed for each slot. In the end, a total of twenty-one participants were therefore counted, divided in six slots, and visits both in Italian and in English (following the demand of foreign visitors asking to participate) were offered.

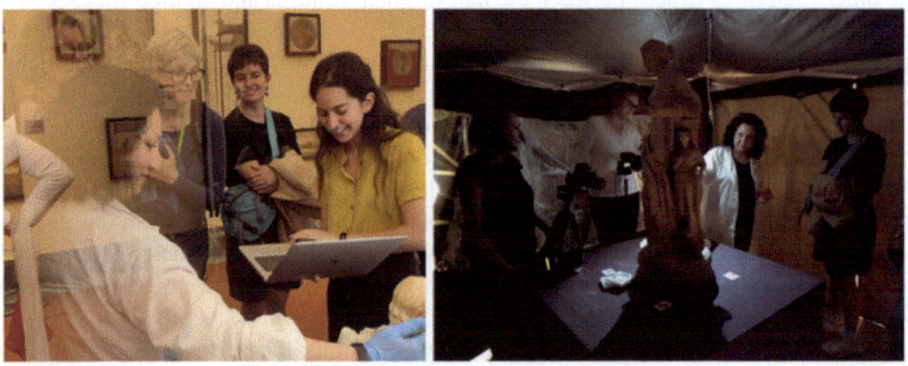

Fig. 9. Diagnostic analyses during the Expert Room at MANN © Museo Archeologico Nazionale di Napoli

After the visits all participants were asked to fill in a very brief questionnaire (English version[1]), that was created to gain data firstly on their previous knowledge on the topic of ancient polychromy and their level of expertise, and secondly to understand what aspects of the visit triggered their curiosity the most, to which level and why, what questions

[1] https://forms.gle/6hh6GuzRnNBup4MU6.

were triggered and concluding by asking what the visitors were planning to do with this newly found curiosity, to understand if the triggered curiosity was enough to evolve in personal interest, and if so, how to exploit it.

During the visits very high levels of curiosity had been assessed, both from visitors completely unaware of the topic of polychromy, and from neophytes and curious onlookers. Evaluating their curiosity on a scale from 1 to 10, 90% of the participants (19/21) marked high scores, between 8 and 10, resulting in an average curiosity level of 9,04/10, slightly higher in participants with a previous knowledge on the topic (9,4/10) rather than in the ones that were completely new to it (8,7/10). Interesting results came from the answers gathered from the third question on the form, regarding what aspect of the research, among the ones presented during the visit, generated the most curiosity. Wanting to group the answers in fixed categories, we could identify mainly four:

1. Diagnostic techniques and tools, archaeometric analysis
2. Digitization and photographic acquisition
3. Coloured reconstruction
4. Challenges of the entire process of identification of polychromy

A curious intake comes up from analysing these results, that is that almost 72% of the total participants claimed that the aspect that mostly triggered their curiosity was either entirely or at least partially the diagnostic techniques and the archaeometric analysis.

> "*Even if I knew about coloured statues, I didn't expect all of the different techniques to go together like this and all the various comparisons that were needed before you can create a reconstruction*"

For the majority of the participants (72%) the stated curiosity was sufficient to develop in interest, either in the form of independent follow-ups, by autonomously looking for more information on the topic on the web or in papers and books, or by looking for more events or exhibition on the topic organised in this or other museums. Lastly, another interesting factor that has been assessed during the event regarded the kind of questions that visitors asked throughout the visit. As question asking is one of the way that people show their curiosity towards something, as a way to gather answers and fill their knowledge gap, it was also worth to record and analyse the questions they made to understand what aspects of the subject matter are the ones that most deserve to be acknowledged during the design of an interactive experience. By analysing them, they could indeed, be subdivided between questions on bindings and way of applying the colour, the meaning this colour may have had, but mainly where traces were found, what are the pigments and how come they disappeared in most scenarios, or on the other hand why in the cases they were saved, was that possible indeed.

2.4 Results and Requirements About Curiosity and XR Experiences

Through the analysis of all the data gathered from the Curiosity Probe Kits and the MANN experiments, it was possible to establish some points that are indeed fundamental for the design of an experience that aims at trigger curiosity in the public towards the topic of the lost polychromy of sculptures. Thanks to the analysis of the data collected from the Curiosity Probe Kits, it was possible to identify the most important triggers for curiosity among the students. In most cases, *novelty* was the main trigger (something completely new), followed by the *unexpected* (they meant situations where they had certain expectations, and their expectations were not met) and *unknown* (curious because they had no prior knowledge about the subject). In this last case, it's the absence of knowledge that triggered curiosity, as if the trigger was related to a sense of *mystery*:

- *"You're not expecting to see it happening every day!" (Curiosity Probe Kit, n°21)*
- *"I wanted to understand how it works, I've always found myself caught by the unknown" (Curiosity Probe Kit, n°4)*

Other identifiable triggers have been the oddness, the weirdness of the subject, (to be aggregate with the sense of unexpected); previous interest (10%) and personal connection (10%). The less quoted motivation was the usefulness.

Addressing the topic of lost polychromy, from the fourth activity of the first section, regarding the two different captions of the Hercules, it emerged that those who reported to be interested in finding more information on it, were motivated by the *novelty* trigger (that they had never seen the statue in person before and would have liked to visit it). Specifically, they reported to be interested to polychromy related aspect.

- *"I think I would look further into its history, who discovered the pigments first and if there are any reproduction of the coloured version" (Curiosity Probe Kit, n°16)*
- *"I would like to look to some documentaries or videos showing how the colours were found, and maybe why they faded away" (Curiosity Probe Kit, n°50)*

The topics that were considered worthy of being deepen were the pigments' identification process, the reason behind their fading and the sources to rely on for possible reconstructions, together with the interest in seeing some of these plausible colour reconstructions.

In the last section it was possible to follow the increasing and motivation of curiosity toward ancient polychromy. Specifically in the second activity, on Athena Parthenos colour reconstruction, carried out at the Boston MFA (link: https://blackmath.com/athena-3d-reconstruction), the reported curiosity reached particularly high levels. When asked what the participants would have added to the scientific narration to make the topic more interesting, the responses included the possibility to show more steps of the diagnostic and recolouring process, the addition of information about the painting techniques in ancient times and the way it is possible to recreate the appearance of lost colours. Other topics of interest emerged from the last question, where the participants were given the chance to hypothetically ask the team of researchers some questions. Here, the data showed that they were intrigued by researchers' expectations of what they would find during the research process and the initial clues that sparked the investigation. In addition, participants asked for more information about the specific archaeometric analyses and the origin and use of different pigments. They confirmed to be interested in the reasons why sculptures were coloured, what the colours were meant to convey and the reasons for the possible loss of colour over time. Participants were also curious about the possibility of ancient painters adopting durable materials or techniques, as well as the methods and general workflow used to recover and identify traces of colour.

Finally, they also wanted insights into the digital component of recreating 3D models of sculptures and virtually reconstructing possible colour maps.

Combining these results with the data collected from the MANN experiments, it was possible to draw up a list of priorities to be included when designing an experience dedicated to lost polychromy:

1. To stimulate curiosity the most efficient triggers to work with and with which the visitors resonated the most were the novelty of the information, the unexpected side of the research, the mystery underlying the loss of colours, and the unknown of the reason why something is visible, and something has been lost.
2. The need to focus on the workflow of the research, to let visitors wear the shoes of the researchers and accompany them step by step along the discovery of lost polychromy.
3. The experimental side of the analyses and their outcome were the most significant aspect to trigger their curiosity, even when not explained in high detail, showing how the invisible becomes information acted as a way to convince them of the authenticity of the message and made them feel closer to the topic of polychromy.
4. Understanding and giving hint on how the colours were made, how they used to stick to the marble and why they were important for the complexity of the sculptures.
5. Once letting them see the reconstruction, making clear that's only a hypothesis and showing both the comparisons to get a fuller image and to recontextualise it inside the colourful environment the statues were located in helped them catch the imagine the biggest picture. By allowing viewers to see the reconstruction and clarifying that it is only a hypothesis, we can provide comparisons that offer a fuller image. This helps recontextualize the statues within the colourful environments where they were originally found, enabling viewers to grasp the bigger picture more effectively.

These aspects combined are the basis to try to efficiently communicate lost polychromy, allowing a dialogue with the public and creating a bridge between the present and what we are now able to see and what it may have looked like in the past.

3 Extending Civic Participation Through Social Cohesion

3.1 A Definition of Social Cohesion

Social cohesion is a multifaceted and complex construct which presupposes stable and trusting relationships between different people, groups, and organizations and it defines the forces which encourage individuals to be part of and function as a joint group. Nowadays, the cohesion policy is one of the main investments of the European Union policies. It aims at strengthening economic, territorial, and social cohesion. Since the early 2000s, the Council of Europe has been engaged in the development of a strategy to promote Social Cohesion, adopting a model of society that may be increasingly capable of guaranteeing the well-being of all its members, minimizing inequalities, and avoiding marginalization (Council of Europe 2008[2]). In the last 20 years, there has been multiple attempts to measure, evaluate, and promote Social Cohesion in European Countries [13;14;15] and, while the research on this front is still ongoing, useful frameworks to

[2] https://rm.coe.int/0900001680747068 (accessed on 18 February 2023).

test the presence of cohesiveness in societies have already been defined [16, 17]. The most recent definition mentions social cohesion as in [16]:

> *"ongoing process of developing well-being, sense of belonging, and voluntary social participation of the members of society, while developing communities that tolerate and promote a multiplicity of values and cultures and granting at the same time equal rights and opportunities in society".*

In recent years there has also been a growing interest in the digitisation and digital preservation of cultural heritage (i.e., the European "Horizon 2020" program) and in encouraging citizen curation for cultural objects (i.e. [18, 19]). Thus, it is essential to understand how to enhance civic participation in public spaces in order to stimulate a general cohesiveness among citizens and cultural institutions.

3.2 Social Cohesion Dimensions

In 2011 the Bertelsmann Stiftung developed the Social Cohesion Radar as an empirical instrument for studying social interactions on national and local level [13]. It is divided into three interconnected domains: social relations, connectedness and focus on the common good; each is subdivided into subdomains. For every subdomain, group cohesiveness indicators are identified (Fig. 10).

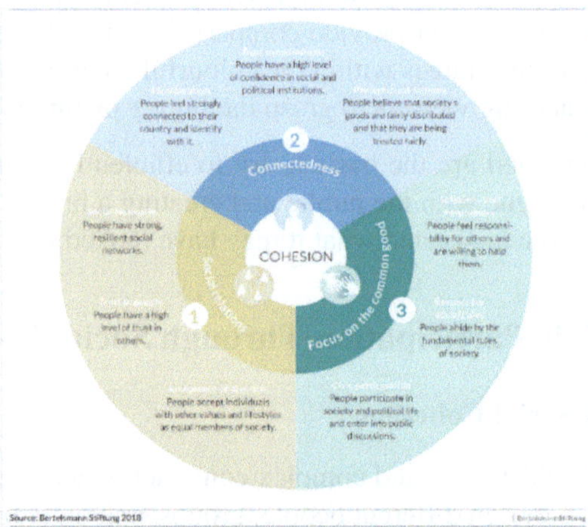

Fig. 10. The Social Cohesion Radar

In particular, the "Focus on the common good" domain includes the "Civic participation" subdomain, that regards the extent to which "people participate in society and political life and enter into public discussions".

3.3 Citizen Engagement and Social Cohesion

The European Commission describes the phenomenon of citizen engagement as "an invited form of citizen participation where public institutions invite citizens to openly discuss matters of concern and care" [17]. At the same time, the concept of "museum" is widely perceived as an *agora* or *forum*: "a gathering place for people and an assembly of minds and ideas – is the privileged understanding". Accordingly, the museum institutions should ideally serve and represent the people visiting them, encouraging the establishment of new relations with the public and inspire **democratic engagement** with cultural heritage [16]. The UK Soundmap project of The British Library (2010) clearly shows the citizens' interest in voluntary participation in Cultural Heritage-related projects [18]. The project concerns a digital collection of pieces of sounds created through *crowdsourcing*. Social media, digital technologies and the Internet are thus useful tools for citizens to access, explore and add to cultural heritage content. In June 2023, the European Commission published a report on civic engagement, explaining how participation in cultural activities strengthens democracy and social cohesion. The report shows that those participating regularly in cultural activities are more likely to vote, volunteer and participate in community activities, projects and organizations. Because of their strong emotional, creative, expressive and collaborative nature, cultural activities can build spaces of dialogue, community gathering, sharing, social interaction and connection between individuals, features which are deeply related to the social cohesion dimensions. The report clearly states that "investing in citizens' cultural participation is essential in any effort to promote civic engagement, democratic vitality and social cohesion in the EU". Cultural activities involving citizens can thus help people bridge social boundaries of ethnicity, religion, gender, age, nationality and occupational status. There is a clear correlation between enticing citizens to vote, making them develop positive social and personal attitudes, nurturing the understanding of different perspectives and facilitating conflict resolution. Citizens who are conscious of their cultural heritage are likely to feel empowered, more active in their community and to perceive a greater sense of cohesiveness.

Acknowledgments. This study was funded by the PERCEIVE project, that received funding from the European Union's Horizon research and innovation programme under grant agreement No 101061157 (Sect. 1 and 2), and by the project Social Cohesion, Participation and Inclusion through Cultural Engagement (SPICE) (Grant Agreement N. 870811) (Sect. 3).

Author Contributions. Arthur Clay and Chiara Gemma Fedon contributed to Sect. 1.; Federica Bonifazi and Sofia Pescarin contributed to Sect. 2; Vanessa Bonanno and Sofia Pescarin contributed to Sect. 3.

Disclosure of Interests. The authors have no competing interests to declare that are relevant to the content of this article.

References

1. Arnone, M.P., Small, R.V., Chauncey, S.A., McKenna, H.P.: Curiosity, interest, and engagement in technology-pervasive learning environments: a new research agenda. Educ. Technol. Res. Dev. Springer Science and Business Media LLC (2011). https://doi.org/10.1007/s11423-011-9190-9
2. Berlyne, D.E.: Conflict, Arousal and Curiosity. McGraw-Hill, New York (1960)

3. Kashdan, B.T.: Curiosity. In: Peterson, C., Seligman, M.E.P. (eds.) Character Strengths and Virtues: A Handbook and Classification, pp. 125–141. Oxford University Press, New York (2004)
4. Loewenstein, G.: The psychology of curiosity: a review and reinterpretation. Psychol. Bullet. Am. Psychol. Assoc. (APA) (1994). https://doi.org/10.1037/0033-2909.116.1.75
5. Peterson, E.G., Cohen, J.: A case for domain-specific curiosity in mathematics. Educ. Psychol. Rev. Springer Science and Business Media LLC (2019). https://doi.org/10.1007/s10648-019-09501-4
6. Shin, D.D., Kim, S.-I.: Homo curious: curious or interested?. Educ. Psychol. Rev. Springer Science and Business Media LLC (2019). https://doi.org/10.1007/s10648-019-09497-x
7. Hidi, S.E., Renninger, K.A.: Interest development and its relation to curiosity: needed neuroscientific research. Educ. Psychol. Rev. Springer Science and Business Media LLC (2019). https://doi.org/10.1007/s10648-019-09491-3
8. Murayama, K., FitzGibbon, L., Sakaki, M.: Process account of curiosity and interest: a reward-learning perspective. Educ. Psychol. Rev. Springer Science and Business Media LLC (2019). https://doi.org/10.1007/s10648-019-09499-9
9. Grossnickle, E.M.: Disentangling curiosity: Dimensionality, definitions, and distinctions from interest in educational contexts. Educ. Psychol. Rev. Springer Science and Business Media LLC (2014). https://doi.org/10.1007/s10648-014-9294-y
10. Thoring, K., Luippold, C., Mueller, M.R.: Opening the cultural probes box: a critical reflection and analysis of the cultural probes method. In: 5th International Congress of International Association of Societies of Design Research, pp. 222–233 (2013)
11. Pescarin, S., Bonanno, V., Marasco, A.: Social cohesion in interactive digital heritage experiences. Multimodal Technol. Interact. **7**(6), 61 (2023)
12. Dragolov, G., Ignacz, Z.S., Lorenz, J., Delhey, J., Boehnke, K., Unzicker, K.: Social Cohesion in the Western World: What Holds Societies Together: Insights from the Social Cohesion Radar, 1st edn. Springer International Publishing, Cham, Switzerland (2016)
13. Larsen, M.M., Boehnke, K.: Measuring social cohesion in the Kyrgyz Republic. SSRN Electr. J. (2016). https://doi.org/10.2139/ssrn.2946732
14. Schiefer, D., Van Der Noll, J., Delhey, J., Boehnke, K.: Cohesion Radar: Measuring Cohesiveness; Bertelsmann Stiftung: Gütersloh, Germany (2012)
15. Fonseca, X., Lukosch, S., Brazier, F.: Social cohesion revisited: a new definition and how to characterize it. Innovation **2018**(32), 231–253 (2018)
16. Boehnke, K., Berrueto, A., Dragolov, G., Ocampo Villegas, P.: Are value preferences and social cohesion interconnected? the case of Mexico. Acta de investigación psicológica, **9**(2), 35–45 (2019)
17. Baggesen, R. H.: Augmenting the agora: Media and civic engagement in museums. MedieKultur: J. Media Commun. Res. **30**(56), 15 (2014)
18. Fargier, N.: Global sound archive. soundmap projects and the perspective of future. Humanit. Future Perspect. Past Present, **21**, 83–97 (2020)
19. Daga, E., et al.: Integrating citizen experiences in cultural heritage archives: requirements, state of the art, and challenges. ACM J. Comput. Cult. Heritage (JOCCH), **15**(1), 1–35 (2022)

Colour Fragility

Cristiana Barandoni[1(✉)], Lucia Burgio[2], Arthur Clay[3], Brenda Doherty[4], Chiara Gemma Fedon[3], Roberta Iannaccone[1], Hartmut Kutzke[5], Dario Lanfranconi[3], Catlin Langford[2], Donata Magrini[1], Letizia Monico[4], Irina Crina Anca Sandu[7], Sophia Sotiropoulou[6], and Giorgio Trumpy[8]

[1] CNR ISPC, Florence, Italy
{cristianabarandoni,robertaiannaccone,donata.magrini}@cnr.it
[2] Victoria and Albert Museum, London, UK
l.burgio@vam.ac.uk
[3] HSLU, Lucerne, Switzerland
{arthur.clay,chiara.fedon,dario.lanfranconi}@hslu.ch
[4] CNR SCITEC, Perugia, Italy
{brenda.doherty,letizia.monico}@cnr.it
[5] Kulturhistorisk Museum Department, Universitetet I Oslo, Oslo, Norway
hartmut.kutzke@khm.uio.no
[6] FORTH IESL, Iraklion, Greece
sophiaso@eap.gr
[7] MUNCH Museum, Oslo, Norway
irina.sandu@munchmuseet.no
[8] NTNU ColourLab, Gjøvik, Norway
giorgio.trumpy@ntnu.no

Abstract. This chapter explores the fragility of colour across five scenarios, focusing on the loss of polychromy in ancient sculptures and architecture, colour variations in paintings, textiles, historical photographs, and digital art. It also emphasizes the need for standardized methodologies to reconstruct, simulate, and restore original colours. For each scenario, the state of the art on colour studies is presented, together with the challenges about colour reconstruction and enviCase studies are discussed, highlighting colour features and characteristics. Modern analytical and diagnostic techniques are finally also discussed.

Keywords: Colour Studies · Colour analysis · Colour reconstruction

To better preserve coloured collection, this chapter focuses on the analysis of colour fragility in five different scenarios, that are considered a priority by projects such as PERCEIVE. These scenarios are:

- Scenario 1: lost polychromy in ancient statues and architecture
- Scenario 2: colour change in paintings and works on paper
- Scenario 3: colour change in textiles (dresses and tapestry collections)
- Scenario 4: colour change in historical photographs/film collections
- Scenario 5: colour rendering and visualization in born digital art.

S. Pescarin et al. (Eds.): *Chromatic Visions*, LNCS 15750, pp. 184–222, 2026.
https://doi.org/10.1007/978-3-032-07792-9_6

Colour studies, research questions and objectives are different in these contexts and, therefore, the authors have defined a "scenario-based" methodology to approach the problem of colour preservation and reconstructions, from data acquisition and analysis, to the evaluation of results, from the definition of needed tools and services, to the design of XR and hybrid experiences.

1 Lost Polychromy in Ancient Statues and Architecture

In recent years, new evidence has emerged about the polychromy of sculptures in classical antiquity. This is largely due to new findings and datasets resulting from analytical studies coming from sculptures and architectural elements held in museums around the world. Thanks to the international scientific community, a standardised methodology for data acquisition has been established, thanks to the coordination of the International Round Table on Polychromy in Ancient Sculpture and Architecture, also known as "Polychromy Round Table"[1] To preserve traces of polychromy and convey a statue's original appearance, it has become increasingly important and interesting to visualise the richness of ancient statuary in terms of colour "pantone" and its original appearance. Polychromy reconstructions can contribute to offering a more correct idea of the original aspect, enabling users to perceive its complexity (i.e. with shades and not homogenous colours applied to certain areas) and semantics. Nevertheless, reconstructing the original colours is a challenging endeavour due to the difficulty in identifying the full extent and original appearance of the painted parts, along with their distribution over the surface, the tools and the painting techniques used. It is particularly difficult, moreover, to reconstruct how primary colours are brought out and enriched by highlights and shadows, given the present preservation state of sculptures. Although we have a good understanding of the pigments used in antiquity, we do not have much information about ancient painting techniques, such as used binders or how the paint was applied. Furthermore, the appearance of the artworks might have been significantly altered over the centuries. Therefore, we should highlight here how this lack of information does not allow a reliable reconstruction of the original colour appearance. For this reason, it is important to follow a scientific workflow and provide a minimum degree of reliability and authenticity, connecting results to analytical data and sources, not only for the scientific community, but also for the general public.

Unfortunately, there is a lack of a universally adopted methodological protocol for polychromy reconstruction, and this has led to inconsistencies in the visualisations (both digital and physical copies) proposed by various research groups. These discrepancies can be attributed not only to varying levels of familiarity and thus fluency of interpretation of historical sources and studies, thereby reducing the comparability and consistency of colour reconstructions, but they are also due to the uncertainty of the research results carried out on the artworks themselves. For instance, not all surveys yield unequivocal results that allow a full understanding; unfortunately, in most cases, the information is not sufficient to clearly understand aspects such as the extent of colour, the designs of decorations, etc.

[1] https://www.polychromyroundtable.com/.

After decades of the neoclassical white "pure form", the replicas and reconstructions of polychrome artworks, often provoke a shocking reaction. For this reason, it is essential that the reconstructions are accompanied by additional explanations and descriptions of the reconstruction process, to make them accessible to the general public, who is often unaware that statues were once coloured. From this perspective, the digital format can be a valuable tool in overcoming public scepticism, once we move from the concept of "definite reconstruction" to that of a "potential appearance", based on source analysis and scientific evidence. The possibilities offered by the use of digital media allow us to provide a wide range of hypotheses and to show the whole process and the studies behind it. State of the Art examples of the reconstruction of polychromy on classical statues are still few represented, due to the long and costly process. To ensure the accuracy of the results, much time should be devoted to the research of coeval sources and historical records and to comparing iconographic models among themselves: for instance, the literary source can serve as a starting point to verify the presence of models both in the realm of painting, such as frescoes, and in their translation into stone.; for the same reason it is of great interest to have the possibility to compare diagnostic investigation results on similar typologies of artifacts. The scientific methodology has been firmly established and results are regularly disseminated through conferences and international journals, although it has never been codified and standardized, while the supporting data often remains inaccessible [1–8]. This wealth of content has immense potential and is of utmost importance to researchers, which underlines the urgent need for greater accessibility. For this reason, a starting point should be the drafting of an "investigation manual", establishing a protocol that will eventually be accepted by the entire scientific community, pointing out general guidelines for research, even addressing the issue of the instrumentation used, as well as the method of communicating and making data accessible in an open repository where containing and accurately displaying all data obtained can be "FAIRly" available for comparative studies and reuse. Since the late 20th century, considerable resources have been devoted to producing reconstructions of ancient sculptural polychromy [9]. Today, polychromy reconstructions are mainly made by painting physical copies of the sculptures [10–12]. However, certain experimentations also occur on digital formats, 2D images or 3D models where polychromy is suggested by colourizing well-defined delineated areas in homogeneous, flat RGB tones [13–16]. More sophisticated reconstructions involve the study of contextual or related, i.e. comparative, artworks (such as other sculptures or mummy portraits) to manipulate (e.g. stretch or deform) their digital images and extract colour textures to simulate layers of paint in the element being reconstructed or features on the sculpture being studied [17]. However, the main problem is still the lack of an established, agreed methodological protocol. Heritage scientists and scholars have relied on the reconstruction of physical copies to disseminate the results of the research on polychromy, although digital reconstruction is becoming increasingly widespread. More sophisticated methods of digital colour reconstruction of archaeological buildings and sites have been developed with the support of software that helps to manage the reliability of the reconstruction [18, 19] but the methodologies used have not been tested on ancient statuary. The current restoration of polychrome statues is particularly deficient in terms of nuance and shading: the colour appears flat, with a total absence of the vibrant tones that we know existed and prevailed in the past. Sculpture was supposed to imitate reality, and it's worth looking at frescoes or analysing the production of Attic ceramics, for example, to understand how detailed the rendering of textiles was and the dedication and care with which artists animated images. Sculpture has certainly not escaped this artistic intention to reproduce reality, but what makes it difficult, if not impossible, to understand lies in its very material nature: what remains today is the substrate and not the finish and we can only work on this layer, which by its very nature does not allow for a better understanding. The efforts of the scientific community are today not only aimed at understanding the possible pictorial characteristics of Greek and Roman sculptures, but also at providing and facilitating physical restoration and virtual reconstruction efforts, to better preserve collections and to raise public awareness, fostering a sense of care.

1.1 Objectives

Our main and ultimate objective is to reconstruct the colour and original appearance of the polychrome statues, improving the quality of rendering, the capacity of available methods and the speeding up of the traditional workflow. The digital reconstruction of the original polychromy is, in fact, a time-consuming process that requires specialised skill (i.e. iconographic comparative analysis takes time). The development of a semi-automatic Artificial Intelligence (AI) protocol to support this research or to suggest possible reconstructive hypotheses can offer a solution to obtain a more reliable colour reconstruction or reconstructions (it would be possible for example to provide more than one potential reconstruction). The aim is to obtain a methodological protocol welcomed by the scientific community and, at the same time, accepted and appreciated by the general audience. If we consider this AI support, there are still not sufficient studies and applications. Although only a few attempts have been made and, when published, widely accepted, there is a clear need to advance the field by developing tools and services that empower researchers to explore and construct multiple reconstruction options, supporting the development of interpretations, fostering dialogue and refining results. Through such an iterative process, researchers can progressively approach a more authentic representation of the object's original colourful appearance.

1.2 Challenges

Although archaeometry techniques enable today to detect minute traces of pigment, even when they are no longer visible to the naked eye, many questions about the appearance of the original polychromy remain unanswered. Some pigments are better preserved than others, and certain pigments change chemically over time. Even when a pigment is scientifically identified, questions often remain as to whether it was applied in its pure state or mixed with other pigments, and whether it played a major or minor role in the overall polychrome scheme [20]. Given these uncertainties, it must be accepted that it is virtually impossible to reconstruct the appearance of the original polychromy. One of the main obstacles to colour reconstruction is the imprecision of the reconstruction itself, since few traces are visible and few comparisons are available, so that the dissemination and presentation of the results is reduced to the strictly scientific data obtained during the measurement campaign. The meticulous analytical protocol can provide a comprehensive physio-chemical characterization of the colour preserved on the surface of the sculpture. Even the tiniest traces of colour are sufficient for experts to determine the nature and composition of the original materials, identifying the pigment or mixture of pigments, albeit with greater complexity, and possibly determining the type of binding medium. However, while this information, although accurate, relates to the colour preserved in a few small, barely visible areas of the surface, the synthesis of a comprehensive understanding of polychromy and the reconstruction of the original appearance of the object at the time of its creation remains a challenge. The preparation of "mock-ups" (pieces of original material treated and painted according to ancient recipes) could overcome the difficulties of reproducing reliable textures and helps to render the appearance of the surfaces, simulating the original aspect and avoiding the "plastic effect" produced by full-tone painting. Another challenge would be to develop tools to digitally visualize the degree of (un)certainty of the polychromy in the different parts of the reconstruction (e.g. objective data such as the location of the visible traces and the description of lost *polychromy* in ancient publications or archival documents and hypothetical data, such as which part of the reconstruction is based only on hypothesis and comparisons with other similar polychrome sculptures).

1.3 Case Studies

To better develop a methodology and supporting tools, we have therefore selected a number of case studies that we are going to describe in the following paragraphs:

- Various versions of Roman statues representing the so-called "Venus Anadyomene";
- The sculptures and frescoes belonging to the Isis complex in Pompeii

The Venus Anadyomene Statues

We have selected a group of marble statues representing the iconography of "**Venus Anadyomene**" (from the Greek, "Venus Rising from the Sea"), the Goddess of love, depicted in the act of rising from the sea and resting her hands on wet hair. The theme of this goddess is a perfect case study, because it enables us to work on a very well represented type and on pretty well preserved and known polychromy remains, such as in the case of the "Lovatelli", "Bikini", and "Marina" Venuses.

We have chosen to study a selected group of "Anadyomene-type" statues with traces of original polychromy, all conserved at the MANN museum in Naples, using a non-invasive analytical protocol, whose results will help to reconstruct the complete pictorial framework. Moreover, this type of statues is also spread in literature, frescoes and mosaics. In particular, the literary sources provide a precise description not only of the genesis of the myth but also of the appearance that is easily recognizable in its artistic rendition. According to the legend recounted by Hesiod, whose Theogony serves as a cornerstone in understanding Greek mythology, detailing the origins and family lineage of the Olympian gods, Titans, and various other divine beings, the goddess of beauty and love in emerged from Uranus's severed genitals, cast into the primordial sea by Cronus. The iconic image of Venus Anadyomene rising from the sea gained renown in antiquity largely due to a revered painting by Apelles, captivating the imagination of ancient writers, who frequently referenced it in Latin literature, from Ovid to Pliny the Elder, Servius, and Apuleius. Apuleius, particularly, offers several exquisite passages on the subject, replete with vivid descriptions and rich chromatic imagery: beyond their refinement and visual allure, Apuleius's depictions stand out for their attention to physical detail. For instance, the goddess's flowing locks are described as "undulating locks of hair overflowing in waves," an image steeped in the aquatic symbolism of dishevelled feminine hair: this is one of the reasons why, emerging from the waters, Anadyomene is always depicted with her hands surrounding two large tufts of hair and squeezing them. Not only Hesiod and Apuleius but also Plato, Strabo, Pliny the Elder, Ovid, and Apuleius, the image of Venus Anadyomene continues to resonate in modern literature. Contemporary writers embrace and expand upon this magnificent creation of GreekRoman culture, further enriching its enduring allure, the same does painters as Titian who in 1520, painted an oil with the same subject, now housed in the National Gallery of Scotland in Edinburgh. The longevity of the myth and, simultaneously, the repetitiveness of the pictorial canon, albeit with minimal variations (such as the presence or absence of jewellery adorning the body), have been the reasons why this specific iconographic theme was selected. This allows comparison and representations, and/or, when feasible, also to speculate on a general pictorial composition: the comparison between different media will certainly help to understand if the use of colour was intended to imitate the reality, if there was any link between different works of art, but also for other purposes such as understanding the way shadows and volumes were rendered on different surfaces (neck decoration in Fig. 1).

On the selected statues, Fig. 2, it is still visible residual polychromy. Moreover, there are several versions of statues with the same iconography around the world with similar characteristics to be compared with, and various examples of frescoes depicting Venus which could serve as an iconographic visual reference as well. Beside a naked Venus, other depictions were documented as fully or partially clothed, very similarly to the statuary examples. In archaeological museums displaying collections of statuary beside frescoes, ceramics, mosaics portraying scenes where statues are shown and depicted with either full or partial colouring, the iconographic comparison is made possible. As shown in the images provided below, the representation of sculptures with painted details not only involves fresco technique but also ceramic production and mosaics, in this specific case, opus sectile. These case studies also adopt important iconographic models onto different media, of which we have evidence in stone artifacts (Fig. 4). Figure 3-left depicts a unique coloured sculpture of Artemis on a base (INV 9301), while Fig. 3-right portrays one of the most famous myths of the classical world, the Callipygian Venus on an Attic krater wearing a dark orange mantle, her body's skin coloured in white, and her hair painted to resemble blonde. Finally, the already mentioned Fig. 4 is another extraordinary example of Venus, in this case fastening her sandal, depicted on a precious *opus sectile*, with decorative details, such as the jewellery, realised with a different stone material, the body, the hairstyle, and the anatomical details highlighted to recreate the folds of the abdomen and arms.

The Isis Complex

The Venus type is also diffused in the Temple of Isis in Pompeii (the myth of Venus is strictly related to the myth of Isis in Campania [22]). For this reason, we have extended

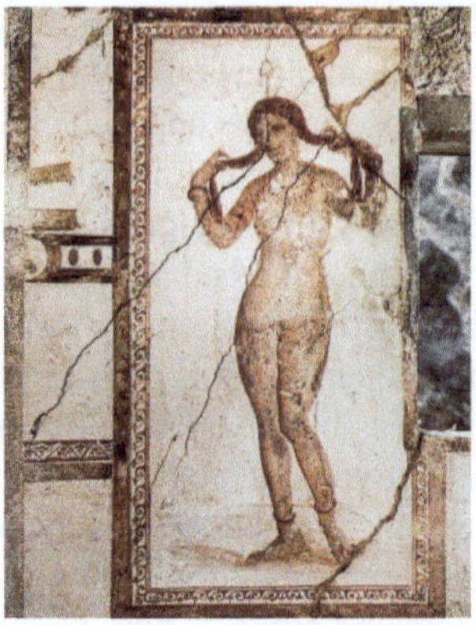

Fig. 1. The life-sized depiction (fresco) of the Anadyomene from the House of the Principe di Napoli, Pompeii VI.15.8. It is located on the walls of the exedra, which forms the central area of the triclinium (Wikipedia).

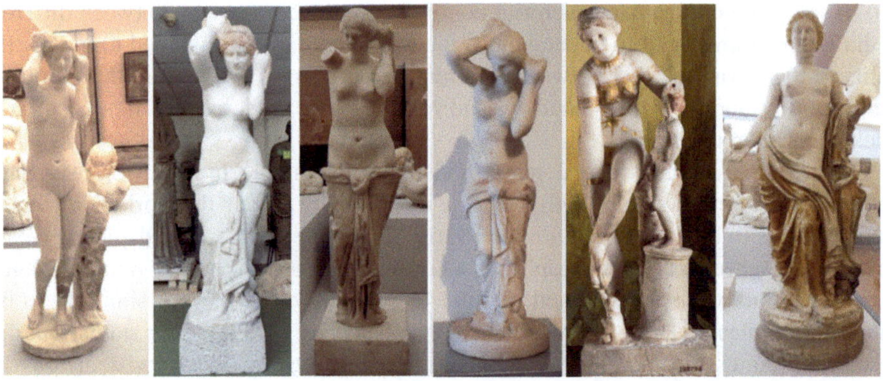

Fig. 2. The selected statues: a) Venus Anadyomene INV 111383, Pompeii, House of Oppius Gratus and Quartilla; b) INV 6292 Venus Anadyomene from Pompeii; c) INV 126248 Venus Anadyomene, from Pompeii; d) INV 6298 Venus Anadyomene from the Iseum of Pompeii; e) INV 152798 Bikini Venus, from the House of the Bikini Venus (I 11, 6–7), Pompeii; f) INV 109608 Lovatelli Venus, from the House of Diomedes, Pompeii (I, 2. 17).

our study to the Iseum [23–25], including not only the sculptures but also the frescoes. This Temple is dedicated to the Egyptian goddess Isis. The walls were all painted in the first and second Pompeian styles, and we may suppose that the statues were also originally painted. Although the site is still visible in Pompeii, even if only partially, the entire decoration, including frescoes, statues and objects linked to the cult, are permanently exhibited at MANN Museum. These remains, excavated in 1764, survived the eruption. They included paintings and fragments of statues, which were scattered across different parts of the temple and later reassembled in the museum. Very few fragments of the frescoes are visible *in situ* today, but the documentation of both the findings and archival sources enables a virtual reconstruction of the entire complex. Additionally, a maquette reconstructing the Iseum complex at the museum serves as a reference for reconstructing this archaeological site.

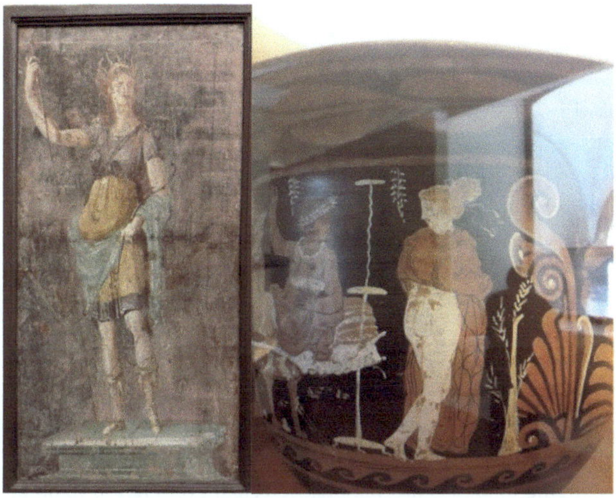

Fig. 3. (left) *Fresco with Artemis, INV 9301from Pompei, VI Insula Occidentalis 10* -(right) Bell crater with red figures from Sant'Agata dei Goti INV. 82570. The vase dates back to the second half of the IV century BC; in the banquet scene is depicted a dancing figure who raises her mantle, robes, faithfully replicating the compositional scheme of the much more famous Marble Venus the pattern of the much more famous marble copy.

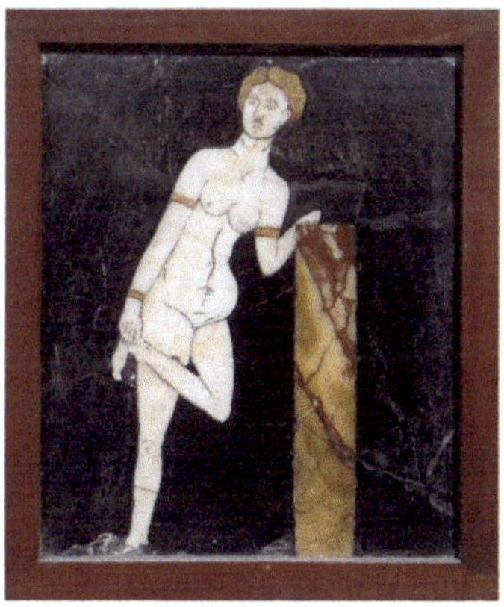

Fig. 4. INV 109678, Venus tying her sandal in *opus sectile*, from Pompeii (I, 2, 10), 41–68 AD

Among the selected statues, we have focused on two particularly meaningful examples: the statue of Isis (Fig. 5b), showcasing significant polychromy and gilding, and another example of Venus Anadyomene (Fig. 5a):

– a) INV 6298 Venus Anadyomene.

- b) INV 976 Isis with sistrum with traces of polychromy
- c) INV 6290 INV Isis head from acrolith inside the temple
- d) INV 6312 INV Dionysus with panther

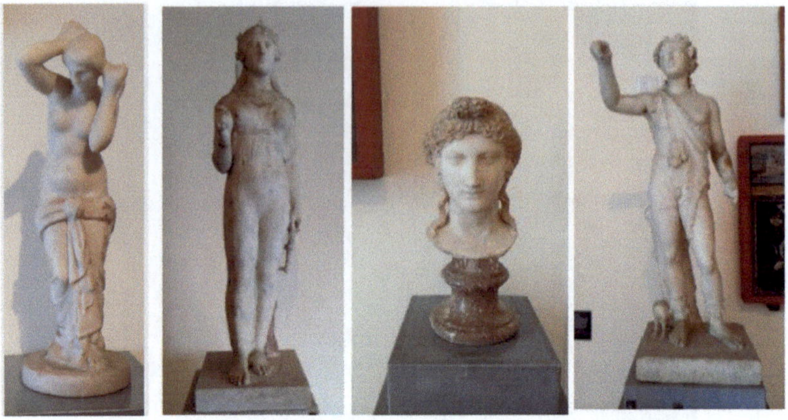

Fig. 5. Statues belonging to the Iseum Complex selected. From left to right: a) INV 6298 Venus Anadyomene; b) INV 976 Isis with sistrum; c) INV 6290 Isis head from acrolith inside the temple; d) INV 6312 Dionysus with panther

According to the plan of the temple and the locations of the findings during the excavation, both the Venus and the Isis were originally placed under the portico (respectively INV 6298 on the corner between the south and the west walls, and INV 976 on the corner between the west and the north walls), whereas the acrolith should have been positioned inside the *naos* and the Dionysus in a niche on the west side of the temple. The reason for choosing these sculptures lies in their polychrome and gilded features, notably visible in certain areas (such as on the statue of Isis) to the naked eye. While the Anadyomene contributes to better understanding the seriality of the iconography, the Egyptian – *Egyptianizing* goddess assumes a significant role in comprehending not just its pictorial but also its iconographic background, being particularly rare, if not unique (Fig. 5b).

Among the frescoes, we have selected those belonging to the portico, partially affected by colour changes problems, partially needing their re-location and reconstruction of missing elements:

- Floor plan sketch of the ancient architecture. The layout includes labeled areas such as the Porticos to the south, Temple in the center and Ekklesiasterion to the east (Fig. 6)
- Frescoes of the eastern wall of the portico (Fig. 7)
- Frescoes of the southern wall of the portico (Fig. 8)
- Frescoes of the western wall of the portico (Fig. 9)
- Frescoes of the northern wall of the portico (Fig. 10)

We decided to specifically select these frescoes, since they represent several interesting clues: they ran all around the porticus, following a repeated composition, interrupted only in correspondence of niches or passages. The decorations depict architectural motives: the *zoccolo* is made of oblong panels in yellow, enlivened by sea monsters and sphinxes, while the main ones are painted in red and decorated with architectural elements, divided by additional panels with fine garland ornamentation and vignettes. Finally, the upper part has a white background with thin *aediculae* separated from the main panels, by a scroll on a black background (Fig. 98).

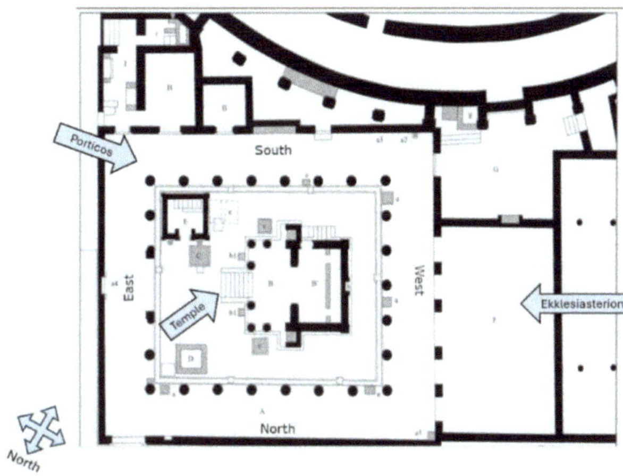

Fig. 6. Floor plan sketch of the ancient architecture. The layout includes labeled areas such as the Porticos to the south, Temple in the center and Ekklesiasterion to the east

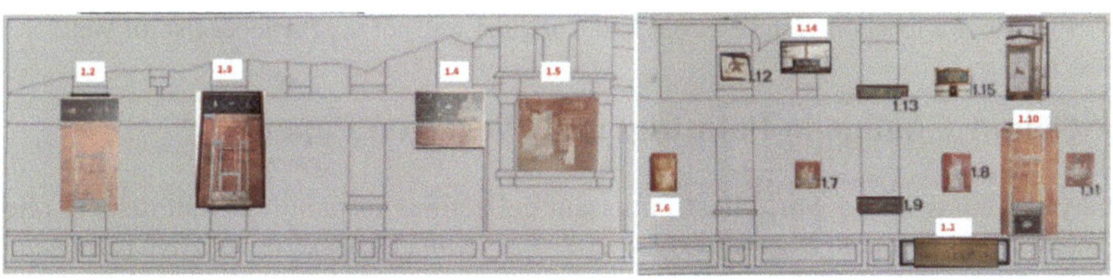

Fig. 7. Re-location of the frescoes in the original walls of the Porticoes (East Wall)

The goal of the project is to reconstruct in 3D the entire complex and to place the statues in their original settings, complemented by colour reconstructions directly linked to diagnostic data. In addition, the frescoes under the portico will be re-located, following De Caro's studies [26] and colour reconstruction will be obtained by the integration with AI and IBR techniques with available historical studies (Fig. 11), while missing parts will be added. In this way, it will be possible to overlap the 3D model of the site, as it is today, with its 3D reconstruction.

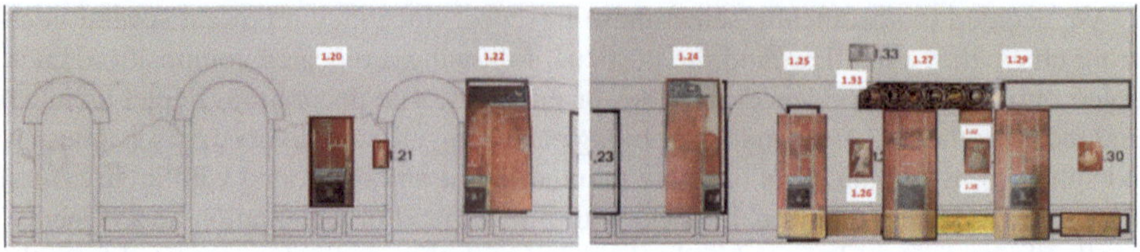

Fig. 8. Re-location of the frescoes in the original walls of the Porticoes (South Wall)

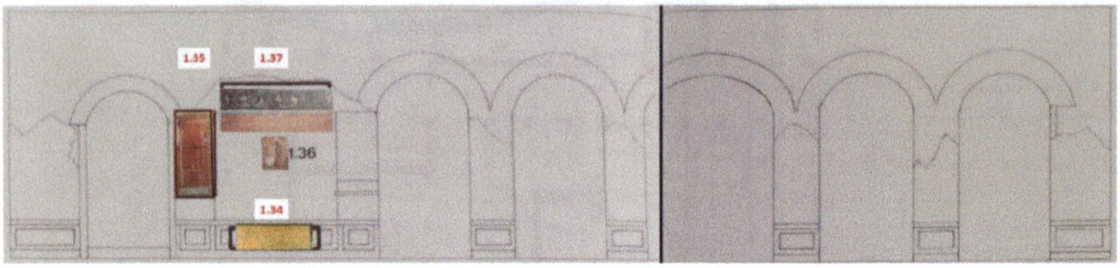

Fig. 9. Re-location of the frescoes in the original walls of the Porticoes (West Wall)

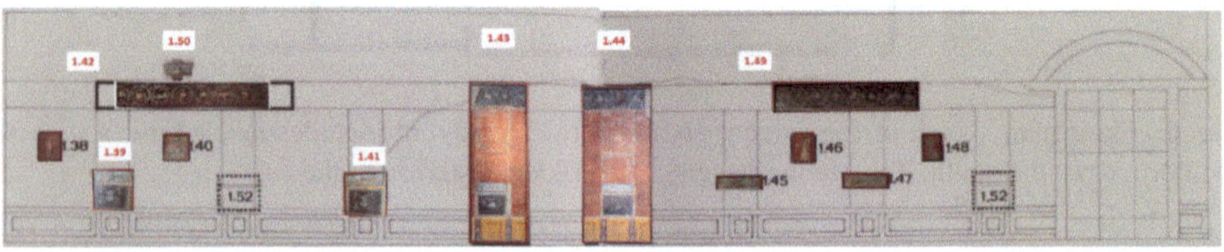

Fig. 10. Re-location of the frescoes in the original walls of the Porticoes (North Wall)

2 Colour Change in Paintings and Works on Paper

Many museums in the world host important collections of paintings that display phenomena of colour change, as is the case of *The Scream* (1910?) from MUNCH museum in Oslo [27], graphic art from the same collection or canvas paintings as those made during the Warne Munde period [28]. We have therefore selected this example to work on the problem of virtual colour reconstruction, and prediction of its potential change over time, based on the use of Colour science and Artificial Intelligence algorithms. Predictive systems together with specific guidelines for the preventive conservation of colours are not fully available and are needed not only for museum professionals, but also for the public in their experience of perception of the art-object as authentic and as a fragile item to be preserved. Scientific literature abounds in studies regarding the use of historically accurate reconstructions for better understanding the original techniques of modern and contemporary artists and the degradation mechanisms involved in the colour change of art materials (e.g., red lakes, cadmium yellow, chrome yellows…) in combination with oil media. Articles published by several authors [29–31], showed

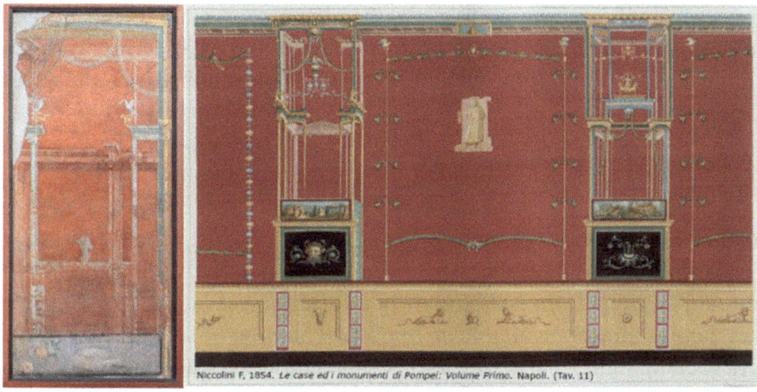

Fig. 11. a) An orthorectified photo of the fresco 1.35 conserved at MANN museum; b) example of manual re-colourisation and reconstruction of a wall with missing parts by Niccolini (19th cent.) Niccolini, F., & Niccolini, F 1854. "Le case ed i monumenti di Pompei disegnati e descritti", 1 (Tav. 11)

the behaviour of 19th century paints when exposed to light or darkness. Specific conditions of lighting, framing or retouching materials can lead to specific phenomena of discolouration phenomena (i.e. fading, yellowing, darkening) in oil paints. Optical characteristics of the paint materials have been measured using colourimetry and Kubelka Munch theory, the latest helping in digitally reconstruct/visualize original colours of discoloured paints. Other authors [32–35] published studies on eosin-based red lakes (e.g. geranium lake) that proved to be highly sensitive to light, leading to rapid fading of many modern paintings where they have been identified and digital reconstructions have been proposed based on the results from several analytical techniques (MA-XRF, micro-Raman, SEM-EDS, colourimetry and micro-fadeometry etc.). Other [36, 37] also published an extensive study on the fading of cadmium yellow paints, observed in many modern and contemporary artworks (e.g. *The Scream,* 1910?), showing the contribution of the manufacturing process and of environmental parameters such as high RH and light in the process of colour change. Spectral pigment mapping using multi and hyperspectral tools has been also reported by several authors [38, 39, 40, 41 pt.1-2-3] on case studies of modern and contemporary artists as source of information for designing a methodology for digital reconstruction of the colour change (fading, darkening, yellowing). Ciortan [42] recently proposed a multi-model approach to model fading effects in the spectral, temporal and spatial dimensions using micro-fading test (MFT) data and hyperspectral imaging on areas of a painting where photodegradation phenomena have been identified. A colourimetric reference of T*he Scream* (1910?) has also been proposed using photographic documentation available in the digital archive of MUNCH and colourimetric data, to approximate the original appearance of the famous icon [43].

2.1 Objectives

The main objective is to design a 3-steps concept and approach for: (I) creation of a sense of embodiment and perception of authenticity; (ii) building a sense of wonder; (iii) developing a sense of care aiming to strengthen the preservation efforts of these fragile collections. This concept design can be further used and explored for applications by curators/art historians, exhibition designers, other creative industries representatives and a general public.

2.2 Challenges

The main challenge regards the availability of the art objects, especially if they are in a restrictive exhibition set-up (as the case of *The Scream*) or in fragile conservation condition (as mural paintings from MANN) or in a location less accessible. Nevertheless, in some cases there is already available an important amount of scientific data (from previous analytical campaigns) that can be used to start colour processing and reconstructions. In addition, there are challenges regarding the references to be used for colour reconstructions, as in most cases the original documentation (coloured photographs) of the art objects at the moment of their creation is missing. In the case of *The Scream* (1910?) there are few black and white images from past publications in 1930 ca. (Fig. 12), but it is difficult to compare them with coloured photographs taken much later (in 1970s and 1980s). Another challenge is the fact that the taste for experimentation of modern artists, as Edvard Munch or Vincent van Gogh, and contemporary as Jackson Pollock, created the premises for further change in colour due to instability or incompatibility of paint materials [30, 44, 45]. Munch painted The Scream at a time when the paint-making industry was undergoing a rapid transition. Prior to the 1880s, paints were mixed in artists' studios with hand-ground materials, but toward the end of the 19th century, progress in the chemical industry brought tubes of premade paint to store shelves. Highly saturated factory-made pigments lent artists' creations vibrancy, but these paints were rarely, if ever, tested for longevity [27, 46]. Thus, in present times we often observe degradation patterns generated by the composition of the paint formulations and interactions with environments, and change in colour are among these patterns, posing also the question on how the future development of the change can be prevented or mitigated. In the art of Munch, the colours become protagonist of his innovative studio experimentation, applied in thin and/or thick brushstroke, squeezed directly from the tube, blended with crayon lines and contours, pure, mixed or overlapped, sometimes even applied with fingers or other tools than brushes. This taste for experimentation led to severe consequences regarding the stability of paint layers and integrity of the representation [47]. Nowadays many of his works present porous, powdery, de-cohesed paint layers, colour change phenomena (fading and darkening), abrasions and cracks, paint layers losses and lacunae. All these degradation forms lead to a loss of the readability of the image depicted but also to many questions related to long-term survival of these artworks. In a few cases, such as The Scream, the integrity of the object has also been affected by theft (2004) and other vandalic acts [27]. Regarding the concepts of authenticity and accurate reproduction of original colours, one main question is if the scientists would be able to reproduce using mock-ups the original appearance of the altered colours and if these model samples can faithfully model the composition of original paints and their degradation/colour change over time. Also, regarding the predictive models to be used, the question is what exactly we can predict, for how long and to which extent these models can be compared to the real processes of fading taking place (as in some cases the process of change reached a certain threshold and slowed down, when for others it might be ongoing). Another question to be discussed is the authenticity concept regarding reconstructions hypotheses and what are the main criteria or variables to be taken into account when proposing and displaying them to the public.

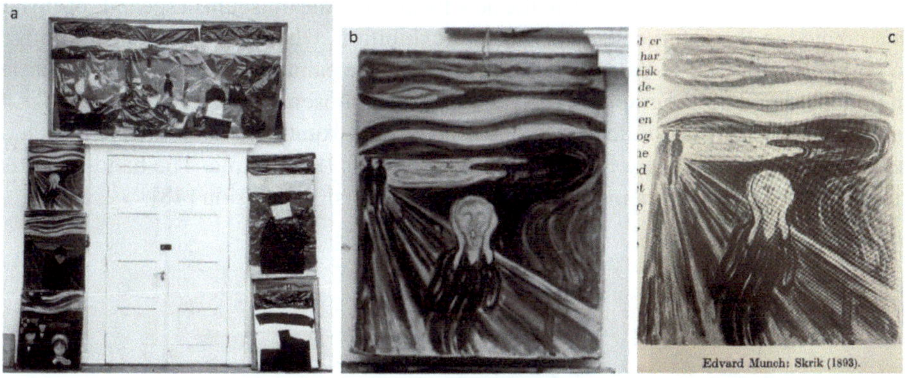

Fig. 12. Black and white images of the Scream around 1930s: a) the studio of Munch at Ekely, where Scream is displayed on a wall together with Anxiety and Despair (photo-archive væring.no/MUNCH, 1938); b) the close-up on Scream (from picture a); c) image published in Østby Leif, Naturalisme til nyromantikk, Oslo 1934, ill. s. 133.

3 Case Studies

We have considered specific case studies:

1. ***The Scream*** (1910?) by Edvard Munch and ***the hand-coloured print*** (1895), with fading phenomenon in areas of yellow and red paint
2. ***Road in Provence (1880–1890),*** a watercolour on paper by Paul Cezanne from the collections of the Art Institute of Chicago with some evidence of colour change and
3. ***Fresco paintings*** from Roman time with changes in colour from yellow to red, housed by MANN in Naples.

The Scream (1910?) by Edvard Munch and the Hand-coloured Print with Fading Phenomenon (1895)

Munch blended various materials such as oil, pastel and tempera to produce "colours that were screaming" as much as possible, to express his own perception of nature's scream when blended with human despair and anxiety (Fig. 13). Unfortunately, over time these experiments challenged the preservation of Munch's artwork. The original yellow brush strokes in the sky in the painting have faded to muted white, or off-white. This also applies to the neck portion of the central screaming. The cadmium sulfide yellow colour is fragile and prone to light induced discolouration as well as some red dye used in the area of the bridge. The thick strips of the sea are also flaking off. A big halo produced by a liquid agent (acting during the theft from 2004 to 2006) generated a deturpated mark in the left corner of the Scream painting, the cardboard also being heavily damaged in this area. To prevent the painting from decaying even more, the Munch Museum has kept the painting in storage, with controlled environmental conditions (low light exposure levels, a constant temperature of ca.18 degrees C, and ca. 50% relative humidity). Before 2021 when the Scream started to be exhibited in the new location in Bjørvika, the main version was barely accessible. Nowadays, in the new location of the museum, a special set-up has been created in the Infinite semi-permanent exhibition, where the Scream room offers the possibility to see 3 of the versions of the motif (the painting on cardboard, the pastel on cardboard and a lithograph). The type of display chosen was selected also thanks to the outcomes of earlier studies [36], thus scheduling a rotation that assures that each of the 3 artworks exhibited are on display not more than 1h at a time. The hand-coloured print is never on display, being also the most light-sensitive version [48].

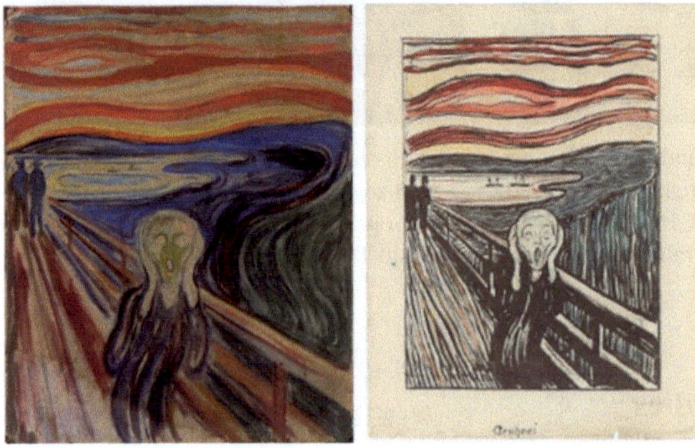

Fig. 13. Two versions of the same motif, The Scream, in different techniques and on different supports: left – main version dated 1910?, on cardboard and right – hand-coloured print on paper dated 1895, MUNCH, Oslo (credits MUNCH)

Road in Provence (1880–1890)

Cezanne's *Road in Provence* was chosen as a case study as it shows possible phenomena related to colour change. An historic B&W photograph of the watercolour – the earliest known image captured around 1934 – will serve as a reference point to determine which areas of the watercolour, and to what extent, may have undergone colour change from 1934 to date. Comparisons between the historic photograph and its current condition will be undertaken with the aid of hyperspectral imaging, X-ray fluorescence spectroscopy and the results of new B&W photography. Research on the spectral sensitivity of historic and contemporary B&W film, lighting conditions and other capture parameters will also inform the reconstruction of the visual appearance of the watercolour, as it appeared in 1934. This watercolour on paper by Paul Cezanne belongs to the collections of the Art Institute of Chicago and presents evidence of colour change (Fig. 14).

Fresco Paintings from the Villa of Papyri (Herculaneum) with Colour Shift.

Some of the Pompei frescoes identified at MANN have been already described. To better define a methodology suited in case of colour change, we have also included new more appropriate examples: a series of frescoes from the Villa of Papyri of Herculaneum.

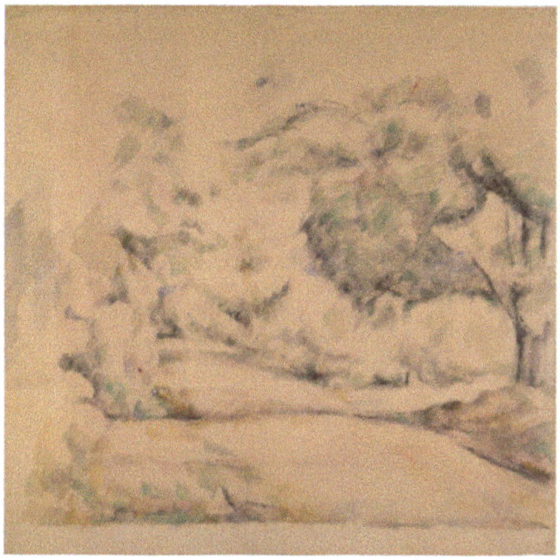

Fig. 14. Road in Provence by Paul Cezanne, watercolour and graphite on tan wove paper (1880–1890), Mr. And Mrs. Martin A. Ryerson Collection, Art Institute of Chicago (Reference number 1933.1221).

They present in fact evident widespread colour toning phenomena caused by the eruption of Vesuvius in 79 AD (Fig. 15).

In a fresco with landscapes from the Villa of Papyri in Herculaneum it is visible the change from yellow to red, widespread over the entire surface; originally it was a yellow monochrome panel but later, due to the heat released by the eruption of Vesuvius [49–51] (around 400 °C), it changed the chemical-physical structure of the pigments on the walls, turning the yellow into red, from goethite to hematite. In a fresco with a duck and deer (Fig. 15 left) the veering occurred in the lower part of the panel, while in the fresco with Cupid (Fig. 15 right), involved the entire surface, leaving intact the upper right part of the painting, between Cupid's wings, where the original red colouring is clearly evident. The same phenomenon of toning is also documented by the pictorial evidence still in situ: an example is in the northernmost room of the first lower level (*basis villae*) of the Villa which has been recently excavated in 2007–8: this room, at the moment of discovery, was totally filled with pyroclastic material, causing part of the stuccoes on the vault to become detached contact, at the time of the eruption and arrival of the pyroclastic flow, between the enormous heat and the paint significantly altered the appearance of the yellow, which turned red. Of particular interest is the impression that the cloud left on the walls, thanks to which the path of access is clearly distinguishable.

This colour toning is exemplified by other known cases in literature from Herculaneum, as those in the House of the Deer in the north wall of corridor A [52]. The frescoes featuring the background decoration painted with yellow ochre (goethite) have experienced oxidation, leading to a change into red ochre (hematite). The chosen panels will be examined with non-invasive archaeometric investigations to obtain a comprehensive diagnostic overview that can later be compared with other documented examples of toning, not only from Herculaneum but also from Pompeii, although the phenomenon is less documented here [53].

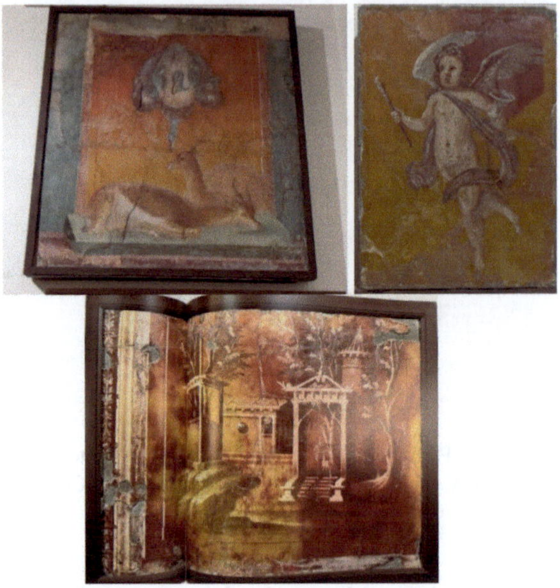

Fig. 15. (Left) Fresco with Duck and Deers, plaster and pigment, atrium, MANN-INV 8759 (Photo CNR ISPC) – (Right) Fresco with Cupid, plaster and pigment. MANN-INV 9319 (Photo CNR ISPC)

4 Colour Change in Textiles

Humankind has been pursuing materials for colouring, painting and dyeing since prehistoric times. There is an astoundingly large number of both inorganic and organic colourant sources to be found in nature worldwide where their identification is currently an intriguing subject because colour is significant to all cultures. Throughout history, organic dyes have been derived from plants and animals and used for colouring purposes; in this context historical textiles reflect the unique artistry and craftsmanship of weavers, artists and craftspeople through time [54, 55]. However, not all colours have resisted the test of time, since fading and degradation occur naturally, and ultimately this has changed our perception and appreciation of the objects as they were originally intended. The colour reconstruction of textiles has therefore generated considerable scientific interest in recent years. Colour, its perception and change in textiles resonate deeply with the public, and the potential for the accessibility of this type of collection is great, including items which remain in archives. Many issues make it difficult for scientists to characterise the origin to dyes in textiles, namely, the large number of chemical classes they belong to and the wide range of potential dye sources; the small amount of sample that can be available for analysis and the low amount of coloured compounds (chromophore-containing molecules) they contain. Furthermore, the complex breakdown processes of chromophore-containing molecules, which are also linked to matrix degradation processes, are highly influenced by the objects' exposure to light, temperature, relative humidity and atmospheric pollutants [56, 57].

4.1 State of the Art

In recent decades, academia and industry have pursued investigations into online digitisation of textiles that have ultimately supported and supplemented the multidisciplinary field regarding art historical interpretations and visual examinations. The identification of natural dyestuffs in ancient objects is a difficult process because of their complex chemical composition and the possibility of having both chromophores and degradation products concurrently [58, 59]. Due to this, new analytical methods and techniques for the identification and characterization of organic dyes in microsamples have been developed and refined during the past few decades. Furthermore, owing to increasing demand of minimally invasive/non-invasive techniques, more recent developments of novel techniques have resulted in the availability of many alternative/complementary methods and protocols to reveal features leading to the identification of fibre substrates and weaves, original dyestuffs, products and markers of degradation and further contaminants. A combination of multispectral imaging, chromatographic and spectroscopic techniques constitute the modern analytical toolbox [60–64]. These techniques are completely non-destructive and provide the possibility to give a rapid overview of used materials (dyestuffs, but potentially also traces of past conservation treatments) in a larger area. Vibrational spectroscopy, i.e. Raman and infrared spectroscopy are easily accessible methods for dyestuff analyses – most dyestuffs are sensitive in the infrared and visible range of wavelengths, however, in the case of dyed textiles the signals caused by the fabric is dominant and overshadow all bands originated by dyes. Dye characterization in textiles by Surface Enhanced Raman Spectroscopy (SERS) alternatively exploits metallic nanoparticles to enhance Raman signals, enabling the identification of trace dyes with high sensitivity and minimal sampling

The most accurate analyses are however still performed by HPLC techniques due to ther widely available databases. Coupled with an infrared spectrometer or a mass spectrometer, components of the dyestuffs can securely be identified. Results can be improved by adding a second mass spectrometer: after separation of the components by LC, a first mass spectrometer would separate all ions; the second allows to select one ion and fragment it. Secondary ion mass spectrometry techniques which allow surface analyses on microscopic samples become more and more used in dyestuff analyses, however, these analyses require specific instruments which are not accessible for everyone, and specific expertise. To investigate metallic components such as metal threads or mordants, XRF based techniques can be used. The approaches range from the use of a simple handheld XRF instrument to the use of synchrotron-based techniques which provide very high sensitivity and spatial resolution in the micro and nanometre scale. According to the choice of protocol, samples may or may not be consumed during analysis depending on the destructiveness of the analytical procedures employed. It is frequently more efficient these days to combine the two strategies, using non-invasive techniques to inform object sampling. In any case, whatever the analytical methodology, the analytical data obtained for unknown samples are typically assessed against reference in order to identify the dyeing sources. Therefore, studying reference materials that may have undergone artificial aging is also extremely important. The appearance and mechanical characteristics of textile artifacts are significantly impacted by the fading of coloured fibres as well as the degradation of the textile substrate. For this reason, understanding photo-oxidation chemical routes and fading mechanisms is essential to inform effective preventive conservation measures. To study fibre degradation, one parameter is the degree of crystallinity. This can be determined by polarised infrared spectroscopy and X-ray diffraction. The crystallinity studies may be complemented by chemical analyses, for example of amino acid sequences, and NMR studies revealing changes in the molecular structure of the fibre [65]. The knowledge afforded by the combination of analytical techniques provide a better understanding of the objects' complex genesis, their materials, techniques and recipes, which is a prerequisite for a more accurate reconstructions of the objects' original appearance.

4.2 Objectives

Identify the composition of selected objects and reconstruct their original appearance, creating opportunities for the general public to experience the digitally recreated appearance of the objects. Behind the scenes, conservators and scientists need to be able to visualise how an object was originally intended to appear; this knowledge, combined with research into the interaction of dyed textiles with the environment, will guide conservation and restoration efforts and provide information on how a textile may continue to age.

4.3 Challenges

The reconstruction of the original colour of textile objects is a relatively new field of investigation; although there are studies that can be used to predict degradation, these are of limited scope; the reliability of results could be reinforced by investigating better preserved areas in a textile (such as less exposed or hidden areas of dresses). The study of complex ageing processes and the characterisation of dyeing mixtures is only possible thanks to the increasing availability of highly specific and sensitive analytical procedures, which can be applied to the research on organic dyes in cultural heritage artefacts. However it could be difficult to infer the precise chemical degradation pathway of the materials in unknown environmental settings and therefore reconstruct the appearance of the object reliably. Historical recipes vary widely, and can easily be altered and adapted by individual artists and craftpeople, so it is extremely challenging to pin-point precise materials and quantities for any construction so that accelerated ageing procedures can be carried out with a degree of certainty to evaluate fading rates and determine the lightfastness of components.

4.4 Case Studies

Four case studies have been identified and are presented below. The objects were chosen depending on their availability and accessibility (the V&A recently underwent the biggest storage move since the second world war and many of the objects were inaccessible), but more importantly depending on the type and extent of fading and degradation exhibited. The contemporary kimono is a very good example of how even recent objects can be ephemeral in their appearance, while the Victorian dress shows how the balance of the various colours can be altered dramatically depending on susceptibility of different dyes to fading. This in turn affects the perception and appreciation of them by the general public. The two 16th century objects with tarnished metal embroidery were chosen to provide the general public with an idea of the dazzling appearance that this type of objects was intended to have, to provide a sense of surprise and awe generated by the appreciation of the difference, and engender a sense of care for the changed objects.

Kimono FE.422-1992

Reconstruction of original composition and appearance of a faded silk kimono, using a mixture of micro-destructive and non-destructive techniques (Fig. 16). Many of the green areas have faded since acquisition. Samples from faded and unfaded areas were analysed using spectroscopic methods to understand the degradation pathway, and the results were compared with the artist's original dye recipe. Photogrammetry will also be undertaken to capture the object in its 3D state, to facilitate the creation of digital replicas.

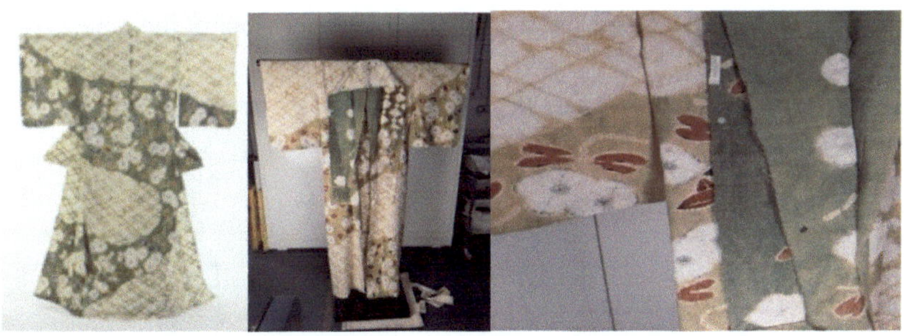

Fig. 16. Kimono (FE.422–1992); left: kimono seemingly unfaded; centre: unfolded to display original green with faded areas, right-higher detail of colour changes in areas of the kimono at V&A

Victorian Dress T.7-1926

Reconstruction of original composition and appearance of a faded cotton dress, using a mixture of micro-destructive and non-destructive techniques (Fig. 17). The bright blue background has faded in all the exposed areas. Threads were removed from significant areas and analysed using chromatographic and spectroscopic methods to identify the dye mixtures used. Colourimetry was used, and photogrammetry will also be undertaken to capture the object in its 3D state and facilitate the creation of digital replicas.

Early 17th Century Purse and Pincushion T.52-1954

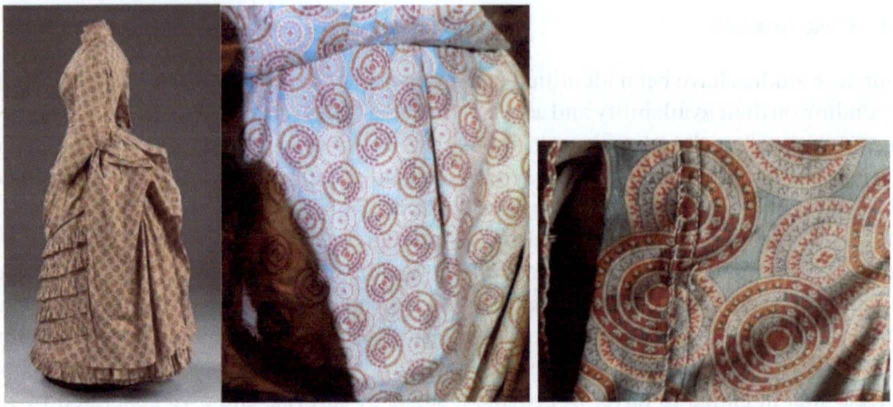

Fig. 17. Dress (T.7–1926); at top left the dress; top right: under a fold showing unfaded blue areas; bottom: detail of faded and unfaded areas on the cotton fabric. V&A©

Reconstruction of original composition and appearance of a faded embroidery and tarnished metal threads on a Elizabethan purse and pincushion, using a mixture of micro-destructive and non-destructive techniques. The objects were analysed with mapping XRF to identify the different types of metal threads used and reconstruct the original appearance of the embroidery. Digital microscopy was also employed to evaluate the morphology of the embroidery and provide a record of the current state of the decorated surface. Non-destructive, in situ dye analysis was also undertaken to narrow down the identity of the natural dyes used and identify the original colour of the objects (Fig. 18).

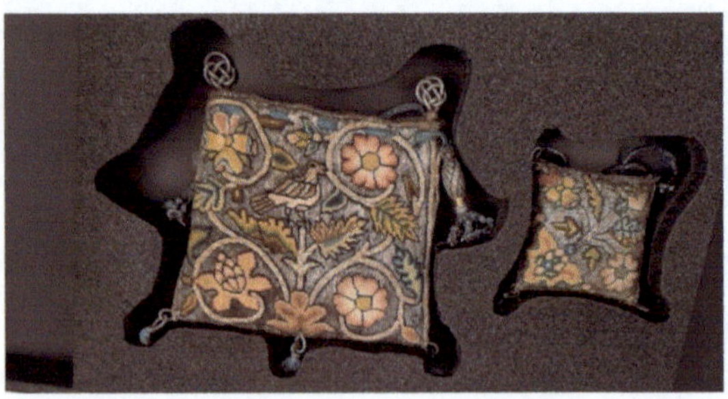

Fig. 18. Details of T.52–1954; left purse, and right – pin cushion ©V&A

16th Century Burse T.40-1986.

Reconstruction of original composition and appearance of tarnished metal thread on an Elizabethan velvet burse used for the Great Seal of England, using a mixture of microdestructive and non-destructive techniques. The burse was analysed with mapping XRF to identify the different types of metal threads used and reconstruct the original appearance of the metal embroidery and identify later repairs and interventions the XRF results. Digital microscopy was also employed to evaluate the morphology of the embroidery and the state of the underlying textile and provide a record of the current state of the object's surface. Non-destructive, in situ dye analysis was also undertaken to narrow down the identity of the natural dyes used for the textile fabric used underneath the embroidery (blue and red colours can be currently discerned upon close inspection) (Fig. 19).

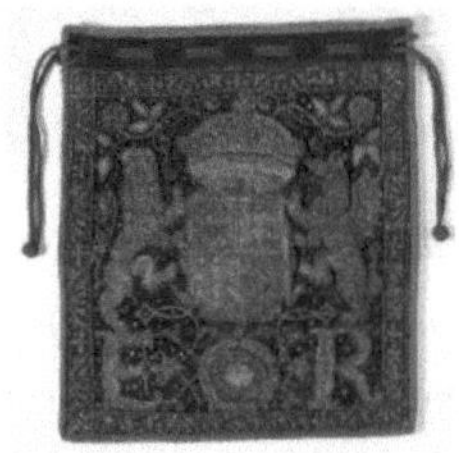

Fig. 19. Image of an Elizabethan burse (T.40–1986) burse at V&A©

5 Colour Change in Photo/Film Collections

Two early photographic and cinematography processes were chosen for investigation: the **autochrome** and **lenticular Kodacolor**.

5.1 Autochrome

The autochrome is considered the first commercially successful colour photography process. Patented in 1903 and released in 1907, the process was developed by the French Lumière brothers. The process comprises a glass plate, on top of which sits a colour filter made of microscopic starch granules dyed red, green, and blue and a panchromatic silver gelatin emulsion (Fig. 20). In the early twentieth century, the colour process was lauded for its faithful and natural production of colour [70].

Today, autochromes are held in a range of museum collections internationally. Among notable collections is the Victoria and Albert Museum, London, which holds over two thousand plates by noted practitioners. While rare and valuable, autochromes pose a number of conservation considerations for the contemporary museum as they are extremely light sensitive [67], and consequently, colours are prone to fading. In addition, autochromes can be damaged by a range of external factors, including humidity and moisture [68]. Some examples of degradation are reported in Fig. 21.

Fig. 21.Degradation phenomena affecting autochromes plates. Top row: Unknown, Broadstairs Beach, c.1910, autochrome, RPS.699–2020; F A Paneth, 'Vierwaldstättersee', c.1927–8, autochrome, RPS.722–2020. Lower row: Hugh C. Knowles, Still life with flowers, c.1910, autochrome, RPS.529–2020; John Cimon Warburg, 'Melisande in the Wood', 1909, autochrome, RPS.1201–2020; Bernad Alfieri Senior, Portrait of artist's wife and son, c.1910, autochrome, RPS.697–2020. All© Victoria and Albert Museum. The Royal Photographic Society Collection at the V&A, acquired with the generous assistance of the National Lottery Heritage Fund and Art Fund.

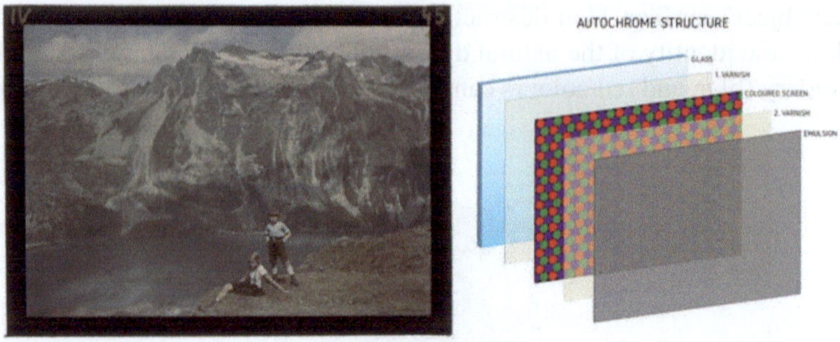

Fig. 20. Left: F. A. Paneth, *Sils: Eva and Heinz at Muotathal in the Engadine,* 1927, Victoria and Albert Museum, RPS.853–2020 © Victoria and Albert Museum, London. Right: layered structure of the autochrome [66].

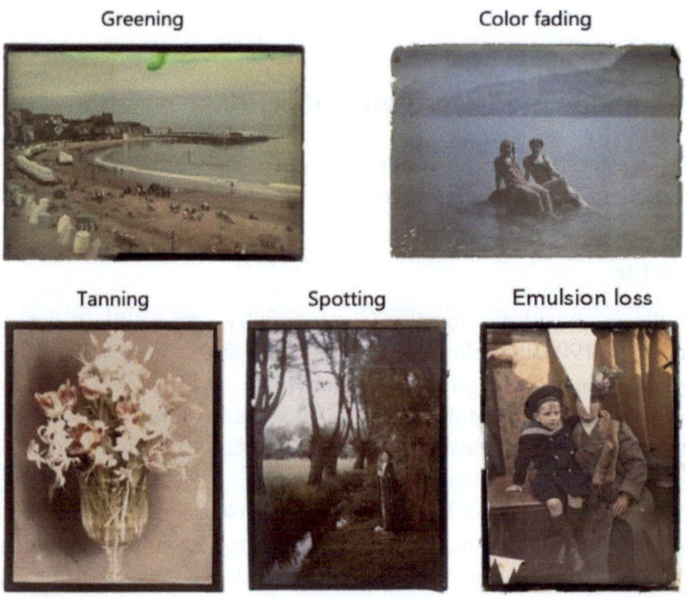

Fig. 21. Degradation phenomena affecting autochromes plates. Top row: Unknown, Broadstairs Beach, c.1910, autochrome, RPS.699–2020; F A Paneth, 'Vierwaldstättersee', c.1927–8, autochrome, RPS.722–2020. Lower row: Hugh C. Knowles, Still life with flowers, c.1910, autochrome, RPS.529–2020; John Cimon Warburg, 'Melisande in the Wood', 1909, autochrome, RPS.1201–2020; Bernad Alfieri Senior, Portrait of artist's wife and son, c.1910, autochrome, RPS.697–2020. All © Victoria and Albert Museum. The Royal Photographic Society Collection at the V&A, acquired with the generous assistance of the National Lottery Heritage Fund and Art Fund.

5.2 Colour Fading

Dye fading is a typical type of degradation affecting autochrome plates [68]. It has been determined that the colour mosaic is light-fast while the photographic emulsion is comparably more stable. As a result, prolonged exposure to intense illumination causes a loss of colour vibrancy in autochrome plates, sometimes shown as complete fading, or a browning of the dyes, known as tanning'. While the dye particles alter, the silver particles remain preserved. An approach to the digital restoration of faded autochromes is presented. The restoration approach adopts spectral imaging techniques aiming to separate the information corresponding to the two overlapping elements of the plates as depicted in Fig. 22. The dyes constituting the colour mosaic are transparent in the near infrared (NIR), while the absorbance of the silver-based photographic image extends in this electromagnetic region. Therefore, capturing an image of the autochrome plate in the infrared allows for the preservation of the photographic image without the colour mosaic absorption [BHJ∗].

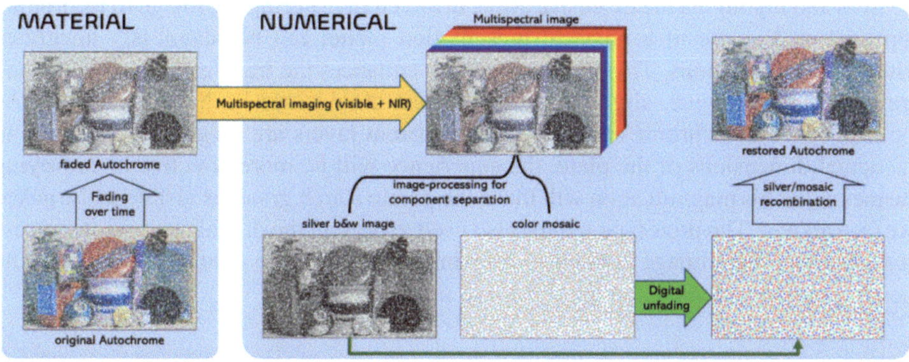

Fig. 22. Illustration depicting the proposed approach for the digital restoration of faded Autochromes

A hyperspectral scan of the plate is conceived to capture the full transmittance information of the colour image in the visible range [69]. An image-processing technique will be developed that combines the information of the NIR image and the hyperspectral cube thus obtaining the transmittance information of the sole colour mosaic. The absorbance of the faded colour mosaic can then be numerically increased to reinstate its supposed original colour. By recombining the restored mosaic and the photographic image a digital representation of the unfaded autochrome can be displayed. PERCEIVE envisions prototypes that introduce a novel approach to exhibiting historical artworks affected by the passage of time. These prototypes provide viewers with a deeper understanding of the restoration process, particularly when it remains faithful to the original. The aim is to emphasise that restoration is an act of care, intended to preserve the artist's original intentions without misrepresentation. This is particularly important because recent projects in early colour photography often involve digital alterations to the plates, such as smoothing or saturating images and colours, which can result in a departure from the faithful representation of the original objects. The use of transparency and transparent materials was favoured by several artists in the early 20th century, such as László Moholy-Nagy (1895–1946), Marcel Duchamp (1887–1968), and John Cage (1932–1992), Fig. 23, who used transparency to create counterpoints between diverse layers of content. As a layering scenario is inherent in the autochromes, it seemed natural to pick up on the use of layered transparency in the history of art and apply that knowledge to the presentation of autochromes, which would allow the visitor to differentiate between the present and the restored state of an autochrome (Fig. 24).

This presentation also highlights the role of technology in the creation of artworks and how they are affected by the passing of time, allowing viewers to see the condition they were in at the time of their creation. In case of colour fading this concept would translate in creating a layer replicating the silver black-and-white image, a layer replicating the faded mosaic, and a third one replicating the restored mosaic. To convey this, a physical exhibition prototype will arrange the layers in a stack, allowing visitors to observe their see-through combination from a specific viewing point. Substituting the faded mosaic with the restored mosaic creates an experience that vividly demonstrates the layer structure of an autochrome. This tangible representation showcases the fragility of the dyes and highlights the extensive effort involved in restoring the image to its original appearance. The use of a physical presentation model can be seamlessly integrated with digital interactions. This hybrid approach enhances the transfer of knowledge and prevents virtual fatigue, aligning with the latest innovations in dissemination. If the replicas of the autochrome elements on transparent layers are magnified compared to the actual dimensions of the plate, the experience will be more effective in conveying the message This magnification will make the potato starch granules visible to the naked eye (see above). The prototype will be also used to present the digital restoration process that generated new image content to impaint parts where the emulsion was lost. The viewing setting is depicted in the following image (Fig. 25).

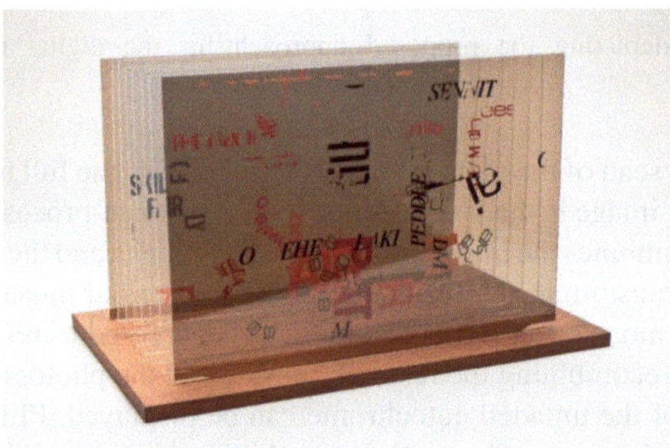

Fig. 23. An image of John Cage's artwork "Not wanting to say something about Marcel Duchamp".

Kodacolor

Some of the first home movies shot in colour used a 16mm lenticular film produced by Kodak from 1928 to the late 1930s. This very special film stock called Kodacolor is embossed with an array of hundreds of vertical cylindrical lenses that allowed recording colour scenes on a black-and-white panchromatic silver emulsion (Fig. 26). A considerable amount of surviving lenticular films is today hidden. As a matter of fact, lenticular film appears like a 'normal' black-and-white film at first quick look, so many lenticular

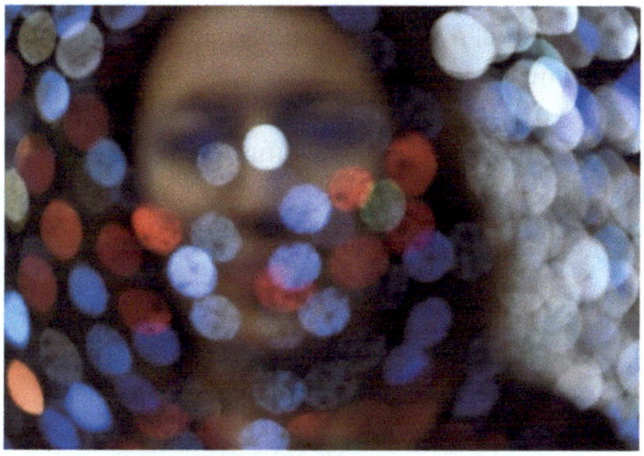

Fig. 24. Image created with Adobe Firefly using the prompt "Museum visitor in a dim environment looks through a transparency that contains a sharp mosaic of red, green and blue circular colour elements randomly arranged. Let's look at the visitor from the other side of the transparency."

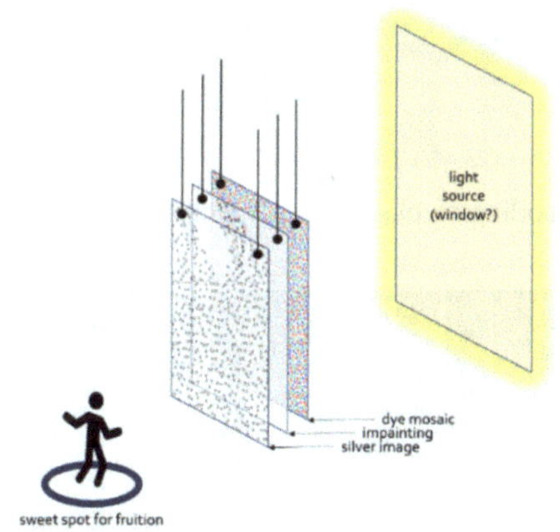

Fig. 25. Viewing setting for autochrome physical exhibition prototype

films have most likely been misclassified by collectors who will need further training to identify them correctly. It is therefore important to revive awareness of the lenticular colour processes thus making these precious historical colour movies available again to the public and securing them for posterity.

The original way to display lenticular Kodacolor consisted of special projection equipment with a tripartite red-green-blue filter that allowed the magnified colour picture to appear on the screen. Today this type of display is outdated. We are currently addressing this issue of technological obsolescence for Kodacolor. There are multiple possible methods to extract the colour information from the film images. Scanning the silver emulsion in high-resolution and letting software extract the encoded colour information represents a modern, efficient method to obtain digital colour images from these historical motion pictures (Fig. 27).

In this context, a new approach based on artificial intelligence has been demonstrated to be more efficient for the localization of the lenticular screen than other previous methods [70]. The results of the state-of-the-art solutions for the colour reconstruction of lenticular film are good. However, some improvements are still necessary. PERCEIVE aims to remove the residual artefacts by improving the quality and dimension of the labelled database, including lenticule warping/curvatures in the data augmentation, treating data as video sequences instead of individual frames, designing a "feedback loop" for data labelling using IQ metrics.

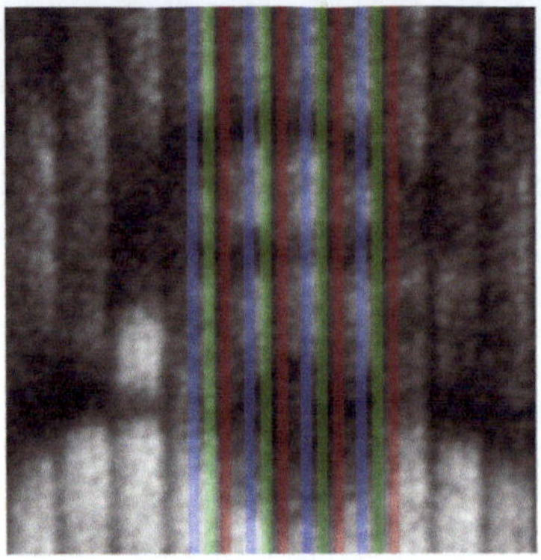

Fig. 26. Spatial colour encoding on black-and-white photographic emulsion in lenticular film

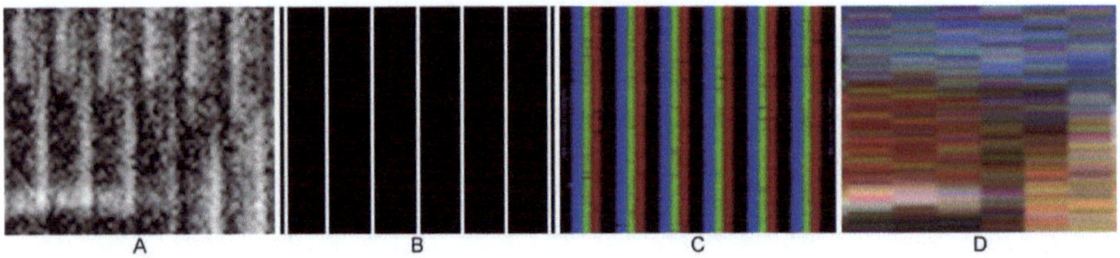

Fig. 27. Digital colour reconstruction process for a tiny fraction of a lenticular image – A: The input black-and-white image

6 Born Digital Art

Born Digital Art represents a transformative shift in contemporary artistic creation. Artists now employ advanced software and coding tools to conceive and realize works entirely within the digital realm. This emerging art form encompasses a wide range of media, including digital painting, three-dimensional computer-generated imagery, virtual and augmented reality experiences, and interactive installations. Defying traditional boundaries, Born Digital Art offers a dynamic and participatory canvas—one that is amplified by its global accessibility. A notable subset of this field, Augmented Reality (AR) art, seamlessly integrates digital content with the physical world, providing audiences with immersive and interactive experiences through AR-enabled devices.

However, the preservation of born-digital artworks poses a significant challenge due to the rapid pace of technological change. Ensuring their long-term accessibility and authenticity requires innovative conservation strategies and adaptive methodologies. Within this context, the PERCEIVE project, part of the European Horizon initiative, is devoted to advancing our perceptual understanding of colour in cultural heritage through the use of emerging technologies. Its primary aim is to develop new approaches for perceiving, preserving, curating, exhibiting, and understanding both historical coloured collections and contemporary digital artworks.

Nonetheless, the preservation of born-digital art remains a formidable challenge, given the rapid pace of technological evolution in the digital sphere. Addressing this issue requires innovative strategies to ensure the longevity and accessibility of these ever-evolving digital masterpieces. The PERCEIVE project, under the European Horizon initiative, is dedicated to advancing our perceptual understanding of colour in cultural heritage through the use of emerging technologies. Its primary goal is to pioneer new approaches for perceiving, safeguarding, curating, exhibiting, and interpreting both historical coloured collections and contemporary digital artworks, thereby fostering renewed public engagement.

By presenting case studies that exemplify artworks conceived within the digital realm and intrinsically linked to the technologies of their creation, PERCEIVE expands its scope to encompass current forms of cultural expression. This perspective invites a deeper reflection on the evolution and continuity of modern art, highlighting the new and dynamic roles that artists and conservators must assume to preserve born-digital works, sustain a living cultural narrative, and safeguard historical continuity.

Shaw 54th Memorial: At Home

The AR Work "Shaw 54th Memorial: At Home" is an innovative Augmented Reality (AR) experience that brings this iconic public artwork into people's homes, enabling remote accessibility through the Hoverlay app. The experience features a three-dimensional reproduction of the memorial accompanied by four holographic narratives that enhance public engagement in a novel and educational way. This work presents a groundbreaking approach to AR art by challenging the conventions of traditional exhibition spaces and expanding its reach to a broader audience. Through its interactive and immersive storytelling, Shaw 54th Memorial: At Home redefines the use of AR as an educational tool, transforming historical narratives into participatory experiences. In times when physical museum visits may be limited, it enriches the home-based encounter with art, emphasizing its capacity to transcend physical boundaries and adapt to our changing world. (Fig. 28).

MetroNeXt+

MetroNeXt+ is a groundbreaking Augmented Reality (AR) journey that virtually connects the bustling heart of New York City with the charming streets of Zurich (Fig. 29). The project transcends physical boundaries, inviting participants to explore Zurich's architecture, arts, sciences, and lifestyle—all from the iconic Grand Central Terminal. Employing cutting-edge 2D and 3D graphics alongside innovative photogrammetry technology, this immersive experience merges floating Platonic and Archimedean solids with 3D models of Zurich's old town, creating a symbolic and experiential bridge

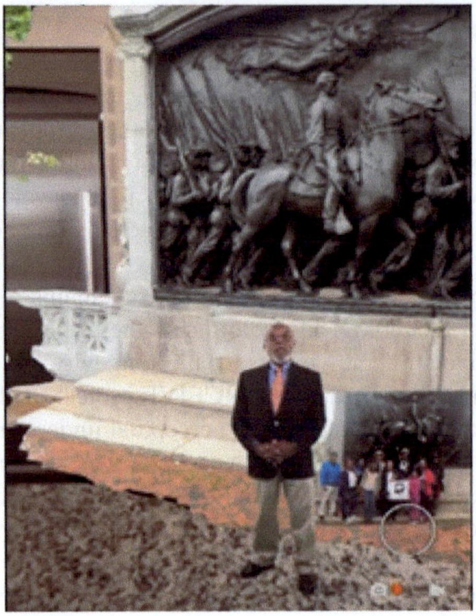

Fig. 28. Shaw & 54th Regiment Memorial: An Augmented Reality Experience – Delve into a digital replica of the iconic Boston Common monument, featuring four accompanying holographic stories. Presented by Hoverlay in collaboration with the National Park Service and the Friends of the Public Garden.

between the two cities. A collaboration between Digital Art Weeks, Virtuale Switzerland, and ETH Zurich, MetroNeXt+ fuses art, technology, and culture to foster a new kind of transatlantic dialogue. It pioneers the concept of virtual transatlantic travel, allowing visitors to move seamlessly from Grand Central Terminal to Zurich's streets within a shared hybrid space that unites the physical and digital realms. By offering an immersive insight into Zurich's culture, the project promotes cross-cultural exchange and dialogue. Its innovative inclusion of a physical subway entrance introduces a "phygital" dimension—blending tangible architecture with virtual experience—to create an exhibition design that transcends the purely digital sphere. Through this approach, MetroNeXt+ redefines how technology, art, and cultural exploration intersect, using photogrammetry to enhance its digital artistry while anchoring the experience in a physical context.

Bring The Greenway Home.

The work "Bring The Greenway Home" is a groundbreaking Augmented Reality (AR) project that transforms your living room or outdoor space into a dynamic AR gallery. Using the Hoverlay app, you can explore historic images of Boston's Greenway corridor while learning about the city's transformation and park's creation (Fig. 30). Contemporary artworks can be experienced at home, and you can share your discoveries through inapp photos and videos. Through a collaboration between the Greenway Conservancy, Boston Cyberarts, and Hoverlay, this initiative reimagines the possibilities of public art and technology. It bridges history, art, and digital innovation, inviting you to explore a century of growth and change along the Rose Kennedy Greenway. The innovation in the "Bring The Greenway Home" project lies in its ability to seamlessly merge art,

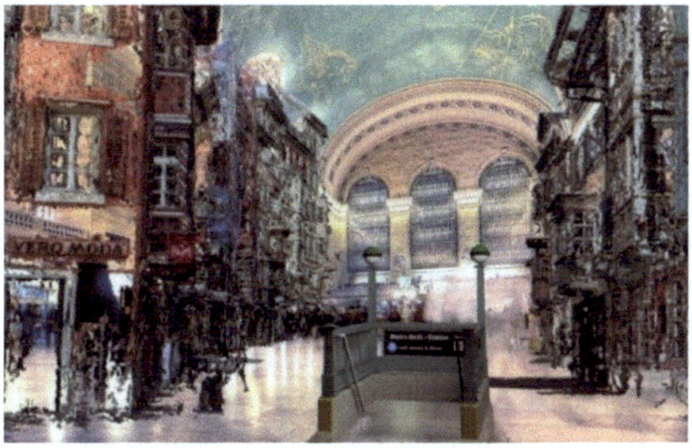

Fig. 29. MetroNeXt +: An AR journey from NYC to Zurich, blending 2D/3D graphics and photogrammetry. Presented by Digital Art Weeks, Virtuale Switzerland, and ETH Zurich, it redefines cultural exchange.

history, and technology into the everyday spaces of the public. This project breaks free from the confines of traditional gallery settings by utilizing Augmented Reality (AR) to transport the iconic Greenway corridor directly into people's homes or backyards. It offers a unique educational experience through interactive holographic stories that delve into historical imagery and contemporary art. Beyond expanding accessibility and engagement, this initiative redefines the use of AR as an educational tool and demonstrates the potential for public art to adapt to the ever-changing world. The project's pioneering spirit is in using technology to bring art, history, and community into our personal spaces.

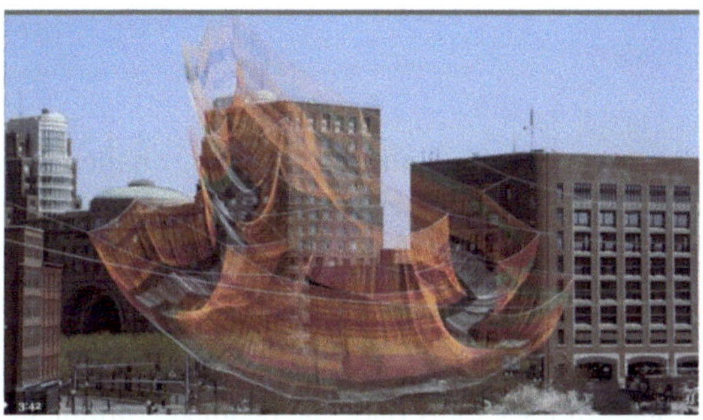

Fig. 30. "Bring The Greenway Home": An AR experience bringing Boston's historic Greenway corridor into your living space. Presented by the Greenway Conservancy, Boston Cyberarts, and Hoverlay, it merges art, history, and technology for an immersive educational journey.

6.1 Objectives

Many colour calibration tools and methods are designed for either analogue (print) or digital (screen) applications. Achieving a perfect match between analogue and digital colours remains challenging due to the inherent differences between these two media. By developing exhibition scenarios that juxtapose analogue and digital colour, it becomes possible to raise awareness of the challenges involved in achieving accurate colour calibration between real and virtual assets that share common colour palettes. This approach also supports the implementation of a colour-managed workflow—one that maintains visual consistency between the perceived colours of physical objects and their digital representations on AR-enabled devices.

To further enhance the AR experience and enable material and texture adjustments on 3D models, one of the project's objectives is to integrate support for the open-source glTF KHR_materials_variants specification. This integration allows 3D models to store multiple colour renderings and lighting conditions, enabling creators to explore how time, context, and restoration processes affect the visual appearance of cultural artefacts.Once this specification is incorporated into the Hoverlay AR Framework, artists will be able to employ dynamic texture adaptation, ensuring visual coherence in real-world settings where light intensity and quality change over time. This technology benefits not only artists but also museums, offering new possibilities for curatorial practice and the presentation of AR artworks in indoor environments.

Because many AR artworks are site-specific and typically displayed outdoors, there is strong interest in "bringing the outside inside." To achieve this, the project proposes the creation of a Reality Reflectance Bubble—a digital environment that replicates the external site where an artwork is intended to be viewed. This feature would allow artists to test or present AR artworks within a lifelike simulation while enabling museums to exhibit site-specific works within their own spaces. The Reality Reflectance Bubble concept would also rely on the glTF KHR_materials_variants specification, ensuring that the virtual textures dynamically respond to the lighting conditions of the outside world, thereby making indoor exhibitions of site-specific AR artworks possible.

6.2 Challenges

The preservation of digital art has long posed a recurring challenge across artistic disciplines. In the case of electronic and digital media, a central question arises: can a viable long-term conservation strategy truly exist? The answer is complex—while some works can be preserved, others inevitably cannot. This dilemma stems from the dependence of electronic art on the continued availability of the technologies used for its creation and display, or on the artist's provision of sufficient documentation to enable faithful re-creation with contemporary tools once the original equipment becomes obsolete. Whereas an artwork carved in stone can endure for millennia and a painting for centuries, digital and electronic works have a far shorter lifespan due to their reliance on fragile, fast-evolving technologies that often become irreparable or unsupported. This reality raises a profound question about the meaning and feasibility of preserving digital art. Yet, the alternative—allowing such works to disappear—is equally untenable. Maintaining a historical record of each generation's digital cultural expression is essential to prevent enduring gaps in the collective memory, ensuring that future audiences can access and understand the creative heritage of the digital age.

6.3 Case Studies

The following case studies exemplify the impact of colour and transparency in Augmented Reality (AR) artworks, each illustrating how digital aesthetics, technological innovation, and conceptual meaning intersect within the contemporary art context. Together, they highlight the transformative power of AR in shaping perception and in redefining the relationship between the physical and virtual worlds.

"Dante's Inferno" by Mark Skwarek.

Crystal Coffin reinterprets the historical display of Mao Zedong's crystal coffin on Tiananmen Square, transforming it into a virtual meditation on power, history, and ideology. Within the virtual "China Pavilion," designed to reflect state-sanctioned architectural forms, the artists invite viewers to question how political symbolism persists in the digital era. The AR installation employs high-transparency glass textures that simulate realism—viewers can perceive Mao's body through a convincingly translucent surface, while light and reflection reveal the illusory nature of authority itself. By merging simulated realism and historical critique, *Crystal Coffin* exemplifies how transparency functions both as a visual and ideological device.

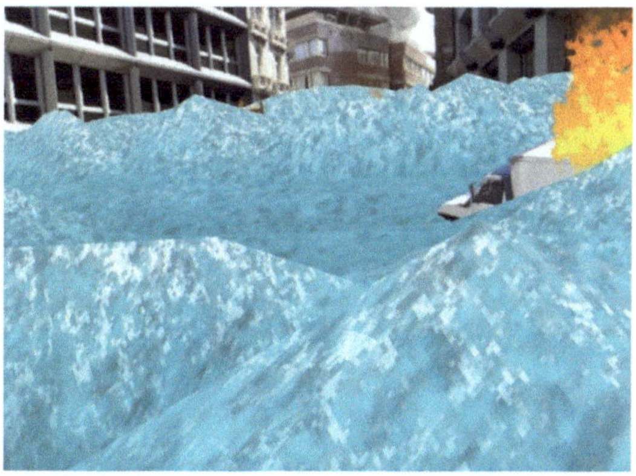

Fig. 31. Dante's Inferno; Artist: Mark Skwarek; Date of Creation: 2013; Transparency Use: Environmental Masking; Intended Setting: Any public square

The AR Artwork "Crystal Coffin".

Crystal Coffin reinterprets the historical display of Mao Zedong's crystal coffin on Tiananmen Square, transforming it into a virtual meditation on power, history, and ideology. Within the virtual "China Pavilion," designed to reflect state-sanctioned architectural forms, the artists invite viewers to question how political symbolism persists in the digital era. The AR installation employs high-transparency glass textures that simulate realism—viewers can perceive Mao's body through a convincingly translucent surface, while light and reflection reveal the illusory nature of authority itself. By merging simulated realism and historical critique, *Crystal Coffin* exemplifies how transparency functions both as a visual and ideological device (Fig. 32).

Simulated Realism; Intended Setting: Any public square

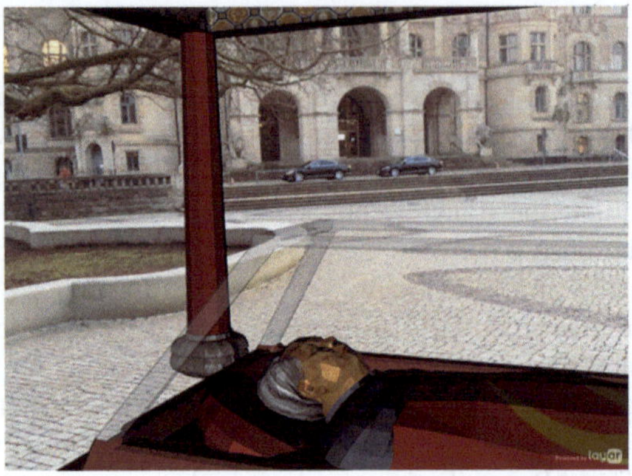

Fig. 32. Crystal Coffin; Artist: Lily & Honglei; Date of Creation: 2014; Transparency Use:

"Ecce Homo".

Ecce Homo is an AR artwork that merges self-portraiture, colour theory, and cosmology. The piece presents a floating self-portrait of the artist divided into eight cubes arranged as a three-dimensional cross, with intersecting red and blue lines symbolizing the connection between humanity and the cosmos. Through the use of transparency dimensionalization, overlapping colours and textures generate depth and motion, extending the artwork into a fourth-dimensional experience. The work challenges the boundaries between digital representation and metaphysical exploration, prompting viewers to contemplate human identity within the digital continuum. (Fig. 33).

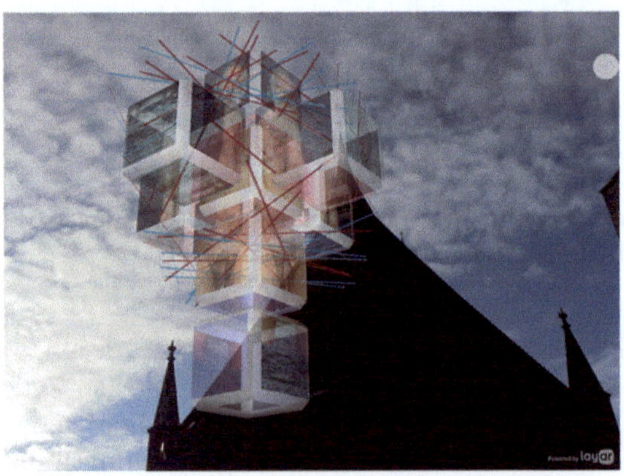

Fig. 33. Ecce Homo; Artist: Arthur Clay & Ingo Lie; Date of Creation: 2017; Transparency Use: Dimensionalization; Intended Setting: Am Markte, Hannover, Germany

The AR Artwork "Ascension of Cod".

Ascension of Cod reflects on the ecological and historical significance of the Massachusetts Bay cod, once central to the region's economy and culture. In this AR experience, a virtual school of codfish ascends through the air like a column of light, symbolizing both reverence and loss. The artwork employs transparency fading to blend digital fish seamlessly with the real environment, allowing natural light to affect their coloration and movement. This dynamic interaction between virtual and ambient light deepens the conceptual duality of the piece—celebration and mourning—while emphasizing the fragility of ecological balance.(Fig. 34).

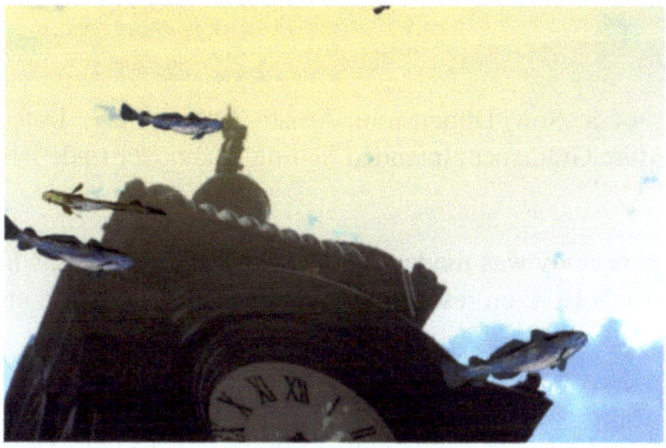

Fig. 34. Ascensions of Cod; Artist: Will Pappenheimer; Date of Creation: 2014; Transparency Use; Transparency Fading; Intended Setting: Any public square

The AR artwork "The Coming of a New Dimension".

The Coming of a New Dimension draws inspiration from Kurt Vonnegut's fictional *Tralfamadorians*, beings capable of perceiving reality in four dimensions. In collaboration with Virtuale Switzerland, the artist translates this concept into an AR "time capsule" that transitions between day and night, immersing viewers in a temporal journey. Using *transparency layering*, the capsule's textures shift through three stages: 90% transparency during daylight, 50% during transition, and 0% at night. This technique visualizes the passage of time and space travel from Earth to deep space, transforming the experience into a poetic meditation on perception, existence, and the limits of human understanding. (Fig.35)

Conclusion

Collectively, these case studies demonstrate the potential of AR art to merge material and immaterial realities. Through transparency, reflection, and light adaptation, each work redefines how digital art engages perception and meaning. The integration of these principles within the PERCEIVE project highlights how technological innovation can serve both artistic expression and cultural preservation, reinforcing the dialogue between contemporary creation and the enduring challenge of authenticity in the digital age.

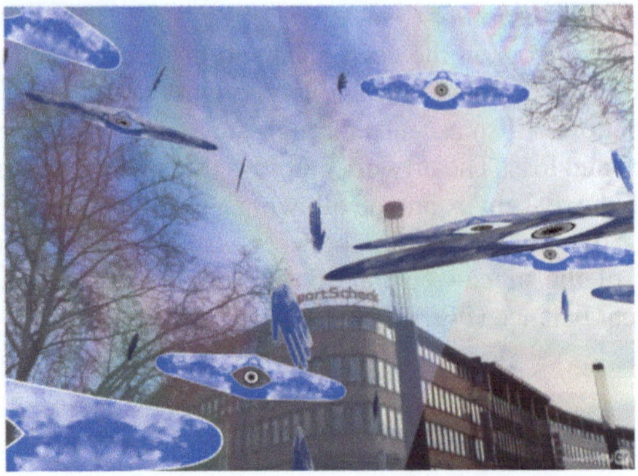

Fig. 35. The Coming of a New Dimension; Artist: Arthur Clay; Date of Creation: 2014; Transparency Use: Texture Gradience; Intended Setting: Hannover trade fair site

Acknowledgments. This study was funded by the PERCEIVE project, which received funding from the European Union's Horizon research and innovation programme under grant agreement No. 101061157.

Authors' Contribution. Cristiana Barandoni, Roberta Iannaccone, Donata Magrini, Sophia Sotiropoulou contributed to Section 1; Irina Crina Anca Sandu, Cristiana Barandoni, Letizia Monico contributed to Section 2; Lucia Burgio, Brenda Doherty, Hartmut Kutzke contributed to Section 3; Giorgia Trumpy; Catlin Langford, Lucia Burgio, Arthur Clay contributed to Section 4; Arthur Clay, Chiara Gemma Fedon and Dario Lanfranconi contributed to Section 5.

Disclosure of Interests. The authors have no competing interests to declare that are relevant to the content of this article.

References

1. Brøns, C., et al.: Tracking colour through time: polychromy on Etruscan urns from ancient creation to modern intervention. Am. J. Archaeol. **128**(2), 167–197 (2024). https://doi.org/10.1086/728694
2. Verri, G., Granger-Taylor, H., Jenkins, I., Sweek, T., Weglowska, K., Wootton, W.T.: The goddess' new clothes: the carving and polychromy of the Parthenon sculptures. Antiquity **97**(395), 1173–1192 (2023). https://doi.org/10.15184/aqy.2023.130
3. Kremer, G., et al.: Minerva in colours: first results on a polychrome Roman sculpture from Carnuntum (Pannonia). Heritage **6**, 5213–5241 (2023). https://doi.org/10.3390/heritage6070277
4. Liverani, P., Bracci, S., Iannaccone, R., Lenzi, S., Magrini, D.: New evidence about the polychromy of early Imperial cycle from the Augusteum of Rusellae (Tuscany). Heritage **6**, 3385–3401 (2023). https://doi.org/10.3390/heritage604017
5. Basso, E., Carò, F., Abramitis, D.H.: Polychromy in ancient Greek sculpture: new scientific research on an Attic funerary stele at the Metropolitan Museum of Art. Appl. Sci. **13**, 3102 (2023). https://doi.org/10.3390/app13053102

6. Campbell, L., Smith, M.: Multi-technique analysis of pigments on sandstone sculptures: renaissance repainting of a Roman relief. Herit. Sci. **10**, 156 (2022). https://doi.org/10.1186/s40494-022-00790-7
7. Brøns, C., Stenger, J., Bredal-Jørgensen, J., Di Gianvincenzo, F., Brandt, L.Ø.: Palmyrene polychromy: investigations of funerary portraits from Palmyra in the collections of the Ny Carlsberg Glyptotek, Copenhagen. Heritage **5**, 1199–1239 (2022). https://doi.org/10.3390/heritage5020063
8. Magrini, D., Bracci, S., Bartolozzi, G., Iannaccone, R., Lenzi, S., Liverani, P.: Revealing Mithras' colour with the ICVBC mobile lab in the museum. Heritage **2**, 2160–2170 (2019). https://doi.org/10.3390/heritage2030130
9. Østergaard, J.S.: "Reconstruction" of the polychromy of ancient sculpture: a necessary evil? Technè **48**, 110–119 (2019). https://doi.org/10.4000/techne.2656
10. Brinkmann, V., Koch-Brinkmann, U.: The Riace Bronzes Experiment. Aesthetics and Narrative. In: Fuchs, M. (ed.) Ahoros. Gedenkschrift für Hugo Meyer, pp. 15–34. Phoibos, Wien (2018)
11. Brinkmann, V., Dreyfus, R., Koch-Brinkmann, U., Camp, J., Piening, H., Primavesi, O.: Fine arts museums of San Francisco. In: Skulpturensammlung, L. (eds.) Gods in Colour: Polychromy in the Ancient World. Prestel, München, London, New York (2017)
12. Brinkmann, V., Wünsche, R. (eds.): Bunte Götter. Die Farbigkeit antiker Skulptur. Hirmer, Mainz (2003)
13. Siotto, E., Callieri, M., Dellepiane, M., Scopigno, R.: The Ulpia Domnina's sarcophagus: preliminary report about the use of digital 3D model for the study and reconstruction of the polychromy. In: Pensabene, P., Gasparini, E. (eds.) ASMOSIA X – Association for the Study of Marble and Other Stones in Antiquity, Rome May 21th–26th 2012, pp. 911–922. L'Erma di Bretschneider, Roma (2015)
14. Siotto, E., et al.: A multidisciplinary approach for the study and the virtual reconstruction of the ancient polychromy of Roman sarcophagi. J. Cult. Herit. **16**(3), 307–314 (2015). https://doi.org/10.1016/j.culher.2014.05.011
15. Skovmøller, A.: Facing the Colours of Roman Portraiture: Exploring the Materiality of Ancient Polychrome Forms, vol. 19. Walter de Gruyter GmbH & Co KG (2020). https://doi.org/10.1515/9783110585520
16. Campbell, L.: Tracking trajectories: projecting polychromy onto a roman relief from a Scottish castle. Heritage **6**, 3722–3744 (2023). https://doi.org/10.3390/heritage6040197
17. Verri, G., Opper, Th., Lazzarini, L.: In picturae modum variata circumlitio?: the reconstruction of the polychromy of a Roman female head ('Treu Head'). In: Liverani, P., Santamaria, M. (eds.), pp. 149–183 (2014)
18. Demetrescu, E., Ferdani, D., Dell'Unto, N., Touati, A.M.L., Lindgren, S.: SCIRES-IT – scientific research and information. J. Cult. Herit. **6**, 11 (2016). https://doi.org/10.2423/i22394303v6n1p51
19. Trizio, I., Demetrescu, E., Ferdani, D. (eds.): Digital Restoration and Virtual Reconstructions: Case Studies and Compared Experiences for Cultural Heritage, Digital Innovations in Architecture, Engineering and Construction Series. Springer, Cham. https://doi.org/10.1007/978-3-031-15321-1
20. Kiilerich, B.: Towards a 'Polychrome History' of Greek and Roman sculpture. J. Art Historiogr. **15**, 1–18 (2016)
21. Niccolini, F., Niccolini, F.: Le case ed i monumenti di Pompei disegnati e descritti. Vol. 1. Naples: NN (1854). https://doi.org/10.11588/DIGLIT.3926
22. Heyob, S.K.: The Cult of Isis Among Women in the Graeco-Roman World (1975)
23. Bragantini, I.: Il culto di Iside e l'egittomania antica in Campania. In: Egittomania. Iside e il mistero, pp. 159–167. Electa, Milano (2006)

24. Bricault, L., Versluys, M.J., Meyboom, P.G. (eds.): Nile into Tiber. Egypt in the Roman world. In: Proceedings of the IIIrd International Conference of Isis Studies, Faculty of Archaeology, Leiden University, Brill (2007)
25. Brysbaert, A., Klinkenberg, V., Gutiérrez Garcia-M.A., Vikatou, I. (eds.): Constructing Monuments, Perceiving Monumentality and the Economics of Building. Theoretical and Methodological Approaches to the Built Environment. Sidestone Press (2018)
26. De Caro, S. Alla ricerca di Iside: analisi, studi e restauri dell'Iseo pompeiano nel Museo di Napoli, ARTI, Roma (1992)
27. Ydstie, I. (ed.): The Scream, Munch Museum. Vigmostad & Bjørke AS (2008)
28. Ferrer, J.S., Sandu, I.C.A., Syversen T., Cardoso A.M., Candeias, A., Borca C.N.: Investigating Colour Changes in Red and Blue Paints – A Preliminary Study of Art Materials and Techniques in Edvard Munch's Old Man in Warnemünde (1907), pp. 209–219 (2019)
29. Geldof, M., Proaño Gaibor, A.N., Ligterink, F., Hendriks, E., Kirchner, E.: Reconstructing Van Gogh's palette to determine the optical characteristics of his paints. Herit. Sci. **6**(1), 17 (2018). https://doi.org/10.1186/s40494-018-0181-6
30. Geldof, M., et al.: Recreating Van Gogh's original colours on museum displays. Electronic Imaging, vol. 2019, Burlingame (CA) (2019)
31. Townsend, J., Carlyle, L., Cho, J.-H., Campos, M.F.: The Yellowing/Bleaching Behaviour of Oil Paint; Further Investigation into Significant Colour Change in Response to Dark Storage Followed by Light Exposure. ICOM-CC Triennial Meeting Postprints, Lisbon (2011)
32. Fieberg, J.E., Knutås, P., Hostettler, K., Smith, G.D.: Paintings fade like flowers: pigment analysis and digital reconstruction of a faded pink lake pigment in Vincent van Gogh's Undergrowth with Two Figures. Appl. Spectrosc. **71**(5), 794–808 (2017)
33. Centeno, S.A., et al.: Van Gogh's Irises and Roses: the contribution of chemical analyses and imaging to the assessment of colour changes in the red lake pigments. Herit. Sci. **5**, 18 (2017). https://doi.org/10.1186/s40494-017-0131-8
34. Monico, L., et al.: Chemical alteration and colour changes in the Amsterdam sunflowers. In: Geldof, M., van Bommel, M., Vellekoop, M., Hendriks, E. (eds.) Van Gogh's Sunflowers Illuminated: Art Meets Science, pp. 125–158. Amsterdam University Press (2019). https://doi.org/10.1021/acs.analchem.0c03251
35. Chieli, A., Miliani, C., Degano, I., Sabatini, F., Tognotti, P., Romani, A.: New insights into the fading mechanism of Geranium Lake in painting matrix. Dyes Pigm. **181**, 108600 (2020)
36. Monico, L., et al.: Probing the chemistry of CdS paints in The Scream by in situ non-invasive spectroscopies and synchrotron-radiation X-ray techniques. Sci. Adv. **6**(20) (2020). https://doi.org/10.1126/sciadv.aay3514
37. Mayda, S., et al.: A combined experimental and computational approach to understanding CdS pigment oxidation in a renowned early 20th-century painting. Chem. Mater. **35**(24), 10403–10415 (2023)
38. Berns, R.S.: Rejuvenating the appearance of cultural heritage using colour and imaging science techniques. In: Proceedings of the 10th Congress of the International Colour Association, pp. 369–374. Granada (2005)
39. Berns, R.S., et al.: Rejuvenating the colour palette of Georges Seurat's A Sunday on La Grande Jatte – 1884: a simulation. Colour Res. Appl. **31**(4), 278–293 (2006). https://doi.org/10.1002/col.20223
40. Deborah, H., Sony, G., Hardeberg, J.Y.: Spectral-divergence based pigment discrimination and mapping: a case study on the Scream (1893) by Edvard Munch. J. Am. Inst. Conserv. **58**(1–2), 90–107 (2019). https://doi.org/10.1080/01971360.2018.1560756
41. Kirchner, E., et al.: Digitally reconstructing Van Gogh's Field with Irises near Arles: Part 1,2,3. Colour Res. Appl. **43**(2), 150–157 (2018). https://doi.org/10.1002/col.22162

42. Ciortan, I.M., Poulsson, T.G., George, S., Hardeberg, J.Y.: Tensor decomposition for painting analysis. Part 2: Spatio-temporal simulation. Herit. Sci. **11**(1), 84 (2023). https://doi.org/10.1186/s40494-023-00913-8
43. Ciortan, B.I.M., Trumpy, G., Bjørngard, H., Sandu, I.C.A.: The Scream (ca. 1910) throughout the years: from photographic documentation to spatio-temporal modelling. In: EUROGRAPHICS Workshop on Graphics and Cultural Heritage, Catania (2023). https://doi.org/10.2312/gch.20231182
44. Sandu, I.C.A., Van Den Berg, K.J., Candeias, A., Godhe Sandbakken, E., Storevik Tveit, E., Van Keulen, H.: Chemical Analysis in Cultural Heritage. Part II. Physicochemical approaches in art and archaeology investigation (chapter 5 – Multi technique and multiscale approaches to the study of ancient and modern art objects on wooden and canvas support). In: Van Den Werft, I., Sabbatini, L. (eds.) Physical Sciences Review. De Gruyter Publ. (2019). https://doi.org/10.1515/psr-2018-0016
45. Monico, L., et al.: Deeper insights into the photoluminescence properties and (photo)chemical reactivity of cadmium red (CdS\(_{1−x}\)Se\(_x\)) paints in renowned twentieth-century paintings by state-of-the-art investigations at multiple length scales. Eur. Phys. J. Plus **137**(3), 1–21 (2022). https://doi.org/10.1140/epjp/s13360-022-02447-7
46. Pettersen, P.: Edvard Munch – the scream. IJCS Spec. Issue **13**(Special Issue 1), 1405–1420 (2022). https://ijcs.ro/public/IJCS-22-104_Pettersen.pdf
47. Frøysaker, T., Streeton, N.L.W., Kutzke, H., Hanssen-Bauer, F., Topalova-Casadiego, B. (eds.): Public Paintings by Edvard Munch and his Contemporaries: Change and Conservation Challenges. Archetype Publications, London (UK) (2015)
48. Chan, E., Aambø, M., Godzimirska, M., Grimstad, I., Lojewski, T., Sandu, I.C.A.: Light-induced colour changes on the scream versions in the Munch Museum collection. In: 1st MUNCH2022 International Conference, Book of Abstracts, p. 93, Oslo, Norway (2022)
49. Allroggen-Bedel, A.: Ein Malerei-Fragment aus der Villa dei Papiri. CRERC **6**, 85–88 (1976)
50. Allroggen-Bedel, A.: Un frammento di dipinto. In AA.VV. La Villa dei Papiri, CRON Ercol **13**(Suppl. 2), 65–68 (1983)
51. Esposito, D. Le Pitture della Villa dei Papiri ad Ercolano. In: Verbanck-Pierard, A., Barbet, A. (eds.), pp. 211–225 (2013)
52. Baraldi, P., Minghelli, S., Scagliarini Corlaita, D.: Viraggi di coloure nelle pitture parietali di Ercolano: un'indagine archeometrica. In: DHER Domus Herculanensis Rationes, Antequem, pp. 299–308 (2011)
53. Baraldi, P., Bensi, P.: Alterazioni delle materie coloranti nelle pitture murali prodotte dalle alte temperature: fonti storiche ed indagini scientifiche. In: Jean, G. (ed.), Salvati dalle fiamme, Atti della giornata di Studio, Lugano, pp. 15–29 (2006)
54. Cardon, D.: Natural Dyes: Sources, Tradition, Technology and Science. Archetype Publications Ltd, London, UK (2007)
55. De Graaff, J.H.H.: The Colourful Past: Origins, Chemistry and Identification of Natural Dyestuffs. Abbeg-Stiftung, Riggisberg & Archetype Publications Ltd., London, UK (2004)
56. Machnowski, W., Wąs-Gubała, J.: Evaluation of selected thermal changes in textile materials arising in the wake of the impact of heat radiation. Appl. Sci. **11**, 6989 (2021). https://doi.org/10.3390/app11156989
57. Iqbal, M., Sohail.: Textile environmental conditioning: effect of relative humidity variation on the tensile properties of different fabrics. J. Anal. Sci. Methods Instrum. **2**(2) (2012). https://doi.org/10.4236/jasmi.2012.22017
58. Degani, L., Riedo, C., Gulmini, M., Chiantore, O.: From plant extracts to historical textiles: characterization of dyestuffs by GC–MS. Chromatographia **77**, 1683–1696 (2014)
59. Ferreira et al.: Chem. Soc. Rev. **33**(6), 329–336 (2004)

60. Degano, I., Ribechini, E., Modugno, F., Colombini, M.P.: Analytical methods for the characterization of organic dyes in artworks and in historical textiles. Appl. Spectrosc. Rev. **44**(5), 363–410 (2009). https://doi.org/10.1080/05704920902937876
61. Dyer, J., Tamburini, D., O'Connell, E.R., Harrison, A.: A multispectral imaging approach integrated into the study of Late Antique textiles from Egypt. PLoS ONE **13**(10) (2018). https://doi.org/10.1371/journal.pone.0204699
62. Houck, M.M.: The use of spectroscopy for textile fiber identification. In Houck, M.M. (ed.), Identification of Textile Fibers, pp. 158–164. Woodhead Publishing (2009). https://doi.org/10.1533/9781845695651.2.158
63. Lee, J., Kim, M.-H., Lee, K.B., Van Elslande, E., Walter, P., Lee, Y.: Analysis of natural dyes in archaeological textiles using TOF-SIMS and other analytical techniques. Surf. Interface Anal. **46**, 312–316 (2014). https://doi.org/10.1002/sia.5594
64. Vanden Berghe, I., Van Bos, M., Vandorpe, M., et al.: Non-invasive analysis of heritage textiles with MA-XRF mapping – exploring the possibilities. Herit. Sci. **11**, 183 (2023). https://doi.org/10.1186/s40494-023-00977-6
65. Gong, X., Zhang, Y.: Insight into the crystallinity of Chinese ancient silk by synchrotron radiation-based and conventional X-ray diffraction methods. J. Conserv. Sci. **36**, 1–14 (2020). https://doi.org/10.12654/JCS.2020.36.1.01
66. Langford, C.: 'Startling', 'extraordinarily beautiful', and 'obtrusive': reactions and responses to the stereo-autochrome. Stereo Immers. Media **6**(1), 90–103 (2022)
67. Pénichon, S.: Twentieth-Century Colour Photographs: Identification and Care. Getty Conservation Institute (2013). ISBN 978-1-60606-156-5
68. Lavédrine, B., Gandolfo, J.-P., Capderou, C., Guinée, R., Mcelhone, J.P.: The Lumière Autochrome: History, Technology, and Preservation. Getty Conservation Institute (2013)
69. Cucci, C., Casini, A., Stefani, L., Cattaneo, B., Picollo, M.: A Novel transmittance VIS–NIR hyperspectral imaging scanner for analysis of photographic negatives: a potential tool for photography conservation. Sensors (2023). https://doi.org/10.3390/s23073562
70. D'Aronco, S., Trumpy, G., Pfluger, D., Wegner, J.D.: A Deep Learning Approach for Digital Colour Reconstruction of Lenticular Films. https://doi.org/10.48550/arXiv.2202.05270

Designing Coloured Experiences

Sofia Pescarin[1](✉), Cristiana Barandoni[1], Vanessa Bonnano[1], Ivana Cerato[1], Daniele Ferdani[1], Donata Magrini[1], Marcello Massidda[1], Irina Crina Anca Sandu[2], Petros Stavroulakis[3], and Laura Travaglini[1]

[1] CNR ISPC, Florence, Italy
{sofia.pescarin,cristianabarandoni,ivana.cerato,daniele.ferdani, donata.magrini,marcello.massidda,lauratravaglini}@cnr.it
[2] MUNCH Museum, Oslo, Norway
irina.sandu@munchmuseet.no
[3] FORTH IESL, Heraklion, Crete, Greece
pstavroulakis@iesl.forth.gr

Abstract. This chapter focuses on the design methodology adopted with the goal of developing interactive experiences for the visitors aimed at engaging them with coloured collections. The methodological approach is described and a detailed description, following the PACT framework, is included. Key stakeholders are drafted, while an extensive space is given to the main nine activities that characterise colour reconstruction. Finally, a Co-Design tool is described.

Keywords: Interactive Media Design · PACT framework · Co-Design tool

1 Design Methodological Approach

To be able to develop appropriate User Experiences and applications for coloured collections, we have started from the analysis of user's needs and requirements, adopting a specific design approach known as **"PACT" framework** [1]. PACT stands for People Activity Context Technology, and it is a way to enquire and let emerge relevant elements to be used in the development of ideas and in the creation of a design briefs. To better adopt this approach, we have developed a question-oriented support for PACT. Questions in fact help to focus and to find answers to common problems within the CH domain. Examples of such questions are reported below in Table 1.

In Volume 2 we describe the design tools, specifically created to support the PACT framework and included in the PERCEIVE design toolbox.

A second approach that we have adopted is users' behaviours exploration designing specific **Cultural Probe Kits (CPK)**. As already described in Chaps. 3 and 4, Probes were originally developed as a means for designers to detect useful information, ideas, and opinions in specific contexts and in a non-intrusive way [2]. Cultural probes typically are a mix of objects included in a physical box that is given to a number of potential stakeholders for a certain amount of time. The objects are hybrid, therefore they might include fully digital items and other tangible items. They are carefully designed to

S. Pescarin et al. (Eds.): *Chromatic Visions*, LNCS 15750, pp. 223–252, 2026.
https://doi.org/10.1007/978-3-032-07792-9_7

Table 1. Examples of question-oriented reference cards

Frame	Methods / Focus	Questions
PEOPLE	– Interviews – Questionnaires – Observations – Social media – Cultural Probe Kits – Focus Groups and Workshops	• Who is the main audience? • Which are the main needs and characteristics?
ACTIVITY	– Scope and Goal	• What is the main problem to solve? • What are main needs and requirements?
	– Temporal Aspects	• How long will the activity last? Regularly? In calm or busy moment?
	– Cooperation	• Is activity meant to be done together or alone? Does it require cooperation?
	– Complexity	• Is the activity simple and well defined with little choices? Or complex and need guidance step by step?
	– Security	• Has the activity potential impact on users' security?
	– Content	• Which input data are needed? Output? Media? Requirements related to those data?
CONTEXT	– Physical Space	• How is the physical space? Climate? Noisy? Echo? Dirty? Wire/Wireless? Open air? Modern/historical? Dark? Electric plugs?
	– Social Context	• Alone or together? Friends or unknown? Language? Religious? Disability? Is feedback sound acceptable?
	– Organisational Context	• How's the working environment? Staff? Who do people work for? Role? Management structure?
TECHNOLOGY	– Media	• Which media? Which digital content should be acquired and processed?
	– Hardware	• Which visualisation devices? Which Interaction devices?
	– Network	• On line? Off Line? Local network?

"provoke inspirational responses" [3]. For example, Gaver's experiment included a map, postcards, disposable camera, and booklets. Probes do no specifically aim at gathering requirements, rather at confronting and providing inspiration to the designers. Moreover, Hulkko [4] referred to probes as stories or fragments of understanding: they are seen as means to engage with the lives of others. Following this approach, specific Cultural Probe Kits have been successfully produced for the understanding of users' behaviours referred to the concepts of caring (Chap. 3), of authenticity (Chap. 4) [5] and of social cohesion (Chap. 5) [6]. The result of this work has been included in a project report [7]. Furthermore, in Chap. 5.2 we have reported the results of another exploration carried out on "curiosity and interest triggers" as elements to be considered in the definition of the needs and requirements. The list of the templates are available in Zenodo:

- Cultural Probe Kit for Authenticity [8]
- Cultural Probe Kit for Sense of Care [9]
- Cultural Probe Kit for Social Cohesion [10]
- Cultural Probe Kit for Curiosity [11]

Other adopted strategies include the use of **questionnaires and focus groups** to better understand needs and requirements of specific types of stakeholders, in the perspective of the design and development of the tools, services and prototypes foreseen by the project. To maximise the results, we have adopted an iterative approach.

During the first iteration (July-October 2023) we have collected internal requests for specific questions to be submitted to the stakeholders, in line with the scenarios and expected outcomes of the project; we have then designed 5 questionnaires, oriented to the 5 types of stakeholders (Scientists/Conservators, Museum professionals, Educators, General Public and SMEs), meant to be submitted to a limited and representative potential users (10 for each type), to test the questionnaires and their efficacy (in this phase we have used Google forms for a productive collaborative development); we have then submitted the 5 questionnaires to 10 stakeholders, selected considering their relevance and representative role, introduced by an explanation on the perceive project and its goals; we have analysed the answers and started to identify the needs and requirements for the development of tools, services and prototypes dedicated to the analysis, preservation and communication of coloured collections. In the second iteration (January-June 2024) we have modified the questionnaire including the findings from the first iteration and added more specific questions, deleting questions not understood or needed; we have analysed alternative survey platforms, compliant with GDPR; collected data have been secured stored and analysed; aggregated data were at this point sent to the developers and designers of the applications. Since the second iteration is still on-going, we report here the statistics of the first iteration only. We have collected data from 11 museum professionals (working in institutions such as the MANN museum Italy; Museum of the Imperial Fora Italy; Victoria and Albert Museum London; Acropolis Museum Greece; Museum of art and design Italy; Gallerie Uffizi Italy; Sistema Museale di Ateneo University of Bologna Italy; Vatican Museums); 25 scientists involved in studies in the five reference scenarios (Chap. 6); 6 creative industries engaged in fields connected to virtual museums, digital exhibitions and digital art; 15 citizens (adults and young adults); 10 educators teaching mainly at the university, but also in high schools and primary schools.

1.1 Stakeholder and Audience

When designing interactive media in the field of Cultural Heritage and Art, designers consider different type of stakeholders, each with different characteristics, needs and requirements. Within the PERCEIVE project, we have identified the following relevant potential stakeholders:

1. Museum Professionals
2. Heritage Scientists
3. Creative industries / Designers
4. Citizens / Museum visitors
5. Educators
6. Decision Makers

Museum Professionals. Museum professionals (internal or external to the museum) work in various departments with a potential interest in digital products for the public or for their conservation / exhibition departments. They might be involved in the concept design of an exhibition, in the preparation of the communication/educational program for their visitors, in the design of the exhibition, in the integration of digital applications for the public or in the selection and exhibition of the materials in the exhibit, with certain lighting and museum displays and in integration with various information. The museum professionals involved in the conservation of the materials are often experts in the heritage science field and might work on the monitoring and diagnostic analysis of their collection. Although the professional scenario changes from country to country, many activities such as those mentioned before are often provided by external companies or experts, especially for the creation of interactive exhibit or for the more complex analysis and acquisitions.

Heritage Scientists. A specific group of researchers and scholars can be involved in the study and analysis of colours of museum collections: lost polychromies in ancient sculptures, fading colours in paintings, colour change in textiles, colour degradation in historical photo -film and colour quality in digital artworks.

Creative Industries. Creative Industries are small and medium companies involved in the creation of digital content or digital applications for cultural institutions, art gallery, monuments; they include professionals involved in the creation of virtual exhibitions or virtual museums; they also include companies working in the field of serious gaming, digital tourism and in general on colour visualisation.

General Public. The public is represented by members of the public of all ages that visit museums or could be interested in cultural heritage and digital art via either virtual (internet, webinars, virtual tours) or physical means (visiting museums and galleries).

Educators. Educators are professionals that teach either to students in schools and universities or to other professionals; they might teach history, art history, design, fashion design, archaeology or related topics, but also deal with Computer Graphics aspects, dealing with visualization and interaction involving colour; they prepare their lectures for their classes on the topic and might be interested in using advanced tools and application to better engage students, or they might be interested in creating interactive material with their own content, or they could be also involved in training activities themselves to be updated on the topic.

Decision Makers. These constitute either policy makers who will benefit from the conclusions of the project overall for informing future policy, or financial investors which will gain by the identification of potential areas of niche financial investment.

The analysis of the needs and requirements of the project stakeholders in each scenario and project outcome is carried out by the organization of focus groups, cultural probe kits, interviews and questionnaires. More details are provided in the following sections.

2 Colour Studies Workflows: The Nine Activities

The final objective of the PERCEIVE project is to develop new ways to perceive, preserve, curate, exhibit, understand and access coloured Cultural Heritage collections and Digital Artworks. The starting point for discussion is the analysis of the main workflows used by the professionals and scientists involved in these activities.

Studying, preserving and communicating colour and coloured collections require, in fact, different activities, some iteratively carried out, others in parallel, others in a temporal sequence. 9 types of activities have been identified for the purpose of the project and for their description, an internal focus group with the PERCEIVE experts of each Scenario has been created. Specific questions have been included in the five above-mentioned questionnaires, to validate the relevance of the activities beyond the project's development.

The identified sequence of activities common to the five PERCEIVE scenarios is as follows:

1. Observation and material characterization of colour in the art-objects or collection items;
2. Scientific analysis and digital acquisitions on the colour, material and shape of the art-objects or collection items;
3. Monitoring colour change and assessment of light fastness to better understand how and if art-objects or collection items should be exhibited;
4. Study of the historical and artistic context to better understand how artists have produced their artwork, using which material, for which purposes and in which context;
5. Digital reconstruction the art-object or collection item as it is today (reality-based) with its shape, material and colour, trying to create a digital twin of it;
6. Digital simulation of the art-object (or part of it), using data acquired in the previous steps, to obtain a new version in a different time frame (digital restoration with colour reconstruction, digital reconstruction with the potential original aspect the object might have had, digital simulation of how the item could evolve in the future);
7. Defining the conservation policies to enable a public/restricted access to the art-object or collection item;
8. Designing and developing interactive and not interactive applications (media in general) to involve and communicate the museum's coloured collections and artworks to the public, but also to share results visually and collaboratively with experts;

9. Developing advanced semantic models of the digital museum items and digital artworks, to be able to augment visual 2d and 3d reconstructions, keeping the scientific process and information such as the sources used and the reliability of the results.

In the following paragraphs, these activities are described in detail, together with their main issues, the type of involvement of different professionals, the complexity of the work, the impact on the community, required input data and the produced outputs.

3 Material Characterisation and Colour Change Detection

Scientific analyses covering a whole range of techniques and analytical approaches (from non-invasive punctual analyses as XRF, Reflectance spectroscopy, Raman spectroscopy, or mapping/imaging methods, as Hyperspectral/multispectral analyses, X-ray radiograph, MA-XRF imaging, 3D Hirox mapping and acquisition to micro-destructive tools needing small samples from the object – microFTIR, microRaman, XRD, GC-MS, MALDI-MS or MFT etc.) are the core of the first step of material characterization and colour change detection when speaking about colourful art-objects that can be subject to change [13, 14]. This range of techniques and methods can give compositional information together with mapping the distribution of coloured materials on the surface and in the structure of the art-object, especially for those multilayered structures we often meet in paintings or photographs [15]. Based on this compositional and stratigraphic information further studies around the possible phenomena of colour change can be performed, involving more specific tools as hyperspectral/multispectral and colourimetric techniques (colourimetry CIEL*a*b*, microfaedometry, MFT).

The colour change often is not easy to detect when it happens on small areas, but when affects the whole object or a material that is relevant and visible on the whole object's surface it can impact the readability and material integrity of it. Thus, in the scientific investigation of such phenomena it is important to have a complementary and multiscale approach, where the change is mapped and characterised from a macro up to nano-level, when possible. The type of sample or surface/material that presents the colour change is also relevant for the choice of the best analytical approach and for complementing bulk analysis with surface investigation.

Furthermore, in addition to the real object and eventually samples that can undergo analysis for a better understanding of the change phenomena, laboratory research on model samples might be required to shed light on aspects of the chemical and physical processes that are involved. These model samples (known also as mock-ups) will be subject to similar analyses as the real samples or objects in order to compare compositional and other type of features. These models have the advantage that they can be destroyed during the analytical investigation. One example of how mock-ups can complement the analyses on real samples and give important clues on the colour change phenomena (in this case is the fading of cadmium yellow) is described by Letizia Monico [16], (Fig. 1).

The main issues come from the limitations of the analytical tools themselves (e.g. whole surface vs. point analyses; dimension of the areas to be analyzed; multilayered structures difficult to access/analyze if the tools do not provide enough depth of penetration for the radiation to map the overlapping materials/layers etc.) but also from the immediate availability of these tools in the museum environments. Not many museums

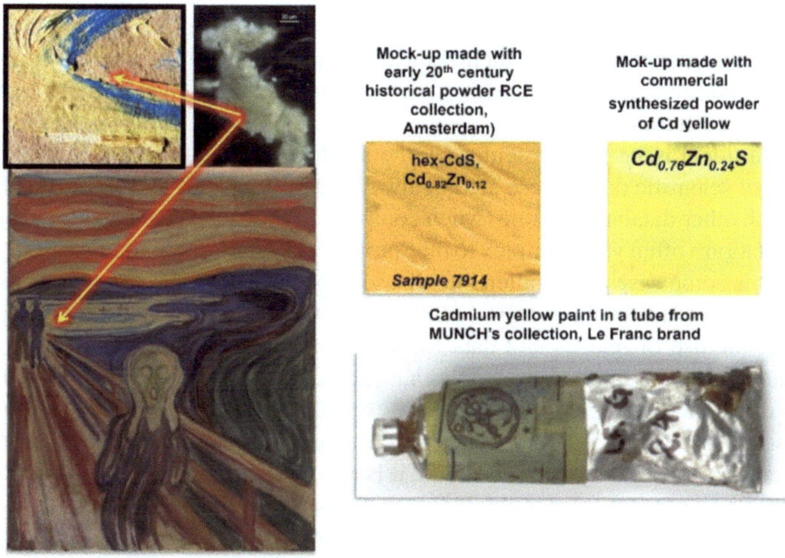

Fig. 1. On left: detail from a yellow area on The Scream where sampling of a Cd yellow flake has been done; On right: 2 images of mock-ups made with a historical powder of the pigment (in its hexagonal form) and also with a commercial powder (with different stoichiometric composition); original paint tube containing Cd yellow (Le Franc) in the MUNCH's collection; credits MUNCH & Letizia Monico (Color figure online)

in the world can afford to have a Scientific Department with such analytical tools. When a Conservation/Heritage scientist is not available sometimes conservators can take over the task of doing some simple analyses (pXRF, MFT or FTIR) or they can ask collaboration to other laboratories where more techniques and resources are available. Some might be accessible through photographic laboratories that are used to document artworks on periodic base for different purposes (conservation state assessment, loans, exhibitions or conservation projects, research projects) but in these cases it is difficult to establish a monitoring schedule for this type of documentation or detection.

Another problem of a nature more specifically linked to the management of collections in museums is the possibility of access to the artwork itself; in museums, not all objects are often on display: those kept in warehouses, for example, require precautions and in most cases, if dealing with marble statuary, preventive conservation is required before applying multispectral diagnostics. In the application of multispectral/hyperspectral investigations, it is necessary to know in detail all the conservation and restoration interventions that the object has undergone over time, and when possible, the factual history from the moment of discovery to its permanence in private collections, up to its arrival at the place of current exhibition.

The complexity of the analytical approach sometimes might require external funding for specific investigations (such as access to MOLAB and FIXLAB facilities within E-RIHS platform[1]), whose results are extremely valuable for conservators and also scholars

[1] https://www.iperionhs.eu, https://www.e-rihs.eu/.

in the field of museum research who need a more detailed and scientific background on the type of colouring materials used, their behaviour and ageing in time, their techniques of application, possible interactions with environmental factors as light, relative humidity, temperature, polluting agents etc.

An additional issue is represented by the availability of repositories, archives and databases of scientific data. These are not always openly accessible or are difficult to be related with other databases online (when comparison of results is needed for example). Archival data are often very complex to interpret because they use outdated language and words that in some cases have fallen into disuse. Hopefully, in most cases the results are easily available (and for free) only if published in Open Access platforms or journals. Regarding the use of images and data from previous investigations on a work of art, many museums have internal protocols that protect the dissemination of the material and sometimes it is not possible to access, except for mere consultation purposes, at previous research. Once all the data necessary to frame the work from the point of view of colour research has been obtained, it is necessary to arrange for its storage: this operation is also not free from problems since, for example, many museums do not have digital repositories that can host a large amount of data.

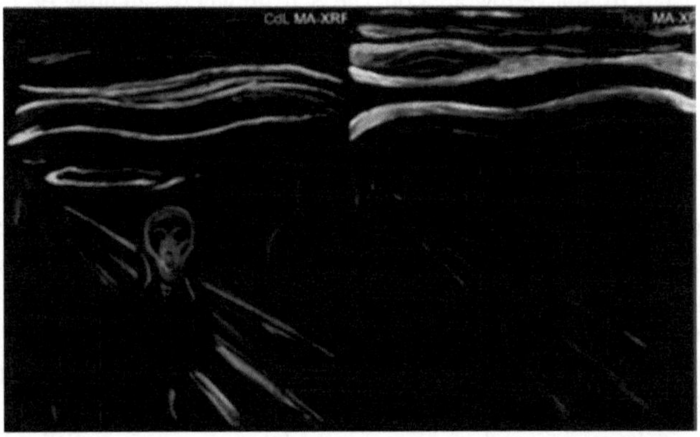

Fig. 2. Visual data from a MA-XRF scan of The Scream (1910?) where the Cd and Hg lines (in white) represent the distribution of cadmium yellow and Vermillion respectively, both sensitive materials but with different behaviours under light exposure (Cd yellow fades, while Vermillion darkens); credits MUNCH (Color figure online)

- *Involvement of different actors/professionals:* Heritage Scientists, Conservators, Curators
- *Complexity of the activity*: The complexity of this activity emerges from the problematics and issues of the material composition of the objects or samples subject to colour change
- *Required Input data*: Data produced, stored and exported by the different analytical equipment and devices
- *Produced Output data*: The output of the analytical approach is also complex and includes a variety of formats, as high-resolution images (jpeg, tiff, png), graphs (compositional maps or spectra), xls sheets with colourimetric and spectral data etc. For

most of these analytical datasets, a certain amount of storage space is needed with specific software tools to visualize, interpretate and elaborate the data (or to extract it when usable on other platforms or for research purposes). An example of such data is shown in Fig. 2.

4 Scientific Analysis for Shape, Colour, Material Acquisitions and Diagnostics

Imaging diagnostics tools offer different applications according to the type of coloured object investigated and their material structure and shape. 2d objects as canvas or cardboard paintings, graphic art on paper (drawing, prints), negatives/positives of old photographs have less depth and tridimensionality but often the surface presents its own 3D profile and can be investigated with three dimensional tools (as video-microscope, OCT tomography or Hirox 3D digital microscope) for topographic, direct and raking light, real colours observations and 2d/3d scans at different magnifications and degrees of resolution. Often the photographic documentation can offer a certain amount of details regarding the condition of the art-object and possible degradation and deterioration patterns, but a more in-depth investigation with 3D equipment (as Hirox 3D microscope) can lead to a multiscale approach in the investigation of surface and multilayered structures and also recording of profilometric data [17] Fig. 3 and Fig. 4 show the Scream (1910?) investigated with different photographic techniques before and after the theft (2004–2006 is the period when the Scream was lost).

Other techniques of imaging that also offer topographic information (3D digital microscopy, profilometry, OCT scans) are a good complement of a normal or raking light photography to document the surface patterns and colours of a painting and also to showcase the difference between several areas with colour changes. In the case of The Scream (1910?) 3D microscopy technology allowed for visualizing at different magnifications (30x, 90x) areas with fading phenomenon (Fig. 5) to be further used for 3D prints as demonstrators in Perceive.

For 3D objects (textile dresses or sculptures in stone), photogrammetry, laser scanners and other 3D acquisition techniques can be used to reproduce their shape, colour and texture and also evaluate eventual degradation.

1. Usually this activity is anticipated by a *discussion with the experts* of the institutions and with curators and by a *survey on site,* to verify the characteristics of the objects to acquire and the logistic.
2. It follows the development of a common *protocol of acquisition*, in which the specific techniques are determined, in accordance with the final goal of a project, together with the logistical – technical aspects and timing.
3. Before starting the acquisition it is often set-up a *digital working environment (often cloud-based, using services such as NextCloud or OwnCloud)* where data will be stored and shared, as to enable all professionals involved to access and process them.
4. In accordance with the type of object, material, dimension, location it is defined the *acquisition methodology*, that usually involve (for shape and RGB acquisition) photogrammetry alone or used in combination with laser scanner. In this case, the chosen type of scanning technique is Structured light projection scanning (SLS) or Time of Flight scanning.

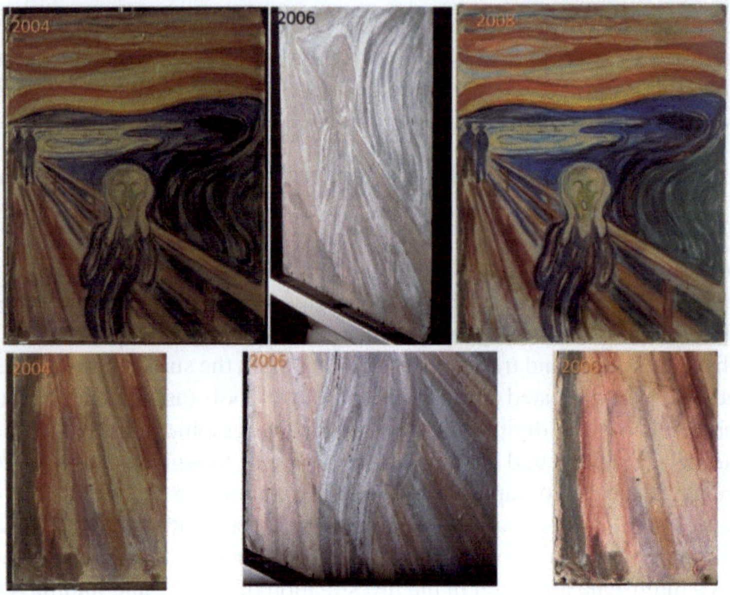

Fig. 3. Upper raw: entire representation of The Scream before and after the theft (from left to right: raking light photo in 2004 and in 2006, RGB photo in 2008); Lower raw: details of the left corner of the painting (where fading occurred) in direct and raking light illumination – 2004, 2006; credits MUNCH.

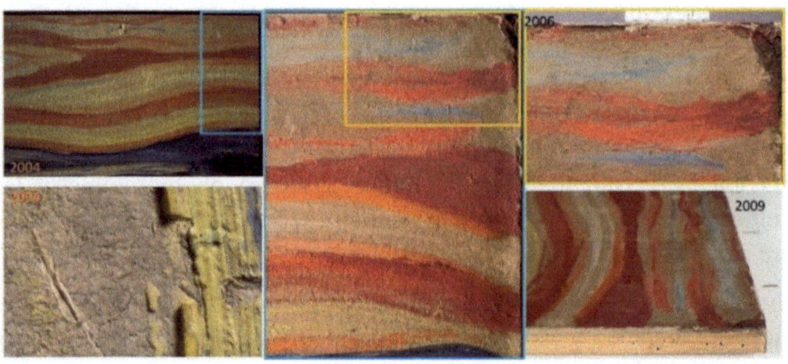

Fig. 4. Details from the right, upper corner of the painting, in different years (2004, 2006, 2009) and using different photographic techniques and angles of observation, credits MUNCH.

5. To better acquire colour and material characteristics and obtain a morphologically accurate models with high-detail textures, it is prepared a *lighting set-up* to allow faster processes and uniform and well-diffused illumination to be performed [18], such as lightbox and turntable, for small-size movable objects, or continuous led lights on stands with photographic studio background for big size or non-movable objects. The lighting set-up provides excellent illumination and enables to set minimal ISO values, reducing image noise [19]. Around the object are placed a grid of *coded*

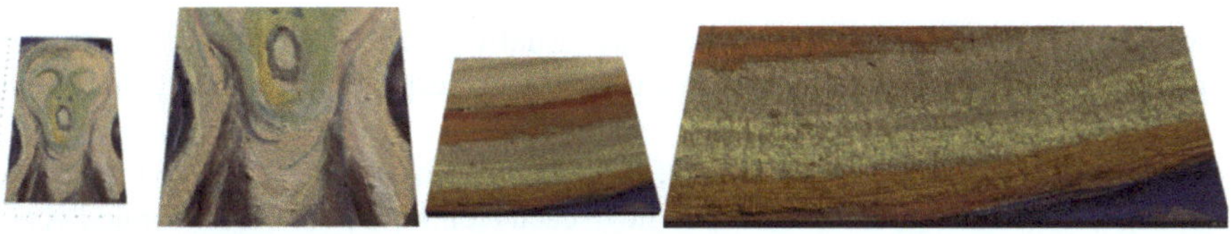

Fig. 5. 3D scans, 90x, of two areas of The Scream surface where colour fading can be perceived; Credits Hirox Europe (Emilien Leonhardt & Vincent Sabatier)

targets to improve camera orientation and model scaling [20–22]. Images are shot in *raw* format and specific colour profiles were created using the ColourChecker Camera Calibration software.

6. This acquisition phase is followed by *software activities*, where data are processed, transformed, and published in their digital version (Fig. 6).

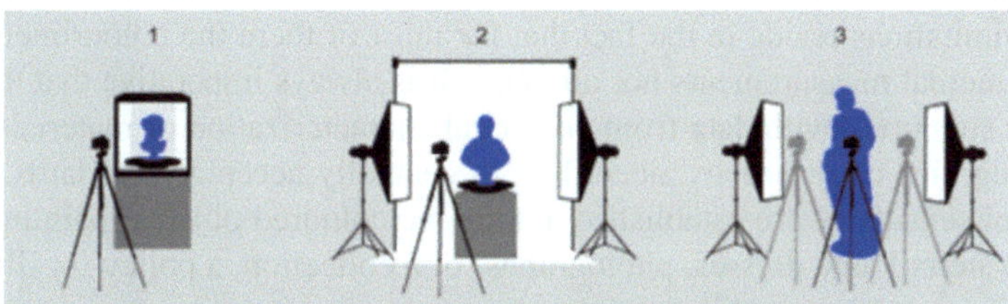

Fig. 6. Example of photogrammetric set-ups

In many cases the main difficulty is connected with the access to the artwork if the scientific laboratories are not part of the museums. Moreover, professionals have to consider objects characteristics, as they might be too fragile, too big or too valuable to be easily transported to external laboratories. For this situation often mobile facilities (e.g. MOLAB - IPERION HS platform[2]) can be accessed according to a specific call for applications subject to evaluation.

Other issues are related with the development of the acquisition equipment and its performance (including the upgrades of software). For photographic techniques for example the illumination conditions and also the use of metric references or colour checker for the photographs are important parameters to take into account for future documentation and reproducibility of colour features in the recorded objects. A consistent and well-established protocol for visual (and analytical) documentation of these objects would be ideal for any institution owning an art-object collection subject to change and degradation.

- *Involvement of different actors/professionals*: These techniques require specialised professionals that are not always available in museums or other institutions owning coloured collections.

[2] https://www.iperionhs.eu/molab/.

- *Required Input data*: The input data are usually those exported directly from the acquisition devices, such as raw images from cameras, point clouds from scanner, etc.
- *Produced Output data*: The results of these techniques can be available immediately as raw data (in different formats specific to each technique) but the elaborated version of them, which can be further used in the next steps of the activities, can take more time and resources to be achieved, stored and also published.

5 Colour Change Monitoring (Sensitivity Detection, Exhibition Policy)

The colour change phenomena in art-objects can be studied and monitored with different tools based on imaging acquisition principles and colourimetric measurements such as MFT (microfadometer testing), CIEL*a*b* colourimetry, Hyperspectral imaging etc. These tools are useful to monitor the degree of colour change and predict the decay rate or value of allowable lux-hour (lux-h) exposure for a sensitive surface or material. Their limitations reside in the fact that for most of them the colourimetric data is based on punctual measurements not on maps. It is always imperative that these kinds of tools are supported with data from analytical characterization of materials to establish the categories of sensitivity, according to generally accepted standards. Once the categories of sensitivity are established among the coloured objects (paintings, prints, drawings, watercolours, dresses, photographs) of a collection, a policy for illumination and exhibition can be created in a museum.

A much more complex matter is that which concerns the monitoring of three-dimensional works, specifically stone ones, since the traces of colour, if they have not previously been documented, are to all intents and purposes non-existent: for this reason, for this specific typology of object, an investigation protocol should be established enabling working in parallel with standardized diagnostics (state of conservation following NORMAL Recommendations for example): where the colour is not found the risk is that of its definitive disappearance or cancellation, due for example to incorrect handling/restoration/conservation activities. A further aspect to evaluate is the impact of environmental pollution, chemical, physical and biological: in both cases the particulate dispersed in the air can attack stone surfaces and alter or destroy what remains of the presence of pigments and their binders. Colour mapping and its state of conservation should be the basis of the choice of whether or not to exhibit each work. There are cases, such as the Venus in Bikini from the MANN in Naples whose continuous loans often due to exhibitions all over the world, represents a very serious risk factor in the conservation of the gilding (Fig. 7).

The colour change monitoring can be performed using a parameter that is Just Noticeable Change (JNC). JNC value perceptible with naked eye shows when a colour has changed. By combining the Blue Wool standards (the responsiveness of material to light) with JNC values (for example for fading of a colour it would be JNF) we can estimate the amount of light a collection item of a particular material can be exposed to light before it fades in a visible way (reference from international standards BS1006 and ISO R105 -Table 2). With a value of 10JNF an item might not be exposed longer because of the amount of fading or discolouration that have occurred. A thumb rule

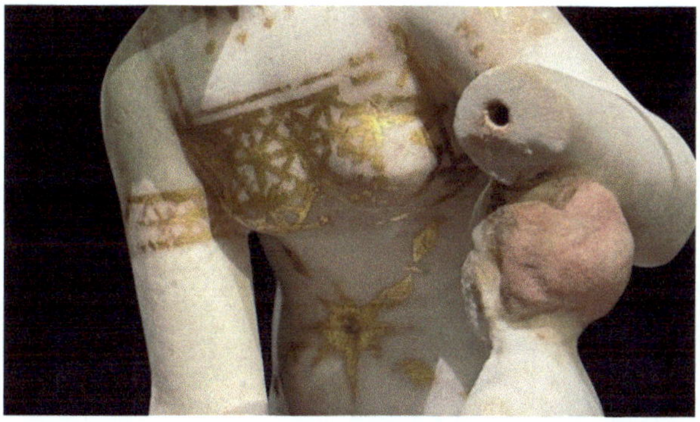

Fig. 7. Marble statuette of Venus in a bikini, found in Pompeii (II, 4, 6), today in the National Archaeological Museum of Naples, Secret Cabinet (inv. 152798) (Credits MANN).

that can apply to these calculations for establishing the period and lux-h exposure for sensitive objects is based on the reciprocity principle of light exposure by varying the intensity/dose of illumination, allowing a longer exposure in an exhibition at low lux-h values being presumably equivalent to a shorter exhibition (in h) with higher illumination intensity (lux) value. When the object is sent on loans and repeatedly exposed, the effect of light is cumulative, so the average illumination conditions have to be correlated to the hours of previous exposure to certain level of lux-h.

Table 2. Reference values from international standards BS1006 and ISO R105, by Blue Wool (BW) category of lightfastness

Category of measure	BW1	BW2	BW3	BW4
Just noticeable fade (JNF)	300.000 lx-h	1.000.000 lx-h	3.000.000 lx-h	30.000.000 lx-h
Years for JNF to occur	100	100	100	100
Recommended exposure limit per year	300 lx-h	10.000 lx-h	30.000 lx-h	100.000 lx-h

As example of such calculations and exhibition policy is the way the different versions of *The Scream* painting have been evaluated in a MFT campaign in 2019 [23]. We know by now that there are 2 versions in MUNCH museum collection that are mostly sensitive to light and humidity (for Cd yellow paints – one of the most sensitive material in the tempera and oil version): *The Scream* (1910?) is in category BW2/BW3, while the hand-coloured print (1895) is in category of higher sensitivity, BW1 (Fig. 8). That means that JNF (Just Noticeable Change) can occur in 50 y if the Scream (1910?) is exposed at 1.000.000 lx-h/year (between 66,7 to 200 days) or in 100 years if the value is 15.000 luxh/year, while for the other version this change can happen in 20 days if the exposure is done at 300.000 lx-h/year (for both, the maximum level of exposure is 25 lx for 1 h).

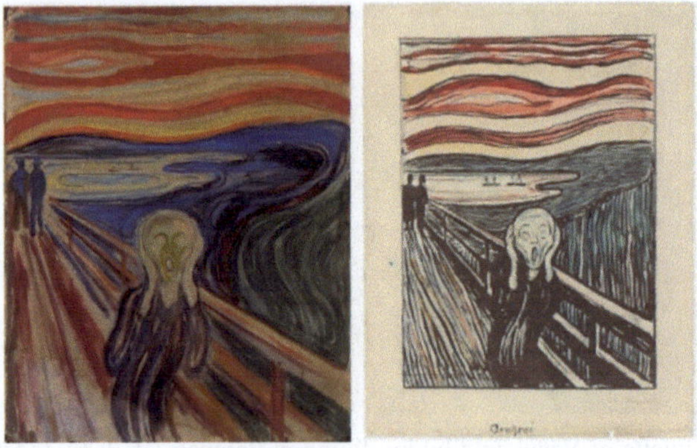

Fig. 8. The 2 versions of The Scream considered mostly sensitive to light and other environmental parameters (relative humidity): BW2/BW3 for the Scream (1910?) on the left; BW1 range of sensitivity for the Scream (1895) on the right (credits MUNCH)

Based on these estimates, the present exhibition set-up (Fig. 9) displays 3 versions of the Scream (not the hand-coloured print but another black and white version, and as there are many copies they can be substituted every 5 months), each exposed for 30 min, in an alternate mode, along a full day of museum opening hours, at 25 lx/h. The level of illumination is thus reduced and the visitors can appreciate each version at a time, never all 3 together, in a space with black walls. Each version is housed in a showcase with automatic doors, that only open when the time to exhibit a version has come. This kind of set-up (Fig. 9) represents a temporary exhibition and has its own limitations regarding the fruition of the whole object and of its details. The information provided in the current set-up does not explain enough the reason behind this choice and the low light levels. A virtual content or augmented reality application able to display all the versions of the Scream (even those that will never be on display) at the same time as the real object with options to also showcase the science behind the colour change using images would very well complement the understanding and appreciation of the full value of these very precious and unique objects.

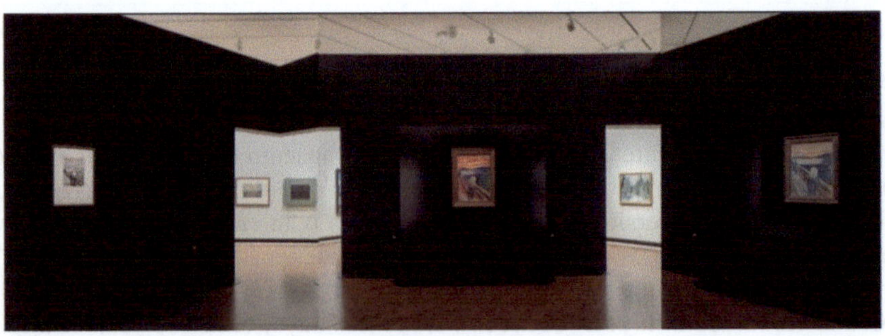

Fig. 9. Set-up of 3 versions of The Scream in Infinite exhibition hall (semi-permanent); Credits MUNCH

A museum is due to carefully and permanently monitor the condition of sensitive items in its collection. An illumination policy can be implemented after establishing a plan for continuous monitoring and mitigation of negative effects of prolonged exposure to light and other environmental factors that can induce colour change. Drafting this policy requires dialogue and common decision making between collection managers, conservation and scientific staff and also exhibition/light designers.

The main output of such a policy is to ensure that the objects are preserved in a stable condition for as long as possible before a JNC occurs. Few online applications are available for some type of objects and materials from museum or conservation institute, but there is no virtual application that shows visually how the change can occur and what the best course of action would be to mitigate the effect of light exposure.

- *Involvement of different actors/professionals*: This activity involves different professionals coming from various scientific and technological fields: conservation science, colour science, material sciences, microscopy, AI engineering, but also light-designers and conservators or registrars.
- *Complexity of the activity*: Main complexity is related to the cumulative effects of long-term exposure and mapping of history of exposure and also potential risks of further damage, real-space and real-time display of sensitive objects.
- *Required Input data*: Scientific publications, Scientific studies, Analysis coming from similar case-studies, MFT monitoring campaigns datasets
- *Produced Output data*: Written and digitalised analytical reports, illumination policy documents, standards of exposure and illumination for collections of sensitive objects, risk assessment methodology and recommendation for risk mitigation in regard to light, relative humidity etc. The main output of an illumination policy is to ensure that the objects are preserved in a stable condition for as long as possible before a JNC occurs. Few online applications are available for some type of objects and materials from museum or conservation institute, but there is no virtual application that shows visually how the change can occur and what the best course of action would be to mitigate the effect of light exposure.
- *Impact on the community*: The main impact is towards the community of museum professionals (from conservators, light designers, registrars, scientists etc.) to the larger public.

6 Interpretation (Multi-disciplinary Study on Colour, Practices, Context and Material)

A crucial activity includes the study of the historical and artistic context to better understand how artists have produced their works, analysing the various materials, purposes and contexts. As an example of such activity we report below the approach followed by the scientists studying lost polychromy on ancient statues.

This branch of studies was born in the 19th century and had its first peak in 1830 thus starting a debate that has not stopped to this day: the prevailing topic, however, contrary to what one might think, did not concern artistic artefacts such as sculptures but rather architecture, therefore the main actors in these interesting debates were architects and not archaeologists, thus giving a very technical imprint to this aspect of art. Analytical field research on the polychromy of Classical marble architecture saw a new beginning in the 1980s with the German research on the temple of Aphaia at Aegina, a building that had been a focus of polychromy research since the early 19th century. Recently, in the last 40 years, the debate has greatly shifted to the artistic and statuary field, launching a second line of research of extreme historical and documentary interest. Although in the beginning there was more discussion on the artistic and aesthetic aspect, the scientific research of materials has managed to carve out a completely different space for itself over time, becoming today the entire fulcrum of the research, the basis on which to hook the documentary and historical part of the process of knowledge. Despite this background, the literary production of the 19th-century full of coloured documentations and excavation data is crucial to deepen to delve into the conservation history of the object and understand how the conservation of colour has changed over time. What do we know about colouring? What can be assumed from archaeological research? For sure we know that the final effect of colouring a surface could be achieved by exploiting the natural colour and texture of a construction material, or by applying paint but more generally the two methods were combined. To reach the ideal final result, there was a specific selection of rocks to be used (or avoided): the preference was quite mandatory since the chosen stone should have some specific characteristics such as being polishable and uniform. There was a basic pigment palette, which was universally applied in both Greek and Etrusco-Roman architecture and also to sculpture, included minerals (iron oxides of local origin and in shades of red, yellow, green, and brown), artificially produced pigments (Egyptian blue, lead white, calcium carbonate white, orpiment), and organic colours (carbon black). More precious and brighter pigments (green malachite, blue azurite, red cinnabar) were usually applied on protected surfaces and architectural sculpture [12]. Not knowing the basics of polychromy standards, with all its subsequent nuances, means setting up the entire research incorrectly: this is why, together with scientific research and analyses of the materials, it is necessary to know in depth the entire history of the object, including the more specifically iconographic aspect. Having a clear iconographic framework of reference is a solid and correct basis for setting up the search for traces of colour in statuary which, unlike painting or frescoes, is sometimes not visible to the naked eye but whose evidence can only be mappable by a specific diagnostic instrumentation.

- *Main issues*: The main issue is to transform scientific studies and publications in information that could be visually recreated
- *Involvement of different actors/professionals*: This activity involves several different professionals coming from the various fields connected to colour and artwork simulation and reconstruction (from art historians to chemists)
- *Complexity of the activity*: Main complexity is related to keeping track of the interpretation process and map this process in the final visual output
- *Required Input data*: Scientific publications, Studies, Analysis coming from similar case studies
- *Produced Output data*: Written reports, visual data to be used as reference in the next activity, semantic data structured in graph databases or knowledge graphs as in the case of the Extended Matrix[3]

[3] https://osiris.itabc.cnr.it/extendedmatrix/index.php/extended-matrix-framework-emf/embt/.

– *Impact on the community*: The main impact is towards the community of scientists and cultural institutions.

7 Reality-Based Reconstruction

In this phase, we digitally reconstruct the museum item or artwork as it is today with its shape, material and colour, trying to create a digital twin of it. We use a reality-based acquisition and reality-based modelling approach adopting photogrammetric and laser scanner approaches, often combined to acquire the geometry and the RGB colour characteristics. Data post processing require an optimization of data, in line with the chose visualization tools. Moreover, the final 3d model is mapped with RGB texture and with the diagnostic analysis previously acquired, especially the imaging (as UV and VIL).uun

– Main issues:

- Data re-use and creation of multiple version (with increased levels of detail);
- Texture and Geometry optimization for webXR;
- Mapping and visualizing imaging diagnostic data;
- Mapping and visualizing punctual diagnostic data (XRF etc.);
- Data integration in digital library
- Interacting with visual data and with semantic information;
- Difficulties in the setup of collaborative and common (aligned) acquisition effort between photogrammetry and diagnostic professionals.

– *Involvement of different actors/professionals*: This activity require the collaboration of the above mentioned professionals. In some cases specific expertise is needed and 3d modelers, 2d/3d artists, computer graphics specialists are involved.
– *Complexity of the activity*: The complexity of this activity is well described by the reported main issues
– *Required Input data*: Photos, Scans, UV and VIL (and other imaging data) images
– *Produced Output data:* 2d orthorectified high resolution images, 3d textured models at various level of details
– *Impact on the community*: still towards the expert and scientists community

8 Colour Simulation (Decay, Original Colour)

With this activity, we create a digital simulation of the item (or part of it), using data acquired in the previous steps, to obtain a new version in a different time frame (digital restauration with colour reconstruction, digital reconstruction with the potential original aspect the item might have had, digital simulation of how the item could evolve in the future). This activity requires to use the reality-based models obtained with Activity 5 and integrate all available sources of information (direct, indirect, comparison with similar cases) to produce a new simulation model. The work of 2d and 3d artists is often required. Colour itself was a medium in ancient times, that's the reason why it is mandatory to recover it. Its relevance is evident for the meaning of a work, as well as the social function of the colour code. The study of colour should embrace other disciplines such as anthropology, linguistics and history without subordinating form to meaning. Colour in ancient times was an extraordinary resource to classify and evaluate society: colours had functions and meanings, were not only mere decorations of stones: they highlighted roles performed in the realms of communication and information. Being able through digital modelling, for example, to access the reconstruction of the image and consequently to its contents and meaning makes available to scholars, as well as the public, a wealth of knowledge that is not guaranteed today. Allowing people to correctly approach the meaning of works of art is of primary importance for the correct diffusion of culture by the relevant bodies; many attempts at physical reconstruction of 1:1 copies of some significant works (such as the Riace Bronzes) have been experimented but the limit between what we can actually deduce through scientific research and what we decide to offer to the public is a delicate border. For this reason, the digital reconstruction/restoration is the only chance we can adopt to correctly convey both the results of scientific, historical-archaeological and finally artistic research. Only by combining these three areas of study is it possible to correctly reconstruct the story of the object and only through the exploration of the potential that the digital world offers we can redefine the concept of culture and sense of care.

- *Main issues*: Acquisition of enough information to enable to develop a digital simulation of the object; Issues in dealing with the knowledge gaps; Complexity in the use and integrating different type of sources (direct analytical sources, historical and iconographic comparison, studies on ancient techniques, context and materials); Issues in the direct interconnection of hypothesis with sources in a way that enable to update results when new data is found; Complexity in keeping and visualising different levels of reliability (see activity nr.9)
- *Involvement of different actors/professionals*: This activity requires the collaboration of photogrammetrists, computer graphics experts, designers, 3d modelers and 2d/3d artists.
- *Complexity of the activity*: The complexity of this activity is well described by the reported main issues
- *Security*: This activity does not have specific impact on security. From a Property Right perspective the produced output usually falls into the work of creativity
- *Required Input data*: 2d and 3d reality-based models, new synthetic images to be used to map the new model
- *Produced Output data*: 2d / 3d simulated models at various level of details
- *Impact on the community*: Main impact is mostly towards the wide community of citizens

9 Conservation vs Access Policy (Identification and Mitigation of Risks)

Every coloured material in the structure of an art-object requires special attention when subject to fading or darkening due to its composition or behaviour to environmental factors. Conservators are the key professionals in a museum that deal with the intimate understanding of these materials and their ageing as they are often called to identify the cause of degradation and propose measures for mitigating the risk of further change or decay or intervene to re-establish the integrity and chromatic wholeness of the object if materials have been lost creating lacunae that disturb the visual appreciation of the object. They eventually work in close contact with scientists and exhibition designers to establish the best conditions of access and usage in different contexts where exposure to light and other environmental parameters needs monitoring and control. Preventive conservation policy [24] is made to establish safe limits for such exposure and calculate the extension and impact of the risk factors (light, humidity, pollution, vibration, pests, vandalism etc.). A magnitude of the risk has a specific calculation formula that implies knowing the following values: object importance, its loss in value, probability of risk occurrence for 100 years and extent of the risk. This will also impact on the display experience, especially for highly fragile and valuable objects, as in the case of The Scream.

Virtual restoration of objects (including the colour reconstruction for areas with change and loss of colour) in exhibition setting is not always possible due to limited space and also restricted exposure levels. In addition, sometimes the accuracy of such a simulation or rejuvenation can be difficult to attain if no original resources or references are available. Nevertheless, the colour science progress using software processing can overcome this challenge and create colourimetric references from scientific results produced with spectral or microfaedometric equipment [25–27].

Virtual or augmented reality applications can be a useful tool to associate with other services or applications for creating educational contents or even prototypes for enhancing the sense of care and authenticity about these objects. The colour and its fading (or darkening) can be protagonists of a unique story to tell the specialised or more general public using images and videos or interactive contents around the creation of the art-object, its fragility and meaning as a material item but also as a representation of emotions generated by the colour perception and its individual or collective interpretation. The colour becomes thus a vehicle of knowledge and experience of the materiality and the fragile impermanence of artistic expression. The colour itself is a descriptive (abstract or realistic) mean, a tool and a process and its understanding can engage all senses, not only the sight. A colour is seen, touched, heard and felt in all its multidimensionality and the experience behind all these ways of bodily engagement (known as embodiment) can be inspired and guided by a science-based approach.

- *Main issues*: Virtual restoration of objects (including the colour reconstruction for areas with change and loss of colour) in exhibition setting is not always possible due to limited space and also restricted exposure levels. In addition, sometimes the accuracy of such a simulation or rejuvenation can be difficult to attain if no original resources or references are available. Nevertheless, the colour science progress using software processing can overcome this challenge and create colourimetric references from scientific results produced with spectral or faedometric equipment.
- *Involvement of different actors/professionals*: This activity involves different professionals from the museum (mainly conservators but also light designers or registrars and conservation scientists) in collaboration with external actors as researchers, scientists or engineers involved in preventive conservation strategy and methodologies or tools design.

- *Complexity of the activity*: Main complexity related to this type of activity resides in finding solutions to match both the limitations of an exposure/illumination policy for the collection/sensitive items and desire to showcase the objects to the larger public as much as possible. Conservators are always concerned with the materiality of the objects and their degradation or deterioration patterns and ongoing processes, while the teams involved in displaying and educating the public about the collection have often a more art historical, aesthetic and design-based approach. Policies for restricted access also need scientific and conservation documentation and procedures and need the approval of the decision making team (usually the leader group of a museum) and these process are time consuming.
- *Required input data*: Archival documents on conservation treatments and history of loans or exposure, scientific publications, and researches, policy and other documents that a museum has to implement for assuring proper conditions for sensitive art-objects in accordance to international guidelines.
- *Produced Output data*: Written and digitalised analytical reports (like for MFT or other campaigns using techniques for colour parameters acquisition), illumination policy and international standards including preventive conservation, risk assessment methodology and recommendations for risk mitigation in correlation to the exhibition/loans programme of the museum.
- *Impact on the community*: The main impact is towards the community of museum professionals (from conservators, light designers, registrars, scientists etc.) to the larger public.

10 Communication of Results (Experts & Public Audience)

In the communication activities, we design and develop interactive and not interactive applications (media in general) to involve and communicate the museums coloured collections and artworks to the public, but also to visually and collaboratively share results with experts. The applications could vary in accordance with the general framework and type of experience (virtual, augmented, mixed, hybrid), and in accordance with specific categories such as: the type of content and the institutional/scientific/audience goals; with the Visualization and Interaction technology, with the type of access and distribution (remote or on site), with the communication style (descriptive vs dramatic), with the level of immersion [28]. Moreover, the user-cantered approach widely adopted in design has recently further evolved into a more concept-oriented approach. This mean the main goal of such applications could be set as to make the users reaching specific cognitive and emotional objectives. PERCEIVE has identified 3 of these objectives:

- making museum visitors and citizens aware of and caring for the fragility and importance of coloured collections
- enabling users to participate to authentic experiences, even in case of fully digital and remote applications
- engaging citizens with art practices through a “provocation, interaction, reflection” loop

This activity starts with the co-design phase, that involve all involved professionals, together with the users. It might use design tools such as the Visitor Card box[4] or other similar tools produced by the GIFT project[5]. It then proceeds with the creation of a User Experience, supported by diagrams, storyboards and scenarios. In parallel one or more narratives and storylines are created; it specifies the User Interfaces and Interaction modalities; It define the framework for its development. The final phase is that of the development of a prototype that is then tested and with the creation of the final product.

- *Main issues*: The main issue is to align institutional and scientific goals with the audience need of being engaged and cognitively-emotionally involved in an experience. A second issue regard the simple and collaborative (possibly web-based) creation of a prototype that could be used to support and test initial ideas and improve the production.
- *Involvement of different actors/professionals*: This activity involves several different professionals coming from the creative industries and cultural institutions (from designers, to programmers, from exhibition experts to light designers, from media artists to writers)
- *Complexity of the activity*: complexity is related to the above-mentioned issues.
- *Required Input data*: Audience data to create personas, Institutional interviews to define goals, Scientific studies and context information to create the narrative; 3d data of the eventual physical spaces where the applications will be installed; digital asset and media to be used in the applications; diagrams; storyboards; written design briefs and scenarios.
- *Produced Output data*: Interactive story describing the user experience (i.e. using tools such as Twine); Web3d or WebXR application, Unity/Unreal projects; final application for PC or Game platforms, mobile App
- *Impact on the community*: The main impact is towards the wide public, although a potential input on cultural institutions, educators and creative industries is foreseen

11 Data Transparency of Processes and Reliability of Results

Nowadays, the 3d documentation and reconstruction of ancient artefacts or historical sites are a common practice in the cultural heritage field [29]. The 3d models are not only visualizing tools but also a way to communicate the work methodology of researchers and experts in the field. Since digital reconstructions always come with a certain degree of interpretation, it is more correct to talk about a "reconstructive hypothesis" [30] which therefore needs to be visualised together with **transparent** data. Transparency is the property of materials that allows us to see things through them. Seemingly, **methodological transparency** is the level at which the user can access and visualise the data history of a specific object, showing what lies behind it. Transparency in research methodology refers to the clear and comprehensive presentation of the research design, data collection, and analysis processes. This notion is essential, since it ensures the validity and reliability of and fosters trust in the research findings; it determines not only how much can be recovered, but helps to understand the data, therefore allowing experts

[4] https://gifting.digital/visitorbox-design-cards/.
[5] https://gifting.digital/design-and-planning-tools/.

and scholars to retrace critical decisions and test conclusions on colleagues' work of and on their own research [31]. The London Charter defines Intellectual transparency as: «The provision of information, presented in any medium or format, to allow users to understand the nature and scope of "knowledge claim" made by a computer-based visualization outcome». What Transparency can describe to researchers and scholars is metadata and paradata. Metadata is useful to describe, manage, preserve, and use the model's data, its content, and its context. On the other hand, paradata refers to the actual use of data: paradata will indicate for example how people use a specific book and its source, while metadata will be the book's title, author, etc. While most of the time metadata and paradata in museums are not *transparent* for the general public, the European project PERCEIVE (www.perceive-horizon.eu), dedicated to the study and reconstruction of coloured museum collections, is engaged in unveiling the results to everybody.

Digital models are not only an "aesthetic" simulation of reality, but they are interactive, manipulable, interrogable, and navigable models which allow new ways of visualisation and access to data and information. Normally, the viewer is presented with the work without commentary about other possible poses in which the object might be realised or reconstructed. Transparency is desirable and entails the presentation of a real-world basis for a virtual reconstruction. One of the main objectives of Kensek [32] is to reply to the question how to use virtual reconstructions in order to enhance scholarship and education (whether of archaeologists, architects, graduate students, third-graders, or tourists) rather than distracting from it. Their work consists in a methodology for making ambiguity, quality of the evidence, and alternative reconstructions dynamically transparent to a user. [34] present a methodological approach to display data-processing which aims to validate 3d modelling practice and to facilitate the exchange of information and collaboration. In particular, the focus is on the visualisation of the uncertainty level of the reconstruction process. Here, the reliability is defined through a gradient colour scale that is related to the different kinds of sources.

In order to add useful visualisations for transparency in these models, Nicolucci [33] suggests the use of a decision tree to represent the modelling process: each terminal node represents a final (alternative) stage, or an alternative reconstruction.

[31] underlines that to visualise data there are two possible ways, the spatial visualisation and multiple windowed mediation of data and information. These could be translated for example with 3d visualisations, texts, graphs, windows with related sources accessible from the model (Fig. 10). Another hint is to use the "rigatino" restoration technique to fill the reconstruction lacunas allowing one to perceive the interventions. One point on which Brusaporci does not compromise is that of avoiding huge amounts of data, which would draw attention from the main object. The author does not mention visualisation tools for the general public. Soto-Martin [35] describes an approach to reconstruct and digitally render the pictorial heritage of St Augustine church in San Cristóbal de La Laguna, an interactive and immersive VR environment «for the enjoyment of all». Users can interact with the environment by activating markers, which will show a text box containing factual information (Fig. 11). The authors hope this environment can provide improvements in terms of student motivation, assimilation of content, and greater autonomy during working reconstruction and conservation processes.

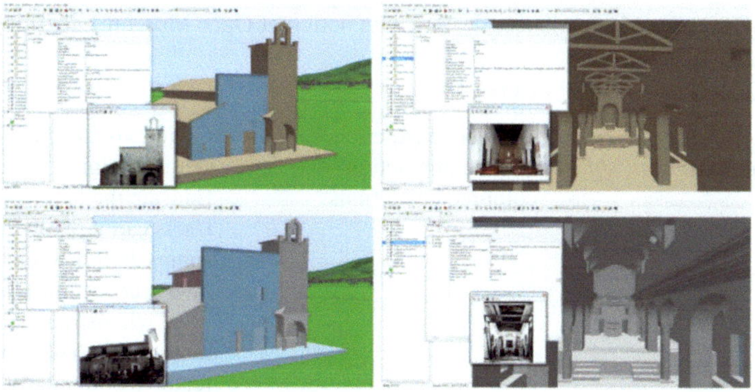

Fig. 10. Church of S. Giustino in Paganica. Integrated informative system, realized with the importation of 3d models in GIS. Correlation of the database with the models of the current and historical building. Spatial visualizations are linked to windowed representation of data.

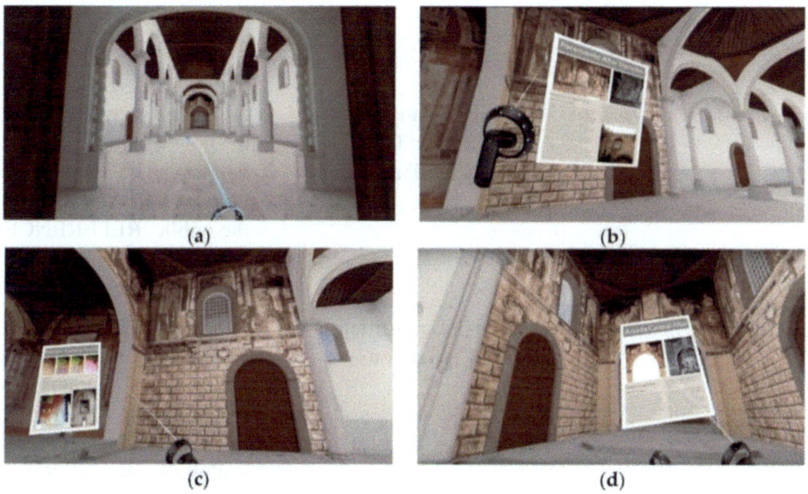

Fig. 11. Virtual interactive actions developed for the program: (a) teleportation, which allows the user to move around the scene, and (b–d), selection by pressing the virtual button, which allows the user to view specific information on the church

- *Main issues*: The greatest issue is that there are very few existing experiences which deal with transparency (see Table 3) and there is a lack of common practices or guidelines on the topic. Nonetheless, it is crucial to communicate the limitations of the 3d reconstructions in relation to the originals [35] thus creating a transparent model accessible for scholars and for the general public and the citizens.
- *Involvement of different actors/professionals*: Boththe 3d modelers who will create the final reliable model and the researchers, scholars, experts in field who studied the origins of the artefact should be involved. Then, it would be necessary to engage the museum professionals.

Table 3. Examples of experiences dealing with notion of Open Museum and Transparency

TITLE	CONCEPT	DESCRIPTION
Chroma: Ancient Sculpture in Colour, New York	Interactive experience, Study and reconstruction of coloured collections	Giving visitors insight into the statue's original appearance. Educate the public. How? Users can virtually recreate the sphinx in full colour, in their own space, comparing the vivid reconstruction and as it looks today. The project features both interpretive and playful components that reflect collaborative work among scientists, imaging specialists, conservators, technologists, art historians, and others. REFERENCE: https://www.metmuseum.org/exhibitions/chroma
Archeologia Invisibile, Torino	Accessibility and Authenticity in Cultural Experiences; Transparency	3d reconstruction of a mummy to show what lays below the bandage. The 3D layers were displayed in the exhibition for the public. REFERENCE: https://museoegizio.it/esplora/mostre/archeologia-invisibile/
Athena Parthenos, Boston	Study and reconstruction of coloured collections	Athena Reveals Her True Colours. A three-minute video explains how the museum uses technology to uncover the appearance of ancient sculptures and shows the statue of Athena Parthenos at the Museum of Fine Arts Boston in its original aspect. REFERENCE: https://www.youtube.com/watch?v=1KIpXsesmSQ

- *Complexity of the activity*: To acquire data and then clearly show the scientific process and information, such as the sources used and the reliability of the results, to the generic public (non-professionals who do not know these topics in depth) is not simple, but fundamental since it gives solidity to the 3d reconstruction.
- *Required Input data*: Metadata, paradata and, in general, every useful data on the dD reconstruction process.
- *Produced Output data*: In general, in [36] point of view:

- o Reconstructions should be accompanied by information making clear the aims and means of the reconstruction
- o Reconstructions may be shown in several valid formats and techniques to reflect a variety of aims and contexts.

  There are different possibilities for the data output in this case. Approaches for visualization include:
- o define reliability with a source-related gradient colour scale [37]
- o show layers of the object, the before and after in a screen near the original artefact [38]
- o Show the process of reconstruction directly projected onto the statue (Chroma exhibition, MET museum, 2023)
- o create projection mapping and natural interaction of colour reconstruction
- o add alternative texturization and personalization multisensoriality and immersiveness [39]
- o make use of AR and natural interaction, allowing visitors to see more details and possible reconstructions by pointing their fingers at an object [39]

– *Impact on the community*: Transparent visualizations will contribute to the visitors' awareness of the presence of colour in ancient sculpture [17]. It is important to:

- o generate understanding of, interest in and support for polychromy research [36]
- o use digital and scientific data that could make understanding damages, how to intervene, to reconstruct and to virtually reconstruct, thus increasing awareness on fragility [39]

At last, the use of "transparent" visualizations will nurture a sense of **cohesion** between general public and restorers, researchers and museum professionals [5], when citizens will truly understand the elaborated process which lies behind the final reconstructions.

12 Designing for Coloured Collections: CoDesignTool

One of the ambitions of the research carried out by a project as PERCEIVE[6] is the development of a user-friendly tool to facilitate and support the design process of digital experiences in alignment with PERCEIVE case studies and scenarios. Furthermore, the intention is to extend the applicability of the tool beyond the scope of the project. Two partners, the Italian National Research Council (CNR) and the German Fraunhofer Institute, have joined forces to pursue this particular endeavour.

Primarily intended to be of benefit to cultural institutions, educators in design courses, and professionals working in the creative industries, the tool offers a structured yet flexible framework for crafting engaging digital experiences. Its systematic approach ensures that all crucial aspects are considered and seamlessly integrated into the final product, while also fostering collaboration and igniting creative thinking throughout the design process. Our tool builds upon an already existing ideation card game, the **VisitorBox**[7]

[6] http://perceive-horizon.eu.

[7] https://gifting.digital/visitorbox-design-cards/.

which emerged from the GIFT Project[8], funded by the EU's Horizon 2020 program. This toolkit was made available as open source, thus allowing for high customisation. Consequently, while maintaining the core elements of the original game, certain alterations were implemented to address the unique needs of our project and to ensure its full support of the PERCEIVE objectives. To this regard, an additional deck was implemented to include a set of cards specifically dedicated to PERCEIVE cognitive goals: Sense of Care and Authenticity. In addition to the physical card game, a digital counterpart in the form of a web application, with the same functionalities, is presently being developed and will be provided at the end of the project. The tool's format notwithstanding, its fundamental nature is that of a card game. Its structure is based on key concepts and components such as cards, boards, and playing rules, with each stage and board of the game being conceived to mirror the phases of a design process. The session begins with the definition of the context, followed by the clarification of institutional and cognitive goals, ultimately converging in the design brief, which outlines the overall parameters of the team's desired outcome. The game then progresses to the ideation and storyboarding steps, after which innovative thinking is encouraged by challenging the ideas just created in detail through a disruptive approach. This initial customised version of the VisitorBox was then employed for a preliminary assessment in a series of dedicated workshops involving all the project's partners with the aim of finalising the design briefs for their respective prototypes. These sessions were held at CNR, the MANN and MUNCH Museums, and the HSLU between November and December 2023. All the valuable information and feedback gathered during these activities allowed the team to proceed with a more thorough redesign of the co-design tool. In February 2024, during a PERCEIVE meetup held in Florence at the CNR Institute, the researchers involved with this task met to brainstorm the new version and discuss its digital implementation. The further developments resulted in the release of **version 0.9**[9] on April 17th of the **PERCEIVE CoDesignTool**. This new version introduced significant alterations to the workflow, the physical arrangement of the boards, canvases and cards, and their layouts and aesthetics overhaul. Thereafter, on the 24th of April, a testing session of this version was conducted at the University of Bologna, involving around 30 students from the Digital Humanities and Digital Knowledge Master's Degree Program (Fig. 12).

The inputs coming from their use of the tool were then translated into further refinements, leading to the release of **version 0.9.1**[10] on the 30th of April. The latter was subsequently employed in two workshops: one held in its entirety remotely by MUNCH on the 7th of May and a hybrid session held by and at MANN on the 14th of May. In both instances, a digital version of the tool was implemented on Miro with the intention of facilitating online partners' interaction and collaboration. During these sessions, only half of the co-design system was utilised, as the teams involved resumed their workflows from their previous meetings held in 2023. As of the time of writing, the CNR team in Florence and the Fraunhofer team are currently engaged in collaborative research and development activities, with the objective of further testing the paper version of the

[8] https://gifting.digital/design-and-planning-tools/.
[9] https://zenodo.org/records/10986215.
[10] https://zenodo.org/records/11091235.

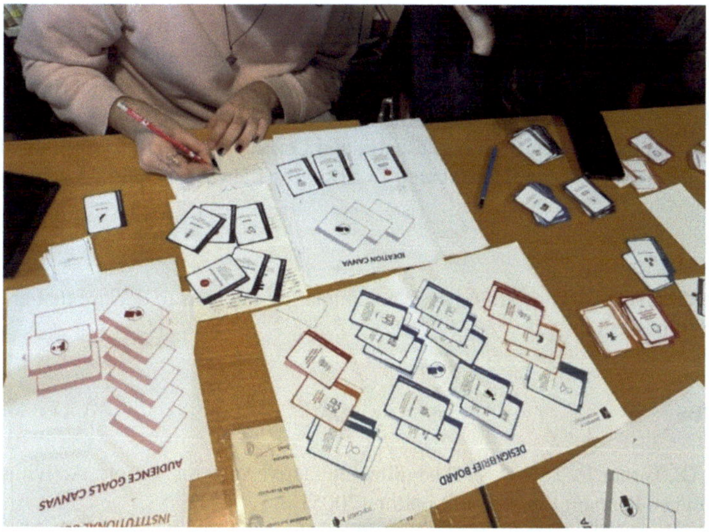

Fig. 12. DHDK students playing with CoDesignTool.

proposed solution while concurrently developing the dedicated web application. The preliminary wireframes are at present available for examination (Fig. 13) while a complete demo is expected to be released at the end of the PERCEIVE Project.

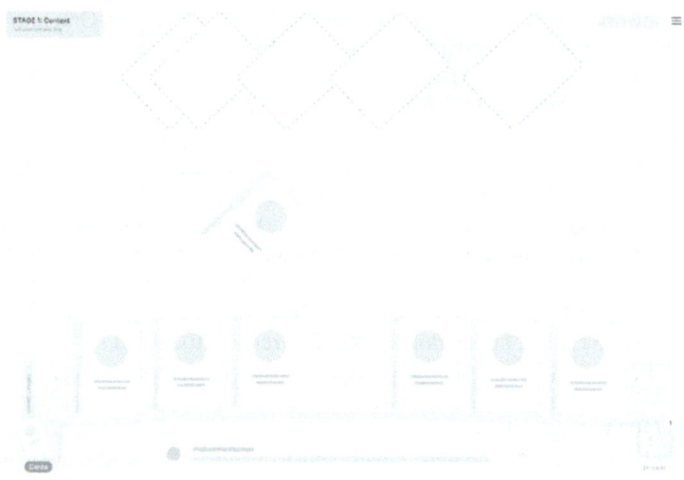

Fig. 13. Example of PERCEIVE CoDesignTool Web Application's Wireframes.

Acknowledgments. This study was funded by the PERCEIVE project that received funding from the European Union's Horizon research and innovation programme under grant agreement No 101061157.

Author Contributions. Sofia Pescarin contributed to Sect. 1, 7, 10; Cristiana Barandoni, Ivana Cerato, Donata Magrini, Sofia Pescarin, Irina Crina Anca Sandu and Petros Stavroulakis contributed to Sect. 2; Cristiana Barandoni, Irina Crina Anca Sandu, Donata Magrini contributed to Sect. 3 and 4; Irina Crina Anca Sandu, Cristiana Barandoni contributed to Sect. 5 and 9; Cristiana Barandoni contributed to Sect. 6; Sofia Pescarin and Cristiana Barandoni contributed to Sect. 8; Vanessa Bonnano, Ivana Cerato contributed to Sect. 11; Laura Travaglini, Marcello Massidda, Sofia Pescarin contributed to Sect. 12.

Disclosure of Interests. The authors have no competing interests to declare that are relevant to the content of this article.

References

1. Benyon, D., Turner, P., Turner, S.: Designing Interactive Systems: People, Activities, Contexts, Technologies. Pearson Education, London (2005)
2. Gaver, B., Dunne, T., Pacenti, E.: Design: cultural probes. interactions **6**(1), 21–29 (1999)
3. Benyon, D.: Designing User Experience. Pearson UK, London (2019)
4. Hulkko, S., et al.: Mobile probes. In: Nordichi '04: Proceedings of the Third Nordic Conference on Human-Computer Interaction (2004). https://doi.org/10.1145/1028014.1028020
5. Pescarin, A., Bonanno, V., Marasco, A.: Social cohesion in interactive digital heritage experiences. Multimodal Technol. Interact. **7**(6), 61 (2023). https://doi.org/10.3390/mti7060061
6. Pescarin, S., Città, G., Gentile, M., Spotti, S.: Authenticity in VR and XR experiences: a conceptual framework for digital heritage. In: Eurographics Workshop on Graphics and Cultural Heritage (2023). https://doi.org/10.2312/gch.20231153
7. Spotti, C., Pescarin, S.: Cultural Probe Kit on Authenticity Triggers in XR Experiences: Activity Book and Diary for Designers to Explore and Develop Authentic Experiences in Digital Heritage (0.9). Zenodo (2023). https://doi.org/10.5281/zenodo.8286449
8. Pescarin, S., Barandoni, C., Sandu, I.: PERCEIVE - What Do You Need and Expect? (Version 1) [Data set]. Zenodo (2024). https://doi.org/10.5281/zenodo.11507479
9. Pescarin, S., Massidda, M., Travaglini, L., Spotti, S., Veggi, M.: PERCEIVE Co-Design Tool for Cultural Heritage Experiences (0.9.1). Zenodo (2024). https://doi.org/10.5281/zenodo.11091235
10. Pescarin, S., Veggi, M., Bonanno, V., Spotti, S.: PERCEIVE Design Cultural Probe Kit Tools [Data set]. Zenodo (2023). https://doi.org/10.5281/zenodo.10260211
11. Bonifazi, F., Pescarin, S.: Cultural Probe Kit on Curiosity Triggers in XR Experiences: Activity Book and Diary for Designers to Explore and Solicit Curiosity in Digital Heritage (Version 1). Zenodo (2023). https://doi.org/10.5281/zenodo.8286038
12. Zink, S.: Polychromy, architectural, Greek and Roman. In: Oxford Classical Dictionary. Oxford, OUP (2019). https://doi.org/10.1093/acrefore/9780199381135.013.8184
13. Sandu, I.C.A., van den Berg, K.J., Candeias, A., Sandbakken, E.G., Tveit, E.S., van Keulen, H.: Chemical analysis in cultural heritage. Part II: physicochemical approaches in art and archaeology investigation. In: van den Werft, I., Sabbatini, L. (eds.) Physical Sciences Review, Chapter 5 - Multi-technique and Multi-scale Approaches to the Study of Ancient and Modern Art Objects on Wooden and Canvas Support, de Gruyter Publ., Published Online 10 Jan 2019. https://doi.org/10.1515/psr-2018-0016

14. Sandu, I.C.A., Joosten, I., Leal, N.: Optical imaging applications for the study of cultural heritage artifacts. In: Arkin, W.T. (ed.) Advances in Laser and Optics Research, vol. 11, pp. 1–46. NOVA Science Publishers Inc. (2015). ISBN: 978-1-63463-494-6
15. Pinna, D., Galleotti, M., Mazzeo, R. (eds.): Scientific Examination for the Investigation of Paintings: A Handbook for Conservators-restorers. Ed. Centro Di, Firenze (2009). ISBN: 978-88-7038-474-1
16. Monico, L., et al.: Probing the chemistry of CdS paints in *The Scream* by in situ non-invasive spectroscopies and synchrotron-radiation X-ray techniques. Sci. Adv. **6**(20) (2020). https://doi.org/10.1126/sciadv.aay3514
17. Sandu, I.C.A., et al.: Multi technique and multiscale approaches to the study of ancient and modern art objects on wooden and canvas support. Phys. Sci. Rev. **4**(5) (2019). https://doi.org/10.1515/psr-2018-0016
18. Webb, E.K., et al.: Quantifying depth of field and sharpness for image-based 3D reconstruction of heritage objects. In: The International Archives of the Photogrammetry, Remote Sensing and Spatial Information Sciences XLIII-B2-2020 (2020). https://doi.org/10.5194/isprs-archives-XLIII-B2-2020-911-2020
19. Verhoeven, G.: Basics of photography for cultural heritage imaging. In: Stylianidis, E., Remondino, F. (eds.) 3D Recording, Documentation and Management of Cultural Heritage, pp. 127–251. Caithness, Whittles Publishing (2016). http://refhub.elsevier.com/S2212-0548(23)00054-1/sref122
20. Guidi, et al.: Acquisizione 3D e modellazione poligonale. McGraw-Hill, Milano (2010)
21. Sapirstein, P.: A high-precision photogrammetric recording system for small artifacts. J. Cult. Herit. **31**, 33–45 (2018). https://doi.org/10.1016/j.culher.2017.10.011
22. Luhmann, T., et al.: Close-Range Photogrammetry and 3D Imaging. Berlin, De Gruyter (2020). ISBN: 978-3110607246
23. Chan, E., Aambø, M., Godzimirska, M., Grimstad, I., Lojewski, T., Sandu, I.C.A.: Light-induced colour changes on the scream versions in the Munch museum collection. In: 1st MUNCH2022 International Conference, vol. 13, special issue 1, pp. 1623–1634 (2022)
24. Staniforth, S. (ed.): Historical Perspectives on Preventive Conservation, vol. 6. Getty Publications, Santa Monica (2013)
25. Ciortan, I.-M., Poulsson, T.G., George, S., Hardeberg, J.Y.: Predicting pigment colour degradation with time series models. In: Colour and Imaging Conference, vol. 30, pp. 250–257 (2022). https://doi.org/10.2352/CIC.2022.30.1.44
26. Ciortan, I.M., Poulsson, T.G., George, S., Hardeberg, J.Y.: Tensor decomposition for painting analysis. Part 1: pigment characterization. Herit. Sci. **11**(1), 76 (2023). https://doi.org/10.1186/s40494-023-00910-x
27. Liang, H., Lange, R., Lucian, A., Hyndes, P., Townsend, J.H., Hackney, S.: Development of portable microfading spectrometers for measurement of light sensitivity of materials. In: ICOM-CC Triennial Conference, Lisbon, Portugal (2011). https://irep.ntu.ac.uk/id/eprint/17689/
28. Ferdani, D., et al.: Handling transparency in 3d reconstructed online environments: Aquae Patavinae VR case study. In: Archaeology in the Digital Era: Papers from the 40th Annual Conference of Computer Applications and Quantitative Methods in Archaeology (CAA), Southampton, 26–29 March 2012, pp. 165–173 (2014)
29. Soler, F., et al.: A complete 3D information system for cultural heritage documentation. J. Cult. Herit. **23** (2017). https://doi.org/10.1016/j.culher.2016.09.008
30. Borra, D.: Sulla verità del modello 3D. Un metodo per comunicare la validità dell'anastilosi virtuale. In: Tecnologie per comunicare l'architettura. Ancona, CLUA Edizioni (2004). ISBN: 9788887965179

31. Brusaporci, S.: The importance of being honest: issues of transparency in digital visualization of architectural heritage. In: Ippolito, A. (ed.) Handbook of Research on Emerging Technologies for Architectural and Archaeological Heritage, pp. 66–93. IGI Global, Hershey - Pennsylvania (2017). https://doi.org/10.4018/978-1-5225-0675-1.ch003.ISBN: 9781522506751
32. Kensek, K.M., et al.: Fantastic reconstructions or reconstructions of the fantastic? Tracking and presenting ambiguity, alternatives, and documentation in virtual worlds. Autom. Constr. **13**(2), 175–186 (2004). https://doi.org/10.1016/j.autcon.2003.09.010
33. Nicolucci, F.: Setting standards for 3D visualization of cultural heritage in Europe and beyond. In: Bentkowska-Kafel, A., Denard, H., Drew, B. (eds.) Paradata and Transparency in Virtual Heritage. Ashgate Publishing, Farnham (2012). ISBN: 978-0754675839
34. Apollonio, F.I., Giovannini, E.C.: A paradata documentation methodology for the Uncertainty Visualization in digital reconstruction of CH artifacts. SCIRES-IT **5**(1), 1–24 (2015)
35. Soto-Martin, O., Fuentes-Porto, A., Martin-Gutierrez, J.: A digital reconstruction of a historical building and virtual reintegration of mural paintings to create an interactive and immersive experience in virtual reality. Appl. Sci. **10**(2), 597 (2020). https://doi.org/10.3390/app10020597
36. Østergaard, J.S.: "Reconstruction" of the polychromy of ancient sculpture: a necessary evil?. Techne **48** (2019). https://doi.org/10.4000/techne.2656
37. Demetrescu, E., Ferdani, D.: From field archaeology to virtual reconstruction: a five steps method using the extended matrix. Appl. Sci. **11**(11) (2020). https://doi.org/10.3390/app11115206
38. Ciccopiedi, C. (ed.): Archeologia Invisibile. Modena, Franco Cosimo Panini Editore (2022). ISBN: 978-88-570-1511-8
39. Pandiani, D.S.M., Pescarin, S.: Beyond static colors: an interactive participatory design perspective on color-centric experiences. Int. J. Conserv. Sci. **13**, 1691–1706 (2022)

Author Index

S. Pescarin et al. (Eds.): *Chromatic Visions*, LNCS 15750, p. 253, 2026.
https://doi.org/10.1007/978-3-032-07792-9

MIX
Papier aus verantwortungsvollen Quellen
Paper from responsible sources
FSC® C105338

If you have any concerns about our products,
you can contact us on
ProductSafety@springernature.com

In case Publisher is established outside the EU,
the EU authorized representative is:
Springer Nature Customer Service Center GmbH
Europaplatz 3, 69115 Heidelberg, Germany

Printed by Libri Plureos GmbH
in Hamburg, Germany